Generalized Structured Component Analysis

A Component-Based Approach to Structural Equation Modeling

T0225578

Chapman & Hall/CRC
Statistics in the Social and Behavioral Sciences Series

Series Editors

Jeff Gill
Washington University, USA

Steven Heeringa
University of Michigan, USA

Wim van der Linden
CTB/McGraw-Hill, USA

J. Scott Long
Indiana University, USA

Tom Snijders
Oxford University, UK
University of Groningen, NL

Aims and scope

Large and complex datasets are becoming prevalent in the social and behavioral sciences and statistical methods are crucial for the analysis and interpretation of such data. This series aims to capture new developments in statistical methodology with particular relevance to applications in the social and behavioral sciences. It seeks to promote appropriate use of statistical, econometric and psychometric methods in these applied sciences by publishing a broad range of reference works, textbooks and handbooks.

The scope of the series is wide, including applications of statistical methodology in sociology, psychology, economics, education, marketing research, political science, criminology, public policy, demography, survey methodology and official statistics. The titles included in the series are designed to appeal to applied statisticians, as well as students, researchers and practitioners from the above disciplines. The inclusion of real examples and case studies is therefore essential.

Published Titles

Analyzing Spatial Models of Choice and Judgment with R
David A. Armstrong II, Ryan Bakker, Royce Carroll, Christopher Hare, Keith T. Poole, and Howard Rosenthal

Analysis of Multivariate Social Science Data, Second Edition
David J. Bartholomew, Fiona Steele, Irini Moustaki, and Jane I. Galbraith

Latent Markov Models for Longitudinal Data
Francesco Bartolucci, Alessio Farcomeni, and Fulvia Pennoni

Statistical Test Theory for the Behavioral Sciences
Dato N. M. de Gruijter and Leo J. Th. van der Kamp

Multivariable Modeling and Multivariate Analysis for the Behavioral Sciences
Brian S. Everitt

Multilevel Modeling Using R
W. Holmes Finch, Jocelyn E. Bolin, and Ken Kelley

Bayesian Methods: A Social and Behavioral Sciences Approach, Second Edition
Jeff Gill

Multiple Correspondence Analysis and Related Methods
Michael Greenacre and Jorg Blasius

Applied Survey Data Analysis
Steven G. Heeringa, Brady T. West, and Patricia A. Berglund

Informative Hypotheses: Theory and Practice for Behavioral and Social Scientists
Herbert Hoijtink

Generalized Structured Component Analysis: A Component-Based Approach to Structural Equation Modeling
Heungsun Hwang and Yoshio Takane

Statistical Studies of Income, Poverty and Inequality in Europe: Computing and Graphics in R Using EU-SILC
Nicholas T. Longford

Foundations of Factor Analysis, Second Edition
Stanley A. Mulaik

Linear Causal Modeling with Structural Equations
Stanley A. Mulaik

Age–Period–Cohort Models: Approaches and Analyses with Aggregate Data
Robert M. O'Brien

Handbook of International Large-Scale Assessment: Background, Technical Issues, and Methods of Data Analysis
Leslie Rutkowski, Matthias von Davier, and David Rutkowski

Generalized Linear Models for Categorical and Continuous Limited Dependent Variables
Michael Smithson and Edgar C. Merkle

Incomplete Categorical Data Design: Non-Randomized Response Techniques for Sensitive Questions in Surveys
Guo-Liang Tian and Man-Lai Tang

Computerized Multistage Testing: Theory and Applications
Duanli Yan, Alina A. von Davier, and Charles Lewis

Chapman & Hall/CRC
Statistics in the Social and Behavioral Sciences Series

Generalized Structured Component Analysis

A Component-Based Approach to Structural Equation Modeling

Heungsun Hwang

McGill University
Montreal, Quebec, Canada

Yoshio Takane

University of Victoria
Victoria, British Columbia, Canada

CRC Press
Taylor & Francis Group
Boca Raton London New York

CRC Press is an imprint of the
Taylor & Francis Group, an **informa** business

A CHAPMAN & HALL BOOK

CRC Press
Taylor & Francis Group
6000 Broken Sound Parkway NW, Suite 300
Boca Raton, FL 33487-2742

First issued in paperback 2020

© 2015 by Taylor & Francis Group, LLC
CRC Press is an imprint of Taylor & Francis Group, an Informa business

No claim to original U.S. Government works

ISBN-13: 978-1-4665-9294-0 (hbk)
ISBN-13: 978-0-367-73875-4 (pbk)

Library of Congress Cataloging-in-Publication Data

Hwang, Heungsun, author.
    Generalized structured component analysis : a component-based approach to structural equation modeling / Heungsun Hwang, Yoshio Takane.
       pages cm
    Includes bibliographical references and index.
    ISBN 978-1-4665-9294-0 (hardcover)
    1. Structural equation modeling. I. Takane, Yoshio, author. II. Title.

QA278.3.H93 2014
519.5'3--dc23   2014028594

Visit the Taylor & Francis Web site at
http://www.taylorandfrancis.com

and the CRC Press Web site at
http://www.crcpress.com

Contents

Preface

This book serves as an introduction to a novel statistical methodology for structural equation modeling, named *generalized structured component analysis*. Generalized structured component analysis was developed by Hwang and Takane (2004a) as an alternative to the two longstanding approaches to structural equation modeling: covariance structure analysis and partial least squares path modeling. As its name explicitly suggests, generalized structured component analysis is a component-based approach akin to partial least squares path modeling. Nonetheless, important differences are apparent from methodological perspectives. Most notably, generalized structured component analysis estimates all model parameters by minimizing a global optimization criterion, whereas partial least squares path modeling does not involve such a global optimization procedure. This seemingly small methodological difference leads, however, to important practical implications.

First, generalized structured component analysis provides measures of overall model fit, which are derived from the global optimization criterion. Such measures are desirable because they allow one to evaluate the adequacy of one's model as a whole and to compare a baseline model to alternative specifications. Conversely, partial least squares path modeling cannot provide measures of overall model fit because it does not involve a global optimization criterion, although some ad hoc indexes are now available. Moreover, because of the use of a global optimization criterion, generalized structured component analysis enables researchers to conduct more complex analyses in a straightforward manner, such as multiple group comparisons involving cross-group equality constraints. In addition, generalized structured component analysis has been extended and refined in various ways over the past decade. In turn, this work has significantly enhanced its data-analytic flexibility and generality. For these reasons, generalized structured component analysis may firmly stand as an appealing alternative to a growing number of researchers and practitioners from various disciplines.

Given the number and scope of these recent advances, it would appear timely to provide a single, comprehensive and rather exhaustive collection of up-to-date developments of this emerging methodology for researchers and practitioners seeking an alternative to traditional approaches to structural equation modeling. Accordingly, this book provides a detailed technical account of generalized structured component analysis and its various extensions. It will, therefore, be useful to quantitative methodologists wishing to gain a good understanding of the theoretical underpinnings of the approach. Simultaneously, this book demonstrates how the approach can be applied to various empirical examples, so that it may be useful to applied researchers and practitioners wishing to grasp the basic concepts behind the approach and to apply it to their own research.

This book can be used at the postgraduate level as a research monograph. It may also serve as a supplementary textbook in an advanced course on structural equation modeling. Throughout this book, conceptual discussions of generalized structured component analysis and of its extensions are placed at the forefront while more technical intricacies (e.g., those on parameter estimation algorithms) are purposely relegated to appendices. To facilitate an understanding of the more technical expositions, readers are expected to hold an intermediate level of knowledge on multivariate statistics and structural equation modeling. If they do not, textbooks such as those by Tabachnick and Fidell (2012) and Bollen (1989) may be of assistance in acquiring this background. Although no prior statistical knowledge is necessary to follow the conceptual discussions presented in this book, it will be useful to the reader to be familiar with basic undergraduate-level matrix algebra (e.g., how to express a system of equations in matrix format).

The organization of this book reflects a hierarchy of increasing technical complexity. Chapter 1 provides a brief overview of traditional approaches to structural equation modeling and discusses why generalized structured component analysis was developed. Chapter 2 exposes the foundations of generalized structured component analysis. Thorough discussions are provided of the issues of model specification, parameter estimation, and model evaluation. This chapter also describes theoretical differences and similarities between generalized structured component analysis and several multivariate statistical methods. A comparative discussion of generalized structured component analysis and partial least squares path modeling at the conceptual and empirical levels is also provided. Chapter 3 introduces relatively simple extensions of generalized structured component analysis. It explains how to conduct constrained single and multiple group analyses, to specify and examine higher-order latent variable structures, to calculate total and indirect effects, and to handle missing data. Familiarity with the three first chapters of this book is sufficient for readers to gain an adequate understanding of the basic characteristics of generalized structured component analysis.

Remaining chapters discuss more advanced developments in generalized structured component analysis. Chapter 4 considers an extension for disaggregate sample analyses under the assumption that observations come from heterogeneous subgroups in the population. This extension combines generalized structured component analysis with fuzzy clustering in a unified framework which is able to obtain cluster memberships of cases and cluster-specific parameter estimates simultaneously. Chapter 5 discusses that a data transformation technique called optimal scaling can be fruitfully integrated into generalized structured component analysis toward the analysis of discrete data measured at nominal or ordinal levels. Chapter 6 focuses on the specification and testing of interactions between latent variables in generalized structured component analysis. Chapter 7 explains how generalized structured component analysis can be extended to deal with multi-level or hierarchical data where individual-level cases are grouped within higher-level units. This extension can take into account such nested structures in both observed and latent variables. Chapters 8 and 9 discuss two different regularized forms of generalized structured component analysis. Specifically, the former considers a ridge-type regularized variant for dealing with the presence of multicollinearity in observed and latent variables, whereas the latter presents a lasso-type regularized variant for selecting subsets of observed and latent variables. Chapter 10 demonstrates how generalized structured component analysis can be extended to the analysis of time series data. Specifically, we show how it is combined with multivariate autoregressive models with an application to the analysis of brain connectivity based on neuroimaging data. Finally, Chapter 11 discusses an extension of generalized structured component analysis which was developed for the analysis of so-called functional data collected in the form of curves, images, or in any other form varying over time, space, or other continua. In each chapter, whenever feasible, we compare generalized structured component analysis to partial least squares path modeling in an effort to show how the two component-based approaches differ when addressing an identical issue.

We are indebted to many colleagues who have contributed to the technical and empirical development of generalized structured component analysis since its inception about a decade ago: Ji Yeh Choi, Wayne DeSarbo, Bill Dillon, Moon-ho Ringo Ho, Sungjin Hong, Kwanghee Jung, Sunho Jung, Youngchan Kim, Aurelie Labbe, Jonathan Lee, Naresh Malhotra, Héla Romdhani, Hye Won Suk, Marc Tomiuk, and Todd Woodward. We especially wish to acknowledge the role of the McGill Quantitative Psychology program as a main vehicle for sustaining and furthering the development of generalized structured component analysis in various directions. We are also grateful to John Abela, Richard Bagozzi, Terry Duncan, Claes Fornell, Jooseop Lim, Thomas Novak, and Liang Wang for generously providing their data, which we have used for demonstrating applications of generalized structured component analysis and its extensions. We wish to express our appreciation to Jae Cha, Vincenzo Esposito Vinzi, Manfred Glang, Jörg Henseler, Giorgio Russolillo, Arthur Tenenhaus, Michel Tenenhaus, and Laura Trinchera, who kindly shared their personal views with us in regards to generalized structured component analysis and partial least squares path modeling, and contributed to widening our understanding and perspectives on component-based approaches to structural equation modeling. We wish to express our gratitude to Emmanuel Jakobowicz and Sunyoung Park for having collaborated in the development of computer software programs for generalized structured component analysis. We also wish to thank Michael Hunter, Marina Takane, and Marc Tomiuk for their editorial assistance with some of the chapters. Lastly, we are deeply indebted to our editor John Kimmel for his patience, support, and encouragement.

As a final note, an online software program named GeSCA is currently available to interested users (www.sem-gesca.org). This program implements the basic features of generalized structured component analysis. A detailed manual and sample data sets are also provided on this website. In addition, an Excel-based software program, called XLSTAT, for generalized structured component analysis will be made available over the summer of 2014.

Heungsun Hwang
Yoshio Takane

MATLAB® and Simulink® are registered trademarks of The MathWorks, Inc. For product information, please contact:

The MathWorks, Inc.
3 Apple Hill Drive
Natick, MA, 01760-2098 USA
Tel: 508-647-7000
Fax: 508-647-7001
E-mail: info@mathworks.com
Web: www.mathworks.com

1

Introduction

We begin this book by discussing why we have developed generalized structured component analysis, which is a relatively new technique for structural equation modeling. We first give an overview of what structural equation modeling is, and then provide a review of existing methods for structural equation modeling, emphasizing their advantages and disadvantages. Finally, we introduce generalized structured component analysis and describe how it maintains the advantages of existing methods while overcoming many of the disadvantages.

1.1 Structural Equation Modeling

Researchers in psychology and other social sciences often collect multivariate data on entities of their interest (cases, subjects, or other sampling units), and want to evaluate the validity of their hypotheses about the relationships among the variables, against the data collected. Structural equation modeling refers to a class of statistical techniques that allow such evaluations. For example, two variables may be measures of essentially the same quantity; one variable may be a cause of another, and so on. These hypotheses are expressed in the form of a model whose empirical validity may be tested using techniques for structural equation modeling.

Structural equation modeling has existed in the form of path analysis in sociology since the 1930s (Wright 1934) and in the form of simultaneous equation modeling, which gained great popularity in econometrics in the 1960s (Amemiya 1985). These techniques were, however, largely restricted to modeling the relationships among observed variables. When the techniques were introduced into psychology in the early 1970s (Jöreskog 1970), an important innovation was added to the methodology. That is, latent variables were introduced to simplify the relationships among observed variables. In psychology and other social sciences, hypothetical constructs, such as intelligence, personality, attitude, and so on, often play important roles, and the latent variables are considered "proxies" of these constructs.

As an example, suppose that we are investigating the relationships between certain foods and diseases. Suppose further that we have a hypothesis that the western European diet (hereafter simply referred to as the western style diet) increases the mortality rate by cancer in the lower digestive system. How can we measure the degree to which people conform to this type of diet? No precise definitions exist. We only have a vague idea that this style of diet is characterized by high calorie and high fat foods. We may begin by collecting information on how many calories people take in on average each day, how much meat and milk products they consume, and so on in different countries. We may then extract a latent variable that captures the most representative variation in these variables, which we may tentatively call the "western style diet." Once the western style diet is measured, its effects on cancer mortality rates can be investigated by regressing the latter on the former. For the cancer mortality rates, we may use the mortality rate by cancer in the large intestine and in the rectum as provided by the World Health Organization. The two cancer variables may be separately regressed onto the western style diet, if there are substantial differences in the pattern of mortality rates between the two variables. If, on the other hand, they are highly correlated, we may first calculate a combined mortality rate, which is then regressed onto the western style diet. Or we might create a second latent variable that captures the most representative variation in the two cancer variables, which we might tentatively call "proneness to cancer in lower digestive system." A regression analysis is then conducted with the proneness to cancer as the criterion variable and the western style diet as the predictor variable. A combination of these analyses constitutes a structural equation model to be tested against the empirical data.

As suggested above, structural equation models consist of two sub-models that researchers must specify, one called the measurement model and the other called the structural model. The former specifies the hypothesized relationships between observed and latent variables. In the example given above, the models that relate the food variables to the western style diet and the cancer variables to cancer proneness represent the measurement model. The structural model, on the other hand, states the hypothesized relationships among latent variables. The structural model is often the main focus of interest in the study. In the above example, the regression model predicting the proneness to cancer from the western style diet represents the structural model.

Structural equation modeling allows a series of analyses such as those described above by combining the two sub-models into a unified framework, thereby enabling the assessment of the adequacy of the entire model against the observed data. If the initial model does not fit the data sufficiently well, it may still serve as a guide as to how the model can be improved. (Note, however, that a model that has been improved in this way must undergo a separate validation using an independent sample.) Structural equation modeling may also provide useful hints for making the model more comprehensive. For instance, it is highly implausible to think that the food variables are the only determinants of cancer mortality in the above example. A fair assessment of the unique contributions of the food variables can only be made when other variables, such as overall wealth, accessibility of health care services, and average lifespan, are also included in the analysis. These confounding variables may explain away at least some of the predictive power of the food variables on cancer mortality.

1.2 Traditional Approaches to Structural Equation Modeling

There have been two general approaches to structural equation modeling characterized by two distinctive ways of conceptualizing latent variables. One is based on common factor analysis, and the other based on component analysis. The former is popular in psychology, while the latter is popular in virtually all other scientific disciplines. In this section, we discuss each approach in turn, highlighting their strengths and weaknesses.

1.2.1 Factor-Based Structural Equation Modeling: Covariance Structure Analysis

In psychology, a collection of techniques called common factor analysis have been used extensively for analyzing multivariate data arising in psychological research (e.g., mental test scores). Factor analysis aims to explain correlations among observed variables by hypothesizing latent variables called common factors. In the example, the food variables are typically highly correlated with each other. This may lead us to suppose that there must be something common underlying the food variables. We hypothesize the existence of a common factor, which can account for the (high) correlations among the food variables. This common factor may again be called the "western style diet." We use the same label for this factor as for the latent variable in the previous section. Note, however, a subtle difference in the nature of a factor in this section and a latent variable used as a more generic term in the previous section. That is, the former is supposed to best account for correlations among observed variables, whereas the latter is supposed to capture the most representative variation in the observed variables. Similarly, the two cancer variables are highly correlated. Again, we may hypothesize that there is a common factor underlying these variables, causing the high correlation among them. This common factor may again be called "proneness to cancer in lower digestive system." We also observe fairly high correlations between the food variables and the cancer variables, which may be explained by further hypothesizing a predictive relationship between the two common factors.

In the factor-based approach, hypothesized relationships among observed variables are translated into a prediction system for correlations among the observed variables under some distributional assumptions on the common factors and error variables. For several technical reasons, it is more convenient to deal with covariances among observed variables than correlations. Firstly, covariances are unaffected by measurement errors, while correlations are. Secondly, the distribution of sample correlation matrices is difficult to derive, whereas that of covariance matrices is straightforward under the assumption of multivariate normality on observed variables. In fact, a covariance matrix is a sufficient statistic for a multivariate normal distribution. This implies that we only need to deal with a covariance matrix rather than individual observation vectors. Correlations are easier to interpret than covariances. However, due to the scale invariance nature of the maximum likelihood (and generalized least-squares) estimators in common factor analysis, estimates obtained from covariance matrices can easily be scaled into those obtained from correlation matrices. The approach discussed in this subsection is often called "covariance structure analysis" because we model and analyze covariance matrices. A more comprehensive account of covariance structure analysis can be found, for example, in Bollen (1989).

The factor-based approach has both desirable and undesirable consequences. On the positive side, this approach is accompanied by powerful parameter estimation methods with nice asymptotic properties. For example, under some conditions, the maximum likelihood estimation in factor-based structural equation modeling yields parameter estimates that are asymptotically normally distributed around true population values with the smallest possible variances. The estimates of the variances (and the standard errors of the estimates) are also available as byproducts of computational algorithms. This allows straightforward tests of the statistical significance of parameter estimates. Maximum likelihood estimation also provides a goodness of fit statistic with an asymptotic chi-square distribution. This statistic can readily be used for evaluating the goodness of fit of the entire model (i.e., measurement and structural models combined).

These properties of the maximum likelihood estimators (or the generalized least squares estimators, which are asymptotically equivalent to the maximum likelihood estimators) are, however, contingent upon several conditions. First of all, the sample size must be large for the asymptotic properties to hold. Secondly, the fitted model must be correct, including the distributional assumption on observed vectors. This implies that the observed vectors are independently sampled from a homogeneous multivariate normal population in which the covariance matrix is generated according to the specified model. In the example given above, these conditions entail that there is exactly one common factor that explains covariances among all food variables, that there is another factor that explains all covariances among the cancer variables, and that the predictive relationship between these two factors explains all between-set covariances. This is too unrealistic an assumption in social science research, in which fitted models are usually only rough approximations of reality. The statistics resulting from covariance structure analysis require large samples to ensure their asymptotic properties. Ironically, by the time the sample size becomes large enough to be able to rely on their asymptotic distributions, virtually all models are rejected, for which no asymptotic results hold. Jöreskog (Sörbom 2001) proposed alternative statistics, such as Goodness of Fit Index (GFI) and modified GFI, to alleviate this situation. While these statistics are more lenient toward accepting fitted models, they are somewhat ad hoc in the sense that "reasonable" fits by these criteria do not guarantee that we can rely on the asymptotic properties of the estimators.

Another caution associated with interpreting the overall goodness of fit statistic is that this statistic only indicates how well a fitted model can approximate observed covariances. A good fit by this statistic does not mean that we have good predictive models among common factors. To see if we have a good predictive model, we need to look at the squared multiple correlation (SMC) coefficient of the regression model assumed between common factors. To illustrate this, look at the covariance matrix given in Table 1.1. This closely emulates the food–cancer example above, where we have two distinct subsets of observed variables, one subset comprising the first two variables pertaining to food variables (designated as F1 and F2 in the table) and the other comprising the last two variables pertaining to cancer variables (designated as C1 and C2 in the table). As before, we are investigating the effects of the food variables on the cancer variables. We assume that within-set correlations (covariances) are all 0.7, while all between-set correlations are 0.3, and that there is one common factor each underlying the food variables and the cancer variables, and that there is a predictive relationship between the two factors. Figure 1.1 depicts the path diagram hypothesized for the data. Covariance structure analysis provides the value of 0 for the chi-square goodness of fit statistic (a perfect fit), but the SMC value of $0.18 = (0.43)^2$. The latter indicates that only 18% of the variance of the cancer factor can be explained by the food factor. That is, we have a perfect fitting model, even though the predictive model in this case is far from perfect. This makes sense because the between-set correlations are all low (0.3).

Suppose now that all within-set correlations are reduced from 0.7 to 0.3 in the above example (see Table 1.2). The result of covariance structure analysis in this case is rather striking. We obtain a perfect fit as before according to the chi-square statistic but also an SMC value of 1 (perfect predictability) as

TABLE 1.1

A Hypothetical Correlation (Covariance) Matrix 1

	Variables			
	F1	F2	C1	C2
F1	1	0.7	0.3	0.3
F2	0.7	1	0.3	0.3
C1	0.3	0.3	1	0.7
C2	0.3	0.3	0.7	1

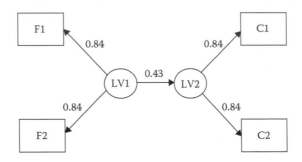

FIGURE 1.1
Path coefficients in the mode for the data in Table 1.1 obtained by factor-based structural equation modeling.

TABLE 1.2

A Hypothetical Correlation (Covariance) Matrix 2

	Variables			
	F1	**F2**	**C1**	**C2**
F1	1	0.3	0.3	0.3
F2	0.3	1	0.3	0.3
C1	0.3	0.3	1	0.3
C2	0.3	0.3	0.3	1

displayed in Figure 1.2. The SMC value indicates that 100% of the variance of the cancer factor can be predicted by the food factor. This result looks counter-intuitive, but is true. This can be easily verified from the fact that the observed covariance (correlation) matrix in this case exhibits a perfect equal variance and equal covariance structure (i.e., compound symmetry), for which the one common factor model fits perfectly. When this factor is split into two factors, they should correlate perfectly with each other. Covariance structure analysis assumes that the low between-set correlations are due to measurement errors in observed variables. If they were adjusted for the measurement errors, they would more highly correlate with each other. This is called the correction for attenuation, which some claim to be an advantage of factor-based techniques over component-based techniques in that the former explicitly take into account the effects of measurement errors. This is a so-called armchair (or utopian) theory. It is of little practical use unless there is a reliable way to measure the variables of concern without measurement errors. Efforts to obtain more reliable factors by increasing the number of high quality indicators are more important than the correction for attenuation. Otherwise, the predictability index (SMC) among factors, no matter how high it may be, is too fictitious to rely on.

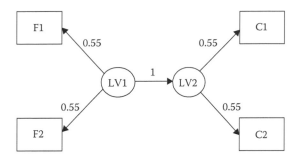

FIGURE 1.2
Path coefficients in the mode for the data in Table 1.2 obtained by factor-based structural equation modeling.

In factor-based structural equation modeling, latent variables (common factors) are regarded as random effects. In psychology, such random effects are often introduced to account for individual differences in ability, preference, and so on. The effects due to individual differences are often needed in the model to construct a realistic model of behavior. However, regarding them as random effects means that we are only interested in their distributions (mostly their variances), but not in individual scores. Consequently, the random effects are often marginalized (integrated) out for the sake of consistent estimation of other structural parameters in the model (Takane and de Leeuw 1987), which is also a standard practice in factor-based structural equation modeling.

One consequence of random-effect common factors is that they are not explicitly defined. Their nature can only be inferred from their correlations with observed variables. The problem is that common factors having prescribed patterns of correlations with the observed variables cannot be determined uniquely. There are infinitely many possible sets of scores that have designated covariances (correlations) with the observed variables. This is known as the problem of factorial indeterminacy in common factor analysis. This is somewhat analogous to describing one's favorite food by its correlations with its constituents. For example, the correlation of my favorite curry and rice with pork is 0.4, the correlation with potato is 0.6, the correlation with carrot is 0.3, and so on. One can make infinitely many varieties of curry and rice with the prescribed profile. A proper recipe, on the other hand, should prescribe exactly how much of each ingredient should be put in to make the designated dish. Harris (1989), in emphasizing the importance of prescriptive instructions, provides an interesting example in which a latent variable is highly correlated with the first observed variable, but is orthogonal to the second, while in fact it is constructed by a simple difference between the two. We may then erroneously conclude that the latent variable is highly correlated with the first variable, but has nothing to do with the second. On the contrary, the second variable plays an essential role in defining the latent variable.

The random common factors may also cause other problems such as improper solutions in factor-based structural equation modeling. This refers to the situation in which logically impossible estimates (negative estimates of variances, correlations whose absolute values are greater than unity) are obtained in estimation. For illustration, assume that between-set correlations are for some reason larger than within-set correlations in Table 1.2. In this case, the standardized path coefficient in Figure 1.2 becomes larger than 1, indicating that more than a 100% of the variance in the cancer factor can be predicted by the food factor, which is logically impossible. Some researchers claim that improper solutions are good indications of misspecified models, so that the fitted models should be carefully examined and modified appropriately. However, identifying the nature of the problem is not always straightforward, nor is determining how to deal with it in practice. In some cases, increasing the sample size may resolve the problem. In other cases, the whole study may have to be redesigned, and entirely new data sets may have to be collected.

In sum, most of the social sciences are not mature enough to take full advantages of factor-based structural equation modeling. This approach is like a razor used to crack a hard rock, where you need a sledge hammer. It is too "fine-pointed" to be useful in modeling complex phenomena in the social sciences.

1.2.2 Component-Based Structural Equation Modeling: Partial Least Squares Path Modeling

In contrast to treating latent variables as common factors, component-based structural equation modeling methods consider latent variables to be components, which are defined as exact linear combinations of observed variables. In this section, we discuss one such approach called partial least squares path modeling as a precursor to generalized structured component analysis, which is the main topic of this monograph. The history of partial least squares path modeling dates back to the early 1980s (Wold 1982; Lohmöller 1984), but it did not gain much popularity until recently (Tenenhaus et al. 2005; Tenenhaus and Tenenhaus 2011), perhaps due to its algorithmic obscurity, as will be briefly described below.

In partial least squares path modeling, path models are estimated in two separate stages. In the first stage, latent variables (components) are estimated as exact linear combinations of observed variables using an iterative procedure. Once the components are estimated, other parameters of interest (e.g., component loadings and structural path coefficients) are estimated by applying regression analyses with the estimated components. Since we already know what regression analysis is, we may focus on the first stage estimation here.

Estimating components in the first stage means estimating the weights applied to the observed variables to derive the components. There are several alternative ways (modes and schemes) to update the weights. We may choose the most suitable one, depending on the purpose of the analysis. We will not discuss these alternatives and how to choose among them here; a detailed explanation of the algorithms will be provided in Chapter 2. Instead, we discuss the general features of the algorithms that they all have in common. Iterative procedures are all simple, and convergence is usually very quick. In contrast to factor-based structural equation modeling, there is no problem of factorial indeterminacy or that of improper solutions, since the components are defined as linear combinations of observed variables. No rigid distributional assumptions are made, nor are the specified models assumed correct in order to establish asymptotic properties of the estimates. Instead, a bootstrap method is used to assess the stability of the parameter estimates. Consequently, the method works with relatively small samples. Even the independence assumption does not need to be made among observed vectors, although in this case, the bootstrap method needs some adjustment for sampling observed vectors. Since a fitted model is not necessarily assumed correct, there is no need to explain every single covariance between the observed variables in order to assess the reliability of the estimated parameters. Consequently, a relatively large number of observed variables can be analyzed simultaneously.

Partial least squares path modeling thus seems quite attractive from a practical viewpoint. There is, however, one definite drawback in this approach. The iterative algorithms are defined only in terms of the updating formulas, which are repeatedly applied until convergence. Convergence is defined as a point of equilibrium, where the values of the weight vectors no longer change significantly between iterations. It is unknown what the algorithms lead to or what the convergence point means. In particular, it is unknown what the algorithms optimize. The lack of optimization criteria means that no global goodness of fit measure is available even for the first stage alone. This is a clear disadvantage in practical data analysis situations, where the evaluation of how well the model fits the data is essential. This also leads to some difficulty in extending the method to address emerging technical or empirical issues because there is no principle to guide algorithm constructions. An ad hoc fit index has been invented to remedy this situation (Tenenhaus et al. 2005). However, there is no direct connection to what is optimized by the algorithms (which is unknown).

Note that this lack of optimization criteria is only in the first stage. Regression analyses in the second stage have obvious optimization criteria. The problem is that these criteria have no relationship with the first stage algorithms. This makes it even more difficult to define a global optimization criterion common to both stages. We previously stated that the idea of structural equation modeling was to enable the evaluation of the global fit of the model by combining both the measurement and structural models into a unified framework. Partial least squares path modeling fails to satisfy this fundamental requirement of the methodology.

Estimates of parameters obtained in partial least squares path modeling are not consistent under the assumption that the fitted model is correct. (This is the usual assumption made in factor-based structural equation modeling.) This means that the estimates of parameters may not approach true population values if the number of observed vectors increases indefinitely. Although this is often regarded as a weakness of partial least squares path modeling, it may not be as serious as it may sound because the assumption of a true fitted model rarely holds in any case. Rather, we contend that consistent estimates under unrealistic conditions do not mean very much. If, for some reasons, we want to obtain consistent estimates (and consequently are willing to make the dubious assumption that the fitted model is correct), there are a number of ways to obtain them (e.g., Croon 2002; Dijkstra 1981; Lu et al. 2011). All these procedures, however, essentially reduce to covariance structure analysis with less efficient criteria (e.g., simple least squares) than maximum likelihood.

1.3 Why Generalized Structured Component Analysis?

Generalized structured component analysis is similar to partial least squares path modeling in many ways. Consequently, it inherits many of the advantages of partial least squares path modeling, which were outlined in the previous section. At the same time, it overcomes some crucial disadvantages. In generalized structured component analysis, a global optimization criterion is explicitly defined and optimized throughout iterations. As in the factor-based approach, both measurement and structural models are separately stated and then combined into a unified framework under a single common optimization criterion.

The optimized criterion value at convergence of an iterative optimization procedure in turn serves as an index of global goodness of fit of the combined model, named FIT and defined in Equation 2.20. The FIT denotes how much variance in the data is accounted for by the combined model. Local fit indices are also available that indicate where misfits may exist. It may be that the measurement model fits well, but not the structural model or vice versa. Or it may be that only a few of the entire set of equations do not fit well. This can be readily checked by examining various local fit indices.

Generalized structured component analysis can avoid anomalous phenomena associated with the correction for attenuation in factor-based structural equation modeling. For the covariance matrix in Table 1.2 (the second example in Section 1.2.1), where observed covariances are all equal to 0.3, both generalized structured component analysis and partial least squares

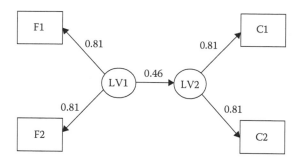

FIGURE 1.3
Path coefficients in the mode for the data in Table 1.2 obtained by component-based structural
equation modeling.

path modeling give an identical result of an SMC value of $0.21 = (0.46)^2$, as
opposed to 1 from covariance structure analysis (see Figure 1.3). The overall fit obtained by generalized structured component analysis is FIT = 0.47
(which is unavailable from partial least squares path modeling). These results
are more realistic than those obtained by factor-based structural equation
modeling.

Thus, we believe that generalized structured component analysis inherits
strengths of existing methods, while overcoming some of their weaknesses.
In the chapters that follow, we present more detailed features of generalized
structured component analysis from various perspectives, as outlined in the
preface.

2

Generalized Structured Component Analysis

In this chapter, we provide a detailed description of generalized structured component analysis. We discuss the issues of model specification, parameter estimation, and model evaluation. We demonstrate the use of generalized structured component analysis for the analysis of real data. We also describe theoretical differences and similarities between generalized structured component analysis and several multivariate statistical techniques that involve components or weighted composites of observed variables. Furthermore, we empirically compare generalized structured component analysis with another well-known component-based approach to structural equation modeling—partial least squares path modeling. We provide technical intricacies related to generalized structured component analysis and partial least squares path modeling in Appendices.

2.1 Model Specification

Generalized structured component analysis involves the specification of three submodels to specify a structural equation model. The three submodels are *measurement, structural,* and *weighted relation* models.

We shall use a prototype structural equation model to introduce notation and model specification in generalized structured component analysis. Figure 2.1 shows a path diagram of the prototype model. In the path diagram, square boxes are used to indicate observed variables or indicators, and circles are used to represent latent variables. Moreover, straight arrows are used to signify that the variable at the base of an arrow affects the variable at the head of the arrow, whereas straight lines are used to represent the weighted relations between indicators and latent variables (i.e., a latent variable is determined as a weighted composite of indicators). An indicator or latent variable is endogenous, if it is affected by one or more variables in the model. On the other hand, an indicator or latent variable whose determinants are not given in the model is exogenous. In this book, we use z and γ to denote an indicator and latent variable, respectively.

In the prototype model, there are eight indicators (z_1, ..., z_8). Each of four latent variables (γ_1, γ_2, γ_3, and γ_4) influences two indicators. In addition, two

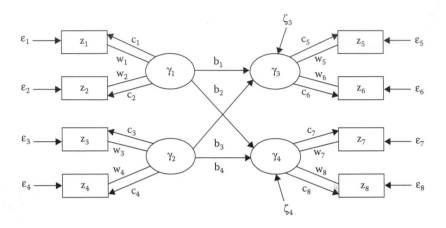

FIGURE 2.1
A path diagram of a prototype recursive structural equation model.

latent variables (γ_1 and γ_2) are specified to influence the other latent variables (γ_3 and γ_4).

2.1.1 Measurement Model

The measurement model specifies the relationships between indicators and latent variables. This model is also called the outer model in the literature of partial least squares path modeling (Fornell and Bookstein 1982). An indicator is considered reflective if it is influenced by the corresponding latent variable, whereas it is considered formative if it forms its latent variable. In generalized structured component analysis, the measurement model is specified only when there exist reflective indicators because formative indicators are dealt with by a different submodel—the weighted relation model. The prototype model depicted in Figure 2.1 requires the specification of the measurement model because indicators are reflective.

Specifically, the measurement model for the prototype model is as follows:

$$z_1 = \gamma_1 c_1 + \varepsilon_1$$
$$z_2 = \gamma_1 c_2 + \varepsilon_2$$
$$z_3 = \gamma_2 c_3 + \varepsilon_3$$
$$z_4 = \gamma_2 c_4 + \varepsilon_4$$
$$z_5 = \gamma_3 c_5 + \varepsilon_5$$
$$z_6 = \gamma_3 c_6 + \varepsilon_6$$
$$z_7 = \gamma_4 c_7 + \varepsilon_7$$
$$z_8 = \gamma_4 c_8 + \varepsilon_8. \tag{2.1}$$

In the above model, c's are loadings relating latent variables to indicators, and ε's are the residuals of indicators left unexplained by their corresponding latent variables. Again, this measurement model shows that all indicators are considered reflective because each of them is assumed to be an effect of the corresponding latent variable. Thus, all the indicators are observed endogenous variables.

We can express the above equations in matrix notation, as follows:

$$
\begin{bmatrix} z_1 \\ z_2 \\ z_3 \\ z_4 \\ z_5 \\ z_6 \\ z_7 \\ z_8 \end{bmatrix} = \begin{bmatrix} c_1 & 0 & 0 & 0 \\ c_2 & 0 & 0 & 0 \\ 0 & c_3 & 0 & 0 \\ 0 & c_4 & 0 & 0 \\ 0 & 0 & c_5 & 0 \\ 0 & 0 & c_6 & 0 \\ 0 & 0 & 0 & c_7 \\ 0 & 0 & 0 & c_8 \end{bmatrix} \begin{bmatrix} \gamma_1 \\ \gamma_2 \\ \gamma_3 \\ \gamma_4 \end{bmatrix} + \begin{bmatrix} \varepsilon_1 \\ \varepsilon_2 \\ \varepsilon_3 \\ \varepsilon_4 \\ \varepsilon_5 \\ \varepsilon_6 \\ \varepsilon_7 \\ \varepsilon_8 \end{bmatrix}. \tag{2.2}
$$

This can be generally written as:

$$
\mathbf{z} = \mathbf{C'}\gamma + \varepsilon. \tag{2.3}
$$

Equation 2.3 is the general form of the measurement model in generalized structured component analysis. In the measurement model, $\mathbf{z}$ is a J by 1 vector of all indicators, γ is a P by 1 vector of all latent variables, $\mathbf{C}$ is a P by J matrix of loadings relating P latent variables to J indicators, and ε is a J by 1 vector of the residuals of all indicators. In the prototype model, J and P are equal to 8 and 4, respectively.

As stated above, generalized structured component analysis requires the measurement model when some indicators are reflective or equivalently some loadings (c's) are nonzeros. When all indicators are formative or equivalently all loadings are equal to zeros, that is, $\mathbf{C} = \mathbf{0}$, the measurement model is not needed. We do not make any distributional assumptions on the residuals (ε's).

2.1.2 Structural Model

The structural model expresses the relationships among latent variables, which is also called the latent variable model (Bollen 1989) or the inner model in partial least squares path modeling.

The structural model for the prototype model is as follows:

$$\gamma_3 = \gamma_1 b_1 + \gamma_2 b_3 + \zeta_3$$

$$\gamma_4 = \gamma_1 b_2 + \gamma_2 b_4 + \zeta_4 \tag{2.4}$$

where b's are path coefficients relating a latent variable to other latent variables and ζ's are the residuals of latent variables left unexplained by the corresponding exogenous latent variables. In the prototype model, two latent variables (γ_1 and γ_2) are exogenous, whereas the others (γ_3 and γ_4) are endogenous.

We can express the above equations in matrix notation, as follows:

$$\begin{bmatrix} \gamma_1 \\ \gamma_2 \\ \gamma_3 \\ \gamma_4 \end{bmatrix} = \begin{bmatrix} 0 & 0 & 0 & 0 \\ 0 & 0 & 0 & 0 \\ b_1 & b_3 & 0 & 0 \\ b_2 & b_4 & 0 & 0 \end{bmatrix} \begin{bmatrix} \gamma_1 \\ \gamma_2 \\ \gamma_3 \\ \gamma_4 \end{bmatrix} + \begin{bmatrix} \gamma_1 \\ \gamma_2 \\ \zeta_3 \\ \zeta_4 \end{bmatrix}. \tag{2.5}$$

The structural model for the prototype model can be considered recursive because there is no feedback loop or reciprocal relationship between latent variables. Generalized structured component analysis can also handle nonrecursive relationships between latent variables. For example, consider a slightly modified version of the prototype model, shown in Figure 2.2. In this model, there is a feedback loop between the two endogenous latent variables (γ_3 and γ_4), indicating that they influence each other.

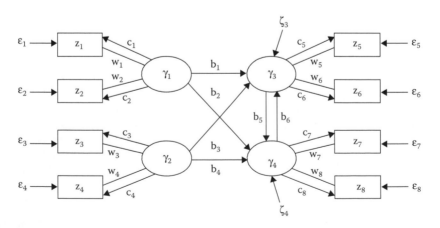

FIGURE 2.2
A path diagram of a prototype nonrecursive structural equation model.

The structural model for this nonrecursive prototype model is as follows:

$$\gamma_3 = \gamma_1 b_1 + \gamma_2 b_3 + \gamma_4 b_6 + \zeta_3$$
$$\gamma_4 = \gamma_1 b_2 + \gamma_2 b_4 + \gamma_3 b_5 + \zeta_4. \tag{2.6}$$

We can express the above equations in matrix notation, as follows:

$$\begin{bmatrix} \gamma_1 \\ \gamma_2 \\ \gamma_3 \\ \gamma_4 \end{bmatrix} = \begin{bmatrix} 0 & 0 & 0 & 0 \\ 0 & 0 & 0 & 0 \\ b_1 & b_3 & 0 & b_6 \\ b_2 & b_4 & b_5 & 0 \end{bmatrix} \begin{bmatrix} \gamma_1 \\ \gamma_2 \\ \gamma_3 \\ \gamma_4 \end{bmatrix} + \begin{bmatrix} \gamma_1 \\ \gamma_2 \\ \zeta_3 \\ \zeta_4 \end{bmatrix}. \tag{2.7}$$

Thus, both recursive and nonrecursive structural models can be rewritten concisely as:

$$\gamma = \mathbf{B}'\gamma + \zeta. \tag{2.8}$$

Equation 2.8 is the general form of the structural model in generalized structured component analysis. In the structural model, $\mathbf{B}$ is a P by P matrix of path coefficients relating P latent variables among themselves, and ζ is a P by 1 vector of the residuals of all latent variables. In recursive structural models, the path coefficients that represent the relationships between latent variables are included in the lower triangular portion of $\mathbf{B}'$. Conversely, in nonrecursive structural models, $\mathbf{B}$ is not lower triangular. We do not make any distributional assumptions with respect to the residuals (ζ's).

2.1.3 Weighed Relation Model

As the name suggests, generalized structured *component* analysis defines a latent variable as a component or weighted composite of indicators. The weighted relation model is used to explicitly express such a relationship between indicators and a latent variable.

The weighted relation model for the prototype model in Figure 2.1 is as follows:

$$\gamma_1 = z_1 w_1 + z_2 w_2$$

$$\gamma_2 = z_3 w_3 + z_4 w_4$$

$$\gamma_3 = z_5 w_5 + z_6 w_6$$

$$\gamma_4 = z_7 w_7 + z_8 w_8. \tag{2.9}$$

We can express the above equations in matrix notation, as follows:

$$\begin{bmatrix} \gamma_1 \\ \gamma_2 \\ \gamma_3 \\ \gamma_4 \end{bmatrix} = \begin{bmatrix} w_1 & w_2 & 0 & 0 & 0 & 0 & 0 & 0 \\ 0 & 0 & w_3 & w_4 & 0 & 0 & 0 & 0 \\ 0 & 0 & 0 & 0 & w_5 & w_6 & 0 & 0 \\ 0 & 0 & 0 & 0 & 0 & 0 & w_7 & w_8 \end{bmatrix} \begin{bmatrix} z_1 \\ z_2 \\ z_3 \\ z_4 \\ z_5 \\ z_6 \\ z_7 \\ z_8 \end{bmatrix}. \tag{2.10}$$

This can be rewritten compactly as:

$$\gamma = \mathbf{W'z}. \tag{2.11}$$

Equation 2.11 denotes the general form of the weighted relation model in generalized structured component analysis. In the weighted relation model, **W** is a J by P matrix of weights assigned to J indicators, which in turn lead to P latent variables. As will be discussed in Section 2.5.3, the weighted relation model makes a distinction in model specification between generalized structured component analysis and partial least squares path modeling. Moreover, Equation 2.11 indicates that in generalized structured component analysis, latent variable scores are calculated uniquely as component scores. When an indicator is formative, it entails no loading in the measurement model, while its weight denotes how the indicator contributes to forming the corresponding latent variable. Note that throughout the book, we will not display weights (w's) in path diagrams that involve reflective indicators in order to make the diagrams concise, unless displaying the weights is necessary for clarification.

In sum, generalized structured component analysis involves three submodels taking the general forms as follows:

Measurement model: $\mathbf{z} = \mathbf{C'}\gamma + \varepsilon$
Structural model: $\gamma = \mathbf{B'}\gamma + \zeta$
Weighted relation model: $\gamma = \mathbf{W'z}$

where

        $\mathbf{z}$ is a J by 1 vector of indicators
        γ is a P by 1 vector of latent variables

C is a *P* by *J* matrix of loadings
B is a *P* by *P* matrix of path coefficients
W is a *J* by *P* matrix of component weights
ε is a *J* by 1 vector of the residuals of indicators
ζ is a *P* by 1 vector of the residuals of latent variables.

2.1.4 The Generalized Structured Component Analysis Model

In generalized structured component analysis, the three submodels are combined into a single, general model. The single model is called the generalized structured component analysis model. Specifically, the three submodels are integrated into a single model in matrix notation, as follows:

$$
\begin{bmatrix} z \\ \gamma \end{bmatrix} = \begin{bmatrix} C' \\ B' \end{bmatrix} \gamma + \begin{bmatrix} \varepsilon \\ \zeta \end{bmatrix}
$$

$$
\begin{bmatrix} z \\ W'z \end{bmatrix} = \begin{bmatrix} C' \\ B' \end{bmatrix} W'z + \begin{bmatrix} \varepsilon \\ \zeta \end{bmatrix}
$$

$$
\begin{bmatrix} I \\ W' \end{bmatrix} z = \begin{bmatrix} C' \\ B' \end{bmatrix} W'z + \begin{bmatrix} \varepsilon \\ \zeta \end{bmatrix} \tag{2.12}
$$

where

I is the identity or unit matrix of size *J*, $\begin{bmatrix} I \\ W' \end{bmatrix}$ is a *T* by *J* matrix, $\begin{bmatrix} C' \\ B' \end{bmatrix}$ is a *T* by *P* matrix, and $\begin{bmatrix} \varepsilon \\ \zeta \end{bmatrix}$ is a *T* by 1 vector, and $T = J + P$.

Let

$$
V = \begin{bmatrix} I \\ W' \end{bmatrix}, \quad A = \begin{bmatrix} C' \\ B' \end{bmatrix}, \quad \text{and} \quad e = \begin{bmatrix} \varepsilon \\ \zeta \end{bmatrix}.
$$

Then, Equation 2.12 is rewritten in matrix notation as follows:

$$
V'z = A'W'z + e. \tag{2.13}
$$

This is the general form of the generalized structured component analysis model. In Equation 2.13, the **V'z** vector consists of all indicators and latent variables, the **A** matrix includes all loadings and path coefficients, and the **e** vector has all residuals.

For example, the prototype model in Figure 2.1 is expressed by Equation 2.13, as follows.

$$
\begin{bmatrix}
1 & 0 & 0 & 0 & 0 & 0 & 0 & 0 \\
0 & 1 & 0 & 0 & 0 & 0 & 0 & 0 \\
0 & 0 & 1 & 0 & 0 & 0 & 0 & 0 \\
0 & 0 & 0 & 1 & 0 & 0 & 0 & 0 \\
0 & 0 & 0 & 0 & 1 & 0 & 0 & 0 \\
0 & 0 & 0 & 0 & 0 & 1 & 0 & 0 \\
0 & 0 & 0 & 0 & 0 & 0 & 1 & 0 \\
0 & 0 & 0 & 0 & 0 & 0 & 0 & 1 \\
w_1 & w_2 & 0 & 0 & 0 & 0 & 0 & 0 \\
0 & 0 & w_3 & w_4 & 0 & 0 & 0 & 0 \\
0 & 0 & 0 & 0 & w_5 & w_6 & 0 & 0 \\
0 & 0 & 0 & 0 & 0 & 0 & w_7 & w_8
\end{bmatrix}
\begin{bmatrix}
z_1 \\ z_2 \\ z_3 \\ z_4 \\ z_5 \\ z_6 \\ z_7 \\ z_8
\end{bmatrix}
$$

$$
=
\begin{bmatrix}
c_1 & 0 & 0 & 0 \\
c_2 & 0 & 0 & 0 \\
0 & c_3 & 0 & 0 \\
0 & c_4 & 0 & 0 \\
0 & 0 & c_5 & 0 \\
0 & 0 & c_6 & 0 \\
0 & 0 & 0 & c_7 \\
0 & 0 & 0 & c_8 \\
0 & 0 & 0 & 0 \\
0 & 0 & 0 & 0 \\
b_1 & b_3 & 0 & 0 \\
b_2 & b_4 & 0 & 0
\end{bmatrix}
\begin{bmatrix}
w_1 & w_2 & 0 & 0 & 0 & 0 & 0 & 0 \\
0 & 0 & w_3 & w_4 & 0 & 0 & 0 & 0 \\
0 & 0 & 0 & 0 & w_5 & w_6 & 0 & 0 \\
0 & 0 & 0 & 0 & 0 & 0 & w_7 & w_8
\end{bmatrix}
\begin{bmatrix}
z_1 \\ z_2 \\ z_3 \\ z_4 \\ z_5 \\ z_6 \\ z_7 \\ z_8
\end{bmatrix}
+
\begin{bmatrix}
\varepsilon_1 \\ \varepsilon_2 \\ \varepsilon_3 \\ \varepsilon_4 \\ \varepsilon_5 \\ \varepsilon_6 \\ \varepsilon_7 \\ \varepsilon_8 \\ \gamma_1 \\ \gamma_2 \\ \zeta_3 \\ \zeta_4
\end{bmatrix}
$$

$$\mathbf{V'z = A'W'z + e}. \tag{2.14}$$

Although it is identical in form, the generalized structured component analysis model (Equation 2.13) is slightly different from the original model proposed by Hwang and Takane (2004a). In the original exposition, **V** was specified to include only endogenous indicators and endogenous latent variables, whereas **W** was to comprise weights associated with latent variables that influence other variables only. On the other hand, in Equation 2.13, **V** is specified to include all indicators and latent variables, so that it always remains a *T* by *J* matrix, while **W** is to include all latent variables, so that it is always a *J* by *P* matrix. The generalized structured component analysis model (Equation 2.13) was introduced by Hwang, DeSarbo, and Takane (2007).

Equation 2.13 helps to simplify model specification and parameter estimation because it is not necessary to distinguish which variables should be included in **W** and **V** based on a given model specification, as described by Hwang and Takane (2004a). As long as the same set of indicators and latent variables is used in model specification, the structures of **W** and **V** remain the same in Equation 2.13. In addition, this formulation makes it easier to understand the relationship between the generalized structured component analysis model and the reticular action model (McArdle and McDonald 1984) in factor-based structural equation modeling or covariance structure analysis. Specifically, Equation 2.13 can be re-expressed as:

$$\mathbf{V'z = A'W'z + e}$$

$$\mathbf{V'z} = \begin{bmatrix} 0 & \mathbf{C'} \\ 0 & \mathbf{B'} \end{bmatrix} \begin{bmatrix} \mathbf{I} \\ \mathbf{W'} \end{bmatrix} \mathbf{z + e}$$

$$\mathbf{V'z = T'V'z + e}$$

$$\mathbf{u = T'u + e,} \tag{2.15}$$

where $\mathbf{u = V'z}$, and

$$\mathbf{T} = \begin{bmatrix} 0 & \mathbf{C'} \\ 0 & \mathbf{B'} \end{bmatrix}.$$

This expression shows that the generalized structured component analysis model is essentially of the same form as the reticular action model. The reticular action model is mathematically the most compact specification amongst various formulations of factor-based structural equation modeling, including the LISREL (Jöreskog 1970, 1973) and the Bentler–Weeks (Bentler and Weeks 1980) models. With respect to the reticular action model, the only difference in model specification is that generalized structured component analysis defines latent variables as components, as shown in Equation 2.11 (i.e., $\gamma = \mathbf{W'z}$).

2.2 Estimation of Model Parameters

The unknown parameters of the generalized structured component analysis model include weights (*w*'s) and path coefficients (*b*'s). In addition, loadings (*c*'s) will be unknown parameters when there are reflective indicators in the model. This indicates that all possible unknown parameters are included in

W and **A**. There exist no parameters associated with the residual terms in **e**, such as their means, variances, or covariances, because no assumptions on the residuals are made in the generalized structured component analysis model.

Let z_i denote a J by 1 vector of indicators measured on a single observation of a sample of N observations ($i = 1, \ldots, N$). The data are typically variable-wise standardized (i.e., each indicator has zero mean and unit variance over N observations). In generalized structured component analysis, the unknown parameters in **W** and **A** are estimated such that the sum of squares of all residuals (e_i) over N observations is as small as possible. This is equivalent to minimizing the following least-squares criterion:

$$\phi = \sum_{i=1}^{N} \mathbf{e}_i{}'\mathbf{e}_i = \sum_{i=1}^{N} (\mathbf{V}'\mathbf{z}_i - \mathbf{A}'\mathbf{W}'\mathbf{z}_i)'(\mathbf{V}'\mathbf{z}_i - \mathbf{A}'\mathbf{W}'\mathbf{z}_i), \tag{2.16}$$

with respect to **W** and **A**. Let **Z** denote an N by J matrix of stacking each observation one below another (i.e., $\mathbf{Z} = [\mathbf{z}_1, \ldots, \mathbf{z}_N]'$). Then, Equation 2.16 can be re-expressed more concisely without summation over N observations, as follows:

$$\phi = SS(\mathbf{ZV} - \mathbf{ZWA}) = SS(\mathbf{\Psi} - \mathbf{\Gamma A}), \tag{2.17}$$

where SS stands for sum of squares, that is, $SS(\mathbf{X}) = \text{trace}(\mathbf{X}'\mathbf{X})$ for any matrix **X**, $\mathbf{\Psi} = \mathbf{ZV}$ denotes an N by T matrix of all indicators and latent variables, and $\mathbf{\Gamma} = \mathbf{ZW}$ denotes an N by P matrix of latent variables. While minimizing the least-squares criterion, each latent variable is constrained to be standardized

$$\sum_{i=1}^{N} \gamma_{ip}^2 = N,$$

so that the scales of latent variables are the same as those of indicators. The standardization of both indicators and latent variables leads to that the least-squares estimates of parameters are standardized ones. The consistency of scaling in indicators and latent variables is of importance in generalized structured component analysis, because if they are scaled differently, the estimates of weights and loadings, which relate indicators and latent variables, cannot be standardized estimates. As Hwang et al. (2010b) explained, in the previous studies with respect to generalized structured component analysis, each indicator was standardized

$$\sum_{i=1}^{N} z_{ij}^2 = N,$$

whereas each latent variable was normalized such that its length is equal to one

$$\sum_{i=1}^{N} \gamma_{ip}^2 = 1.$$

In fact, such scaling inconsistency was dealt with by rescaling the estimates of weights and loadings upon convergence (e.g., by multiplying the weight estimates by $\sqrt{N}$ and dividing the loading estimates by $\sqrt{N}$). However, it is required to maintain the scaling consistency of indicators and latent variables while minimizing the criterion (also see Henseler 2012). By convention, we assume that both indicators and latent variables are standardized. However, from a computational standpoint, it is perfectly acceptable to normalize both indicators and latent variables. What is important is to make the scales of indicators and latent variables consistent. The standardized scores of latent variables can always be obtained by multiplying the normalized scores by $\sqrt{N}$.

The least-squares criterion (Equation 2.16) appears similar to that for multivariate linear regression in the sense that a set of (dependent) variables in Ψ is affected by a set of (predictor) variables in Γ. In multivariate linear regression where all regression coefficients are typically free parameters to be estimated, the least squares estimates of regression coefficients can be obtained in closed form or noniteratively. On the other hand, as illustrated by the prototype models in Figures 2.1 and 2.2, W and A in Equation 2.17 consist of fixed values, such as zeros, which should not be estimated. This makes it difficult to estimate the parameters of W and A in closed form. Instead, an iterative estimation procedure, called an alternating least-squares algorithm (de Leeuw, Young, and Takane 1976; Yates 1933), was developed to minimize the criterion (Hwang and Takane 2004a; Hwang et al. 2007a).

The alternating least-squares algorithm is a general, iterative estimation procedure that divides the entire set of parameters into subsets and *alternately* updates each subset optimally in the *least-squares* manner. More specifically, the algorithm begins by choosing arbitrary initial values for the entire set of parameters. The optimal values of a subset of parameters are obtained by minimizing a least-squares criterion, while the other subsets are temporarily considered constant. Subsequently, another subset of parameters is updated optimally by minimizing the same least-squares criterion, with the remaining subsets fixed. Within each iteration, these steps are carried out until all subsets of parameters are updated. The criterion value based on the updated values of parameter estimates in the current iteration will never be greater than that obtained in the previous iteration, because the updated values cannot be less optimal than their previous values (de Leeuw, Young, and Takane 1976). This procedure can be repeated until the decrease of the criterion value become smaller than a predetermined threshold value (e.g., 0.0001).

In generalized structured component analysis, we divide the entire set of parameters into two subsets (**W** and **A**). We assign arbitrary initial values to **W** and **A**. Then, the following two steps are carried out per iteration. In the first step, we consider **A** to be fixed temporarily and obtain the least-squares estimates of **W** by minimizing the least-squares criterion (Equation 2.17) only with respect to **W**. In the second step, we reverse the previous procedure. We consider **W** constant, and obtain the least-squares estimates of **A** by minimizing the same criterion only with respect to **A**. We provide a detailed description of the alternating least-squares algorithm for generalized structured component analysis in Appendix 2.1.

As stated above, the alternating least-squares algorithm monotonically decreases the value of the criterion (Equation 2.17), which is also bounded from below. This algorithm is therefore convergent. Nevertheless, it is not ensured that the convergence point is the global minimum. This so-called convergence to local minimum issue may be addressed in two ways (e.g., ten Berge 1993, p. 47). One way is to choose good initial values for **W** and **A**, so that the criterion value is likely to start near the global minimum and the likelihood of convergence to the global minimum increases. Hwang and Takane (2004a) suggested applying a constrained component analysis (Takane, Kiers, and de Leeuw 1995), which can be considered a component-based counterpart of confirmatory factor analysis (Kiers, Takane, and ten Berge 1996), to fit a specified measurement model to the data. The resultant component weights can be used as initial values for **W**. The initial values for **A** can then be obtained as the least-squares estimates given **W**, as presented in Appendix 2.1.

Another way is to repeat the estimation procedure with several sets of random initial values for **W** and **A** successively. The alternating least-squares algorithm will result in as many values of the criterion as sets of random initial values used, after convergence. The resultant criterion values are compared and only the solution associated with the smallest value is chosen.

Our finding accumulated to date is that the algorithm seems to converge quickly and to be less vulnerable to local minima with random initial values used. However, it would be ideal to carry out systematic investigations into the behavior of the alternating least-squares algorithm for generalized structured component analysis.

Minimization of the least-squares criterion (Equation 2.17) does not require any distributional assumption such as multivariate normality of indicators. Thus, generalized structured component analysis is a distribution-free or nonparametric approach to structural equation modeling. This will be of use when researchers are not confident about the distribution by which the data are generated. However, as a trade-off of no reliance on a distributional assumption in parameter estimation, generalized structured component analysis cannot estimate the standard errors of parameter estimates based on asymptotic (normal-theory) approximations. Instead, it employs the (nonparametric) bootstrap method (Efron 1979, 1982) to obtain the standard errors of parameter estimates without recourse to a distributional assumption. The bootstrap method is a general tool for statistical inference based on an empirical (data-driven) sampling distribution for an estimator or statistic. In this method, a sample of N observations is treated as the population, from which a large number of resamples of size N are randomly drawn with replacement. An estimator or statistic is calculated from each of resamples and the relative frequency distribution of the estimator over the resamples is considered an empirical approximation of its sampling distribution. The bootstrapped standard errors or confidence intervals derived from the empirical sampling distribution can be used for testing the statistical significance of the parameter estimates or other statistics in generalized structured component analysis. For example, a bootstrapped t statistic can be calculated by dividing a parameter estimate by its bootstrapped standard error. If the bootstrapped t statistic of a parameter estimate is equal to or greater than 1.96 in absolute value, the parameter estimate may be considered statistically significant at the 0.05 level under the assumption that the empirical sampling distribution of the parameter estimate is approximately normal. In addition, the 95% confidence interval (CI) of a parameter estimate can be obtained nonparametrically from the empirical sampling distribution of the estimate.

In the above estimation procedure, the unknown parameters are estimated based on N individual observations of indicators. However, it is also possible to utilize the correlation or covariance matrix of indicators to estimate the parameters. This is derived from the fact that the least-squares criterion (Equation 2.17) can be rewritten as:

$$\phi = SS(\mathbf{Z}(\mathbf{V} - \mathbf{W}\mathbf{A}))$$

$$= \text{trace}((\mathbf{V} - \mathbf{W}\mathbf{A})'\mathbf{Z}'\mathbf{Z}(\mathbf{V} - \mathbf{W}\mathbf{A}))$$

$$= \text{trace}((\mathbf{V} - \mathbf{W}\mathbf{A})'\mathbf{M}(\mathbf{V} - \mathbf{W}\mathbf{A})), \qquad (2.18)$$

where $\mathbf{M} = \mathbf{Z}'\mathbf{Z}$. In generalized structured component analysis, indicators are assumed to be standardized. Consequently, $\mathbf{M}$ in Equation 2.18 can be obtained from the correlation or covariance matrix of indicators as follows: $\mathbf{M} = N\mathbf{R} = N\mathbf{DSD}$, where $\mathbf{R}$ and $\mathbf{S}$ are the correlation and covariance matrices of the indicators, respectively, and $\mathbf{D}$ is a diagonal matrix consisting of the inverses of the standard deviations for indicators as elements. When the number of observations (N) is much greater than that of indicators (J), using the correlation or covariance matrix will be more efficient computationally. Refer to Appendix 2.1 for more technical details.

Once the estimates of weights are obtained, individual latent variable scores are uniquely determined by postmultiplying individual observations by the weight estimates, as shown in the weighted relation model in

Equation 2.11. As stated earlier, the latent variable scores are standard scores with zero means and unit variances. In some situations, however, researchers may be interested in calculating the means of latent variables in the same scales of original indicators. In generalized structured component analysis, the means of latent variables may be computed as follows. Let $\bar{z}$ denote a J by 1 vector of the means of original indicators. Let $\mathbf{W}^{*\prime}$ denote a P by J matrix of the weight estimates, whose elements are divided by the standard deviations of the corresponding original indicators (i.e., $\mathbf{W}^{*\prime} = \mathbf{W'D}$). Let $\bar{\gamma}$ denote a P by 1 vector of the means of latent variables. Then, $\bar{\gamma}$ is calculated by $\bar{\gamma} = \mathbf{W}^{*\prime}\bar{z}$. This computation stems from the fact that standardized latent variables are obtained as follows.

$$\gamma = \mathbf{W'z} = \mathbf{W'D}(\mathbf{z}^* - \bar{z}). \tag{2.19}$$

where $\mathbf{z}^*$ is a vector of original unstandardized indicators. In partial least squares path modeling, the means of latent variables are computed in the same manner (e.g., Tenenhaus et al. 2005).

Another popular way of calculating the means of latent variables adopted in partial least squares path modeling is to rescale $\mathbf{W}^{*\prime}$ such that the sum of weight estimates for each latent variable is equal to one and then to premultiply $\bar{z}$ by the rescaled $\mathbf{W}^{*\prime}$ (Fornell 1992). This approach can be appropriate when all indicators are measured in the same scale and all their weight estimates are positive.

When raw observations are used, the individual scores of latent variables can be simply calculated by postmultiplying the individual observations by weight estimates.

2.3 Model Evaluation

In generalized structured component analysis, a variety of statistical measures of model fit can be used for the assessment of a model. We can divide these measures into overall and local measures.

2.3.1 Overall Model Fit Measures

An overall model fit measure represents a single value that summarizes the discrepancies between the model and the data. This measure can be of use in evaluating how well a model fits to the data as a whole and in comparing a model to alternative models (Bollen 1989, p. 256). Generalized structured component analysis estimates model parameters by consistently minimizing a single least-squares criterion. This enables the provision of measures of overall model fit. Specifically, generalized structured component analysis affords an overall measure of fit, called FIT, given by:

$$\text{FIT} = 1 - [\text{SS}(\mathbf{ZV} - \mathbf{ZWA}) / \text{SS}(\mathbf{ZV})]. \qquad (2.20)$$

FIT shows the proportion of the total variance of all indicators and latent variables explained by a given particular model specification. The values of FIT range from 0 to 1. The larger this value, the more variance in the variables is accounted for by the model specification. The characteristic and interpretation of FIT are comparable to the coefficient of determination or R-squared in linear regression. Actually, this fit measure is equivalent to the average of the R-squared values for all indicators and latent variables, that is,

$$\text{FIT} = \frac{1}{T} \sum_{t=1}^{T} R_t^2,$$

where R_t^2 is the R-squared value of each indicator or latent variable (Henseler 2012).

There is no clear cut-off value of FIT, which is indicative of a good fit. For example, if a model provides FIT = 0.58, it is uncertain that this value shows whether the model fits to the data acceptably well. We can merely conclude that the model explains about 58% of the total variance of all variables. Nevertheless, it is feasible to make a comparison between different models based on FIT. Similar to bootstrapping R-squared in linear regression (Ohtani 2000), we can compute the bootstrapped standard errors or confidence intervals of the difference in FIT between two models so as to examine whether there is a statistically significant difference in the FIT values. This procedure can be regarded as a nonparametric version of the paired t test for two groups of FIT values. However, it may not be sensible to apply the procedure for comparing two models that involve different sets of indicators and latent variables, that is, when the constituents of $\mathbf{\Psi}$ in Equation 2.16 differ. This situation is similar to comparing the R-squared values of two linear regression models whose dependent variables are different.

As with R-squared in linear regression, FIT is also affected by model complexity, that is, the more the parameters the larger the value of FIT. Thus, another index of fit was developed that takes this contingency into account. It is referred to as Adjusted FIT (AFIT) (Hwang, DeSarbo, and Takane 2007), given by:

$$\text{AFIT} = 1 - (1 - \text{FIT})\frac{d_0}{d_1}, \qquad (2.21)$$

where $d_0 = NJ$ is the degrees of freedom for the null model ($\mathbf{W} = 0$ and $\mathbf{A} = 0$) and $d_1 = NJ - \delta$ is the degrees of freedom for the model being tested, where

δ is the number of free parameters. This measure is comparable to Adjusted
R-squared in linear regression. Like Adjusted R-squared, it cannot be interpreted in the same way as FIT (i.e., the proportion of the total variance
explained). The model that maximizes AFIT can be regarded as the most
appropriate among competing models.

Both FIT and AFIT take into account the variance of the data explained
by a model specification. Two additional measures of overall model fit were
developed to reflect the closeness between the sample covariances and the
covariances reproduced by the model parameter estimates upon convergence: (unweighted least-squares) GFI (Jöreskog and Sörbom 1986) and standardized root mean square residual (SRMR) (Hwang 2008).

From the generalized structured component analysis model (Equation 2.13),
we have

$$(\mathbf{V}' - \mathbf{A}'\mathbf{W}')\mathbf{z} = \mathbf{Q}'\mathbf{z} = \mathbf{e}$$

$$\mathbf{z} = (\mathbf{Q}\mathbf{Q}')^{-1}\mathbf{Q}\mathbf{e} = \mathbf{\Omega}\mathbf{e}$$

where $\mathbf{Q}' = \mathbf{V}' - \mathbf{A}'\mathbf{W}'$, and $\mathbf{\Omega} = (\mathbf{Q}\mathbf{Q}')^{-1}\mathbf{Q}$. The reproduced covariance matrix
(also, called the implied covariance matrix) of $\mathbf{z}$, denoted by $\mathbf{\Sigma}$, is then given by:

$$\mathbf{\Sigma} = \mathbf{\Omega}\mathbf{\Xi}\mathbf{\Omega}',$$

where $\mathbf{\Xi} = E(\mathbf{e}\mathbf{e}')$ is the covariance matrix of the residuals (Hwang and
Takane 2004a). Let $\mathbf{S}$ and $\hat{\mathbf{\Sigma}}$ denote the sample covariance matrix and the
reproduced covariance matrix evaluated at the least-squares estimates of
parameters upon convergence. Let s_{jq} and $\hat{\sigma}_{jq}$ denote the jqth element in $\mathbf{S}$
and $\hat{\mathbf{\Sigma}}$, respectively. Then, the GFI and SRMR are calculated by:

$$\text{GFI} = 1 - \frac{\text{trace}\left[(\mathbf{S} - \hat{\mathbf{\Sigma}})^2\right]}{\text{trace}(\mathbf{S}^2)}$$

$$\text{SRMR} = \sqrt{2\sum_{j=1}^{J}\sum_{q=1}^{j}\frac{\left[(s_{jq} - \hat{\sigma}_{jq})/(s_{jj}s_{qq})\right]^2}{J(J+1)}}. \tag{2.22}$$

In general, the GFI values close to 1 and the SRMR values close to 0 may be
taken as indicative of good fit. In factor-based structural equation modeling,
some practical recommendations are available as to how to use these measures to decide whether a model has an acceptable level of fit. For example,
a value of SRMR less than 0.08 may be regarded as a good fit (e.g., Hu and
Bentler 1999), whereas a value of GFI higher than 0.90 may be considered

a good fit (e.g., McDonald and Ho 2002), although GFI seems to become less popular in practice (Sharma et al. 2005). On the other hand, it is not yet known whether the same cut-off values of GFI and SRMR can also be used in generalized structured component analysis. A future study is warranted to investigate the behavior of the two measures in generalized structured component analysis.

Furthermore, it is important to examine the correlation residual matrix between the sample correlations and the reproduced correlations evaluated upon convergence (e.g., Bollen 1989, p. 258; McDonald and Ho 2002). Both sample correlation and reproduced correlation range from −1 to +1. Thus, the correlation residual can theoretically range from −2 to +2. Any correlation residual close to these limits can be indicative of a seriously bad fit (Bollen 1989, p. 258). On the other hand, when all correlation residuals are close to zeros, this can be indicative of an acceptable fit.

2.3.2 Local Model Fit Measures

Although an overall fit measure serves as a useful summary statistic of model fit, it may be of little help in identifying where bad/good fit of a model comes from and where it can be improved. Thus, it is also crucial to examine local model fit measures that focus on individual parameter estimates and other characteristics associated with the measurement and structural models (Bollen 1989, p. 281).

In generalized structured component analysis, separate model fit measures for the measurement and structural models are available. The calculation of these measures is derived from the fact that the optimization criterion (Equation 2.17) can be expressed as the sum of two terms as follows:

$$\phi = SS(\mathbf{ZV} - \mathbf{ZWA})$$

$$= SS(\mathbf{Z} - \mathbf{ZWC}) + SS(\mathbf{ZW} - \mathbf{ZWB}). \tag{2.23}$$

The first term of Equation 2.23 is the sum of squared residuals in the measurement model, whereas the second is the sum of squared residuals in the structural model (Henseler 2012; Tenenhaus 2008a,b). From Equation 2.23, we can calculate the following two measures:

$$FIT_M = 1 - \frac{SS(\mathbf{Z} - \mathbf{ZWC})}{SS(\mathbf{Z})} = 1 - \frac{SS(\mathbf{Z} - \mathbf{ZWC})}{NJ} \tag{2.24}$$

$$FIT_S = 1 - \frac{SS(\mathbf{ZW} - \mathbf{ZWB})}{SS(\mathbf{ZW})} = 1 - \frac{SS(\mathbf{ZW} - \mathbf{ZWB})}{NP}. \tag{2.25}$$

The FIT_M signifies how much the variance of indicators is explained by a measurement model. The FIT_S indicates how much the variance of latent variables is accounted for by a structural model. Both the measures range from 0 to 1 and can be interpreted in a manner similar to the FIT. In situations where no indicators are reflective, the FIT_M will be equal to zero because $C = 0$.

As will be discussed in the subsequent section, partial least squares path modeling is incapable of providing a measure of overall model fit because it does not have a single optimization criterion stemming from a unified formulation of submodels (e.g., Coolen and de Leeuw 1987; McDonald 1996; Henseler, Ringle, and Sinkovics 2009). As such, researchers in the area of partial least squares path modeling have made a great deal of effort to apply and develop a variety of local fit measures for evaluation of the measurement and structural models (e.g., Chin 1998; Esposito Vinzi, Trinchera, and Amato 2010; Henseler, Ringle, and Sinkovics 2009; Tenenhaus et al. 2005). It will also be fruitful to adopt such local fit measures for systematically investigating the quality of the measurement and structural models in generalized structured component analysis.

For the measurement model, the reliability and validity of latent variables should be evaluated. In the reflective measurement model, a block of indicators associated with a latent variable should be homogenous, so they are assumed to measure the same underlying construct. We can use two reliability measures for checking such internal consistency. One measure is Cronbach's alpha (Cronbach 1951) and the other is Dillon–Goldstein's rho or the composite reliability (Werts, Linn, and Jöreskog 1974). Let α_p and ρ_p denote the values of Cronbach's alpha and Dillon–Goldstein's rho, respectively, for the pth latent variable ($p = 1, ..., P$). They are given by:

$$\alpha_p = \frac{J_p \sum_{j \neq j'}^{J_p} \mathrm{corr}(z_{pj}, z_{pj'})}{K_p + (J_p - 1) \sum_{j \neq j'}^{J_p} \mathrm{corr}(z_{pj}, z_{pj'})}$$

$$\rho_p = \frac{\left(\sum_{j=1}^{J_p} c_{pj} \right)^2}{\left(\sum_{j=1}^{J_p} c_{pj} \right)^2 + \sum_{j=1}^{J_p} (1 - c_{pj}^2)}$$

where z_{pj} is an indicator linked to the pth latent variable, c_{pj} is the loading value for z_{pj}, J_p is the number of indicators for the pth latent variable, and corr(z_{pj}, z_{pj}) is the correlation between two different indicators associated with the pth latent variable, and $K_P = J_P(J_P - 1)/2$ is the number of nonredundant correlation coefficients of the J_p indicators. A value of these measures above 0.7 may be indicative of an acceptable level of reliability in early stages of research (Nunnally 1978, p. 245).

As an index of the reliability of each indicator, we can check for its loading value that is equivalent to the correlation between the indicator and its latent variable, as well as its squared multiple correlation that is the squared loading or R-squared for the indicator. For example, it is recommended that each loading value be greater than 0.7 in absolute value (i.e., its squared multiple correlation ≈ 0.50) (Henseler et al. 2009). In addition, each loading estimate should be statistically significantly different from zero.

We may examine the convergent validity of a latent variable by looking at the communality or average variance extracted (AVE) (Fornell and Larcker 1981). Convergent validity is closely related to unidimensionality of a block of indicators, so that a single and identical construct underlies the block. The communality is calculated as the average of squared multiple correlations of a block of indicators for each latent variable. This shows how much the variance of indicators is explained by their corresponding latent variable on average. A value of communality below 0.5 may be indicative of a lack of convergent validity (Götz, Liehr-Gobbers, and Krafft 2010; Henseler et al. 2009). Another way of checking unidimensionality is to apply a principal component analysis to investigate the dimensionality of a block of indicators. According to the Kaiser criterion (Kaiser 1960), if only the largest eigenvalue is greater than one, this can be indicative of unidimensionality for a block of indicators. The discriminant validity of a latent variable may be examined by using the Fornell–Larcker criterion (Fornell and Larcker 1981). That is, the AVE value of a latent variable should be greater than any of its squared correlations with the other latent variables. This is based on the assumption that each latent variable should share more variance with its own block of indicators than with other latent variable underlying a different block of indicators.

A block of formative indicators for a latent variable is assumed to be error free and does not need to be homogenous and unidimensional (Esposito Vinzi, Trinchera, and Amato 2010). Accordingly, it is irrelevant to consider classical test theory and check internal consistency and validity of latent variables formed by indicators (Bollen 1989; Bagozzi 1994). Instead, we may examine the statistical significance of each weight estimate to evaluate whether an indicator makes a statistically significant contribution to forming a latent variable. We can estimate the standard errors or confidence intervals of weight estimates via bootstrapping. An indicator whose weight estimate is statistically insignificant may be removed from the construction of a latent variable. In addition, it is useful to check the presence of multicollinearity or high correlations among formative indicators for each latent variable. Multicollinearity is likely to produce inaccurate parameter estimates far from true parameters and large standard errors, thus often leading to inference errors. Various heuristics and procedures can be used for checking potential multicollinearity inherent to each block of indicators (Belsley, Kuh, and Welsch 1980). We will discuss some of them in Chapter 8.

To gauge the adequacy of the structural model, we can examine the magnitude and statistical significance of the estimates of path coefficients. The statistical significance of individual path coefficient estimates can be tested through the use of bootstrapping. Moreover, the R-squared value of an endogenous latent variable is better to be sufficiently large. For example, Chin (1998) considered the R-squared values of 0.67, 0.33, and 0.19 to be substantial, moderate, and weak, respectively. Furthermore, multicollinearity among latent variables can be examined to see whether some latent variables are highly correlated to each other, resulting in counterintuitive or unreliable estimates of path coefficients.

Besides statistical measures of model fit discussed above, nonstatistical considerations such as model interpretability often play a role in model evaluation, although they are usually more difficult to justify because they can be subjective in nature (Browne and Cudeck 1993).

2.4 Example

In this section, we apply generalized structured component analysis for the analysis of a real dataset collected from a customer satisfaction database. Customer satisfaction is an issue of vital importance in today's competitive market structures. As a result, much research has been directed at developing reliable and succinct indices of customer satisfaction, which can be used at the national, industry, and company levels (see Fornell 1992; Fornell et al. 1996). A well-known, national index of customer satisfaction in the United States is the American Customer Satisfaction Index (ACSI) (Fornell et al. 1996). Since 1994, the ACSI has been used to generate scores on four levels: a national score, seven economic sector scores, 35 specific industry scores, and 200 company/agency scores within those industries (for additional information, see the ACSI website, http://www.theacsi. org). Moreover, the ACSI has been pilot tested in many countries including New Zealand, Taiwan, Austria, and South Korea (Fornell et al. 1996; Johnson et al. 2001).

The ACSI model is a structural equation model that specified customer satisfaction as a latent variable of major interest and hypothesized the interrelationships among antecedent and consequent latent variables of customer satisfaction. This model is depicted in Figure 2.3.

As shown in Figure 2.3, the ACSI model includes 14 indicators: z_1 = customer expectations about overall quality, z_2 = customer expectations about reliability, z_3 = customer expectations about customization, z_4 = overall quality, z_5 = reliability, z_6 = customization, z_7 = price given quality, z_8 = quality given price, z_9 = overall customer satisfaction, z_{10} = confirmation of expectations, z_{11} = distance to ideal product or service, z_{12} = formal or informal complaint behavior, z_{13} = repurchase intention, and z_{14} = price tolerance. The measures and scales of these indicators are presented in the study of Fornell et al. (1996).

The ACSI model also entails six latent variables that underlie the 14 indicators, as follows: γ_1 = customer expectations (CE), γ_2 = perceived quality (PQ), γ_3 = perceived value (PV), γ_4 = customer satisfaction (CS), γ_5 = customer complaints (CC), and γ_6 = customer loyalty (CL). Thus, in the ACSI, J and P are equal to 14 and 6, respectively.

As depicted in Figure 2.3, all indicators in the ACSI are considered reflective. Thus, the measurement model is given as:

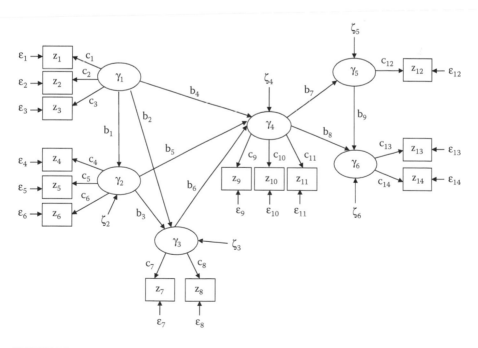

FIGURE 2.3
A path diagram of the American customer satisfaction index model.

$$z_1 = \gamma_1 c_1 + \varepsilon_1$$
$$z_2 = \gamma_1 c_2 + \varepsilon_2$$
$$z_3 = \gamma_1 c_3 + \varepsilon_3$$
$$z_4 = \gamma_2 c_4 + \varepsilon_4$$
$$z_5 = \gamma_2 c_5 + \varepsilon_5$$
$$z_6 = \gamma_2 c_6 + \varepsilon_6$$
$$z_7 = \gamma_3 c_7 + \varepsilon_7$$
$$z_8 = \gamma_3 c_8 + \varepsilon_8$$
$$z_9 = \gamma_4 c_9 + \varepsilon_9$$
$$z_{10} = \gamma_4 c_{10} + \varepsilon_{10}$$
$$z_{11} = \gamma_4 c_{11} + \varepsilon_{11}$$
$$z_{12} = \gamma_5 c_{12} + \varepsilon_{12}$$
$$z_{13} = \gamma_6 c_{13} + \varepsilon_{13}$$
$$z_{14} = \gamma_6 c_{14} + \varepsilon_{14} .$$

(2.26)

We express the above equations in matrix notation, as follows:

$$
\begin{bmatrix} z_1 \\ z_2 \\ z_3 \\ z_4 \\ z_5 \\ z_6 \\ z_7 \\ z_8 \\ z_9 \\ z_{10} \\ z_{11} \\ z_{12} \\ z_{13} \\ z_{14} \end{bmatrix}
=
\begin{bmatrix}
c_1 & 0 & 0 & 0 & 0 & 0 \\
c_2 & 0 & 0 & 0 & 0 & 0 \\
c_3 & 0 & 0 & 0 & 0 & 0 \\
0 & c_4 & 0 & 0 & 0 & 0 \\
0 & c_5 & 0 & 0 & 0 & 0 \\
0 & c_6 & 0 & 0 & 0 & 0 \\
0 & 0 & c_7 & 0 & 0 & 0 \\
0 & 0 & c_8 & 0 & 0 & 0 \\
0 & 0 & 0 & c_9 & 0 & 0 \\
0 & 0 & 0 & c_{10} & 0 & 0 \\
0 & 0 & 0 & c_{11} & 0 & 0 \\
0 & 0 & 0 & 0 & c_{12} & 0 \\
0 & 0 & 0 & 0 & 0 & c_{13} \\
0 & 0 & 0 & 0 & 0 & c_{14}
\end{bmatrix}
\begin{bmatrix} \gamma_1 \\ \gamma_2 \\ \gamma_3 \\ \gamma_4 \\ \gamma_5 \\ \gamma_6 \end{bmatrix}
+
\begin{bmatrix} \varepsilon_1 \\ \varepsilon_2 \\ \varepsilon_3 \\ \varepsilon_4 \\ \varepsilon_5 \\ \varepsilon_6 \\ \varepsilon_7 \\ \varepsilon_8 \\ \varepsilon_9 \\ \varepsilon_{10} \\ \varepsilon_{11} \\ \varepsilon_{12} \\ \varepsilon_{13} \\ \varepsilon_{14} \end{bmatrix}
$$

$$z = C'\gamma + \varepsilon. \tag{2.27}$$

In addition, the structural model for the ACSI is as follows:

$$\gamma_2 = \gamma_1 b_1 + \zeta_2$$

$$\gamma_3 = \gamma_1 b_2 + \gamma_2 b_3 + \zeta_3$$

$$\gamma_4 = \gamma_1 b_4 + \gamma_2 b_5 + \gamma_3 b_6 + \zeta_4$$

$$\gamma_5 = \gamma_4 b_7 + \zeta_5$$

$$\gamma_6 = \gamma_4 b_8 + \gamma_5 b_9 + \zeta_6. \tag{2.28}$$

We express the above equations in matrix notation, as follows:

$$
\begin{bmatrix} \gamma_1 \\ \gamma_2 \\ \gamma_3 \\ \gamma_4 \\ \gamma_5 \\ \gamma_6 \end{bmatrix} =
\begin{bmatrix}
0 & 0 & 0 & 0 & 0 & 0 \\
b_1 & 0 & 0 & 0 & 0 & 0 \\
b_2 & b_3 & 0 & 0 & 0 & 0 \\
b_4 & b_5 & b_6 & 0 & 0 & 0 \\
0 & 0 & 0 & b_7 & 0 & 0 \\
0 & 0 & 0 & b_8 & b_9 & 0
\end{bmatrix}
\begin{bmatrix} \gamma_1 \\ \gamma_2 \\ \gamma_3 \\ \gamma_4 \\ \gamma_5 \\ \gamma_6 \end{bmatrix} +
\begin{bmatrix} \gamma_1 \\ \zeta_2 \\ \zeta_3 \\ \zeta_4 \\ \zeta_5 \\ \zeta_6 \end{bmatrix}
$$

$$\gamma = B'\gamma + \zeta. \tag{2.29}$$

Lastly, the weighted relation model for the ACSI is given as:

$$\gamma_1 = z_1 w_1 + z_2 w_2 + z_3 w_3$$

$$\gamma_2 = z_4 w_4 + z_5 w_5 + z_6 w_6$$

$$\gamma_3 = z_7 w_7 + z_8 w_8$$

$$\gamma_4 = z_9 w_9 + z_{10} w_{10} + z_{11} w_{11}$$

$$\gamma_5 = z_{12} w_{12}$$

$$\gamma_6 = z_{13} w_{13} + z_{14} w_{14}. \tag{2.30}$$

We express the above equations in matrix notation as follows:

$$
\begin{bmatrix} \gamma_1 \\ \gamma_2 \\ \gamma_3 \\ \gamma_4 \\ \gamma_5 \\ \gamma_6 \end{bmatrix}
=
\begin{bmatrix}
w_1 & w_2 & w_3 & 0 & 0 & 0 & 0 & 0 & 0 & 0 & 0 & 0 & 0 & 0 \\
0 & 0 & 0 & w_4 & w_5 & w_6 & 0 & 0 & 0 & 0 & 0 & 0 & 0 & 0 \\
0 & 0 & 0 & 0 & 0 & 0 & w_7 & w_8 & 0 & 0 & 0 & 0 & 0 & 0 \\
0 & 0 & 0 & 0 & 0 & 0 & 0 & 0 & w_9 & w_{10} & w_{11} & 0 & 0 & 0 \\
0 & 0 & 0 & 0 & 0 & 0 & 0 & 0 & 0 & 0 & 0 & w_{12} & 0 & 0 \\
0 & 0 & 0 & 0 & 0 & 0 & 0 & 0 & 0 & 0 & 0 & 0 & w_{13} & w_{14}
\end{bmatrix}
\begin{bmatrix} z_1 \\ z_2 \\ z_3 \\ z_4 \\ z_5 \\ z_6 \\ z_7 \\ z_8 \\ z_9 \\ z_{10} \\ z_{12} \\ z_{13} \\ z_{14} \end{bmatrix}
$$

$$\gamma = \mathbf{W}'\mathbf{z}. \tag{2.31}$$

We combined these three submodels into the generalized structured component analysis model. We fitted the specified model to a consumer-level ACSI dataset collected in 2002. This dataset consists of the responses of 774 consumers to the service units (e.g., police, garbage pick-up services, etc.) within the US sector of public administration. Thus, the sample size is 774. We used 100 bootstrap samples for the estimation of standard errors and confidence intervals.

Generalized structured component analysis provided that FIT = 0.67 (SE = 0.00, 95% CI = 0.66–0.69), AFIT = 0.67 (SE = 0.01, 95% CI = 0.66–0.69), GFI = 0.99 (SE = 0.00, 95% CI = 0.99–1.00), SRMR = 0.07 (SE = 0.00, 95% CI = 0.07–0.08). This indicates that the ACSI model accounted for about 67% of the total variance of all variables. Both FIT and AFIT turned out to be statistically significantly different from zero. GFI was very close to 1, whose 95% confidence interval includes 1. SRMR was quite smaller and close to 0, although it turned out to be statistically significantly different from zero. Table 2.1 presents a matrix of the sample correlations and correlation residuals of the indicators. The sample correlations are in lower triangle and correlation residuals are in upper triangle. Most of the correlation residuals were

TABLE 2.1

Sample Correlations (in Lower Triangle) and Correlation Residuals (in Upper Triangle) of Indicators in the ACSI

	z_1	z_2	z_3	z_4	z_5	z_6	z_7	z_8	z_9	z_{10}	z_{11}	z_{12}	z_{13}	z_{14}
z_1		-0.09	-0.14	0.03	-0.01	-0.10	-0.07	0.03	0.02	-0.01	0.04	0.03	0.06	0.01
z_2	0.65		-0.11	-0.00	0.01	-0.05	-0.14	0.02	-0.01	-0.07	-0.04	0.04	-0.01	-0.09
z_3	0.43	0.47		-0.10	-0.11	-0.01	-0.12	-0.05	-0.10	-0.11	-0.07	0.04	-0.05	-0.07
z_4	0.53	0.50	0.33		-0.04	-0.08	-0.16	0.06	0.03	-0.03	-0.07	-0.06	0.06	-0.04
z_5	0.49	0.51	0.33	0.84		-0.08	-0.20	0.05	0.03	-0.01	-0.02	-0.08	0.06	-0.06
z_6	0.33	0.39	0.37	0.62	0.61		-0.15	-0.02	-0.11	-0.12	-0.15	0.04	-0.04	-0.11
z_7	0.31	0.25	0.22	0.42	0.39	0.36		-0.13	-0.16	-0.16	-0.20	0.10	-0.08	-0.11
z_8	0.47	0.46	0.33	0.72	0.71	0.56	0.53		0.05	0.02	0.02	-0.04	0.03	-0.02
z_9	0.50	0.48	0.33	0.85	0.85	0.60	0.47	0.77		-0.04	-0.07	-0.02	0.02	-0.07
z_{10}	0.45	0.40	0.30	0.75	0.78	0.56	0.44	0.71	0.81		-0.06	0.04	-0.03	-0.13
z_{11}	0.51	0.43	0.35	0.73	0.77	0.54	0.41	0.72	0.78	0.75		0.02	0.03	-0.05
z_{12}	-0.17	-0.17	-0.13	-.40	-0.42	-0.25	-0.16	-0.34	-0.41	-0.34	-0.36		-0.01	0.03
z_{13}	0.37	0.30	0.22	0.58	0.58	0.41	0.32	0.49	0.62	0.54	0.61	-0.33		-0.10
z_{14}	0.31	0.21	0.19	0.47	0.45	0.33	0.28	0.43	0.51	0.43	0.51	-0.29	0.76	

z_1, Customer expectations about overall quality; z_2, Customer expectations about reliability; z_3, Customer expectations about customization; z_4, Overall quality; z_5, Reliability; z_6, Customization; z_7, Price given quality; z_8, Quality given price; z_9, Overall customer satisfaction; z_{10}, Confirmation of expectations; z_{11}, Distance to ideal product or service; z_{12}, Formal or informal complaint behavior; z_{13}, Repurchase intention; z_{14}, Price tolerance.

small (e.g., <0.1), although the correlation residuals associated with a few indicators (e.g., z_6 and z_7) were relatively large.

Moreover, generalized structured component analysis provided that $FIT_M = 0.80$ (SE = 0.01, 95% CI = 0.79–0.82) and $FIT_S = 0.38$ (SE = 0.01, 95% CI = 0.35–0.40). This indicates that the measurement model of the ACSI accounted for about 80% of the total variance of all indicators, whereas the structural model explained about 38% of the total variance of all latent variables. Table 2.2 reports other local fit measures. As measures for internal consistency, we looked at Cronbach's alpha and Dillon–Goldstein's rho for latent variables. Both values for each latent variable were greater than 0.7.

To check convergent validity, we examined the AVE or communality value of each latent variable. All AVE values were greater than 0.5. Moreover, we have checked discriminant validity of a latent variable based on the Fornell–Larcker criterion. The AVE value of each latent variable was greater than any of the squared correlations between the latent variable and the other latent variables. We also applied a principal component analysis to each block of indicators and found that only the largest eigenvalue was greater than one for every block. This served as another means of checking unidimensionality for each block of indicators.

Table 2.2 also provides the estimates of weights and loadings. All weight estimates for each block of indicators turned out to be statistically significant. This indicates that all indicators made significant contributions to building their latent variables. All loading estimates were large and greater than 0.7 in absolute value, and were statistically significant. In addition, their squared multiple correlations were large (i.e., >0.50). Thus, all latent variables seemed to be well constructed in such a way that they accounted for large portions of the variances of their indicators.

Table 2.3 presents the R-squared values of each endogenous latent variable. The R-squared values were moderate to large for all endogenous variables, except for customer complaints ($R^2 = 0.16$), indicating that customer satisfaction had a weak explanation power for customer complaints.

Table 2.4 shows the estimates of path coefficients. In general, the interpretations of the path coefficient estimates are consistent with the relationships among the latent variables hypothesized in the ACSI model. That is, *customer expectations* had statistically significant and positive influences on *perceived quality* ($b_1 = 0.58$, SE = 0.03, 95% CI = 0.52–0.64) and *perceived value* ($b_2 = 0.12$, SE = 0.03, 95% CI = 0.07–0.18). In turn, *perceived quality* had statistically significant and positive effects on *perceived value* ($b_3 = 0.65$, SE = 0.03, 95% CI = 0.60–0.71) and *customer satisfaction* ($b_5 = 0.68$, SE = 0.03, 95% CI = 0.62–0.73). *Perceived value* had a statistically significant and positive effect on *customer satisfaction* ($b_6 = 0.26$, SE = 0.03, 95% CI = 0.21–0.33). *Customer satisfaction* had a statistically significant and positive impact on *customer loyalty* ($b_8 = 0.59$, SE = 0.04, 95% CI = 0.53–0.65), while a statistically significant and negative impact on *consumer complaints* was also apparent ($b_7 = -0.41$, SE = 0.04, 95% CI = −0.48 to −0.32). *Consumer complaints* had a statistically significant and negative

TABLE 2.2

Estimates of Weights and Loadings and Their Standard Errors and 95% Confidence Intervals Obtained from Generalized Structured Component Analysis for the Original ACSI Model

Latent	Indicator	Weights				Loadings			
		Estimate	SE	$|t|$	95% CI	Estimate	SE	$|t|$	95% CI
$AVE = 0.68, a = 0.76, \rho = 0.86$									
CE	z_1	0.43	0.02	28.83	0.41–0.47	0.86	0.01	57.64	0.83–0.89
	z_2	0.44	0.02	25.01	0.41–0.47	0.88	0.01	75.76	0.85–0.90
	z_3	0.34	0.01	25.01	0.31–0.36	0.73	0.03	28.20	0.66–0.77
$AVE = 0.79, a = 0.87, \rho = 0.92$									
PQ	z_4	0.42	0.02	18.63	0.38–0.46	0.94	0.01	105.74	0.92–0.95
	z_5	0.42	0.02	17.66	0.38–0.47	0.94	0.01	136.26	0.92–0.95
	z_6	0.27	0.01	23.13	0.24–0.29	0.79	0.02	35.25	0.73–0.82
$AVE = 0.75, a = 0.70, \rho = 0.86$									
PV	z_7	0.40	0.02	23.71	0.36–0.43	0.79	0.02	36.71	0.75–0.82
	z_8	0.73	0.02	40.63	0.70–0.76	0.94	0.01	167.18	0.93–0.95
$AVE = 0.85, a = 0.91, \rho = 0.94$									
CS	z_9	0.47	0.02	27.63	0.44–0.49	0.95	0.01	151.49	0.94–0.96
	z_{10}	0.27	0.02	16.61	0.24–0.30	0.91	0.01	81.23	0.88–0.93
	z_{11}	0.34	0.01	23.49	0.32–0.37	0.91	0.01	80.34	0.88–0.93
$AVE = 1.00, a = 1.00, \rho = 1.00$									
CC	z_{12}	1	0		1–1	1	0		1–1
$AVE = 0.88, a = 0.87, \rho = 0.94$									
CL	z_{13}	0.61	0.02	40.20	0.58–0.64	0.96	0.00	238.44	0.95–0.96
	z_{14}	0.45	0.02	28.76	0.42–0.49	0.92	0.01	151.69	0.91–0.93

AVE, Average variance extracted; CC, Customer complaints; CE, Customer expectations; CI, Confidence intervals; CL, Customer loyalty; CS, Customer satisfaction; PQ, Perceived quality; PV, Perceived value; SE, Standard errors.

TABLE 2.3

R-Squared Values of Endogenous Latent Variables
Obtained from Generalized Structured Component
Analysis for the Original ACSI Model

	R^2
CE	
PQ	0.34
PV	0.53
CS	0.82
CC	0.16
CL	0.40

CC, Customer complaints; CE, Customer expectations; CL,
Customer loyalty; CS, Customer satisfaction; PQ, Perceived
quality; PV, Perceived value.

TABLE 2.4

Estimates of Path Coefficients and Their Standard Errors and 95%
Confidence Intervals Obtained from Generalized Structured Component
Analysis for the Original ACSI Model

	Estimate	SE	$\lvert t \rvert$	95% CI
CE → PQ (b_1)	0.58	0.03	19.17	0.52–0.64
CE → PV (b_2)	0.12	0.03	3.71	0.07–0.18
CE → CS (b_4)	0.03	0.02	1.35	−0.02–0.07
PQ → PV (b_3)	0.65	0.03	21.98	0.60–0.71
PQ → CS (b_5)	0.68	0.03	21.53	0.62–0.73
PV → CS (b_6)	0.26	0.03	8.47	0.21–0.33
CS → CC (b_7)	−0.41	0.04	10.68	−0.48 to −0.32
CS → CL (b_8)	0.59	0.04	19.18	0.53–0.65
CC → CL (b_9)	−0.09	0.04	2.57	−0.16 to −0.03

CC, Customer complaints; CE, Customer expectations; CI, Confidence intervals; CL,
Customer loyalty; CS, Customer satisfaction; PQ, Perceived quality; PV, Perceived value;
SE, Standard errors.

effect on *consumer loyalty* (b_9 = −0.09, SE = 0.04, 95% CI = −0.16 to −0.03). Yet, *customer expectations* were found to have no statistically significant effect on *customer satisfaction* (b_4 = 0.03, SE = 0.02, 95% CI = −0.02–0.07). This further suggests that the statistically nonsignificant link from *customer expectation*s to *customer satisfaction* may be eliminated from the ACSI model in the current example.

Accordingly, a simpler model was specified for the data, where no path from *customer expectations* to *customer satisfaction* was specified, indicating that there was no direct effect of *customer expectations* on *customer satisfaction*. We decided to compare the overall model fit of this alternative model with that of the original model. We applied generalized structured component analysis to fit the simpler model.

The simpler model involved FIT = 0.67 (SE = 0.01, 95% CI = 0.66–0.69), AFIT = 0.67 (SE = 0.01, 95% CI = 0.66–0.69), GFI = 0.99 (SE = 0.00, 95% CI = 0.99–1.00), and SRMR = 0.07 (SE = 0.00, 95% CI = 0.07–0.08). These overall fit values are almost identical to those for the original model. In particular, despite the omission of one path coefficient, the AFIT value for the simpler model remained virtually the same as that for the original one. Furthermore, we conducted a paired t-test for the FIT values between the original and simpler models. As described earlier, this test was equivalent to testing whether the difference in the FIT values calculated from 100 bootstrap samples for each model was equal to zero. We found the t-statistic nonsignificant statistically [$t(99) = -0.59$, $p = 0.55$, 95% CI = -0.00–0.00], indicating that there was no statistically significant difference in FIT between the original and simpler models. All these suggest that the simpler model was favored to the original model.

As shown in Table 2.5, the weight and loading estimates of the simpler model were almost identical to those of the original model, leading to the same interpretations. Furthermore, as exhibited in Table 2.6, the path coefficient estimates in the simpler model were quite similar to their counterpart estimates in the original model and could be interpreted in the same manner as described above. As expected, the deletion of the path coefficient did not change substantially the R-squared value of *customer satisfaction* ($R^2 = 0.82$).

2.5 Other Related Component-Based Methods

In this section, we discuss extant multivariate statistical techniques that are related to generalized structured component analysis. We may classify these techniques into three groups. The first group of techniques is related closely to the measurement model of generalized structured component analysis in that it involves reflective indicators. This group includes principal component analysis, constrained component analysis, principal covariates regression, and (extended) redundancy analysis. The second group is related more closely to the structural model of generalized structured component analysis, which focuses on associations among latent variables that are formed as weighted composites of indicators. It includes canonical correlation analysis, canonical regression analysis, and Glang's (1988) approach. Generalized structured component analysis can have many of the techniques in the two groups as special cases, particularly when a single latent variable is considered for each block of indicators.

The last group includes partial least squares path modeling that is a well-known approach to component-based structural equation modeling. This

TABLE 2.5

Estimates of Weights and Loadings and Their Standard Errors and 95% Confidence Intervals for a Simpler ACSI Model

Latent	Indicator	Weights				Loadings			
		Estimate	SE	$\lvert t \rvert$	95% CI	Estimate	SE	$\lvert t \rvert$	95% CI
AVE = 0.68, a = 0.76, ρ = 0.86									
CE	z_1	0.43	0.02	24.50	0.40–0.47	0.86	0.01	66.78	0.83–0.89
	z_2	0.44	0.02	25.25	0.40–0.47	0.88	0.01	76.37	0.84–0.90
	z_3	0.34	0.02	22.33	0.31–0.37	0.73	0.03	27.40	0.68–0.77
AVE = 0.79, a = 0.87, ρ = 0.92									
PQ	z_4	0.42	0.02	18.48	0.38–0.47	0.94	0.01	133.91	0.93–0.95
	z_5	0.42	0.02	20.59	0.37–0.45	0.94	0.01	126.24	0.92–0.95
	z_6	0.27	0.01	22.15	0.25–0.29	0.78	0.02	34.69	0.74–0.83
AVE = 0.75, a = 0.70, ρ = 0.86									
PV	z_7	0.40	0.02	25.74	0.37–0.43	0.79	0.02	36.55	0.75–0.82
	z_8	0.73	0.02	45.01	0.70–0.76	0.94	0.01	156.77	0.93–0.95
AVE = 0.85, a = 0.91, ρ = 0.94									
CS	z_9	0.47	0.02	28.16	0.44–0.50	0.95	0.01	141.10	0.94–0.97
	z_{10}	0.27	0.02	17.57	0.24–0.29	0.91	0.01	73.22	0.88–0.93
	z_{11}	0.34	0.02	22.32	0.31–0.37	0.91	0.01	92.44	0.89–0.93
AVE = 1.00, a = 1.00, ρ = 1.00									
CC	z_{12}	1	0		1–1	1	0		1–1
AVE = 0.88, a = 0.87, ρ = 0.94									
CL	z_{13}	0.61	0.02	38.70	0.58–0.64	0.96	0.00	238.31	0.95–0.96
	z_{14}	0.45	0.02	27.84	0.42–0.49	0.92	0.01	147.13	0.91–0.93

AVE, Average variance extracted; CC, Customer complaints; CE, Customer expectations; CI, Confidence intervals; CL, Customer loyalty; CS, Customer satisfaction; PQ, Perceived quality; PV, Perceived value; SE, Standard errors.

TABLE 2.6

Estimates of Path Coefficients and their Standard Errors and 95% Confidence Intervals for a Simpler ACSI Model

	Estimate	SE	$|t|$	95% CI
CE → PQ (b_1)	0.58	0.03	19.76	0.53–0.63
CE → PV (b_2)	0.12	0.03	3.43	0.05–0.19
PQ → PV (b_3)	0.65	0.03	21.57	0.60–0.71
PQ → CS (b_5)	0.69	0.03	27.51	0.65–0.75
PV → CS (b_6)	0.27	0.03	9.21	0.21–0.32
CS → CC (b_7)	−0.41	0.03	12.29	−0.47 to −0.34
CS → CL (b_8)	0.59	0.04	16.07	0.52–0.66
CC → CL (b_9)	−0.09	0.03	2.75	−0.16 to −0.02

CC, Customer complaints; CE, Customer expectations; CL, Customer loyalty; CS, Customer satisfaction; PQ, Perceived quality; PV, Perceived value; SE, Standard errors.

approach is comparable to generalized structured component analysis in data-analytic scope and capability. We thus discuss the theoretical differences and similarities between the two component-based approaches in detail. We also compare their performance empirically, using simulated and real datasets.

2.5.1 Multivariate Techniques Related to the Measurement Model of Generalized Structured Component Analysis

Principal component analysis is a popular data reduction technique (Horst 1936; Hotelling 1933; Pearson 1901). This technique aims to extract a series of components or weighted composites of indicators such that the components are mutually orthogonal and explain the variance of indicators as much as possible. There are different ways of formulating the objective of principal component analysis (Jolliffe 2002). We will focus on a data matrix-based formulation because this is easier to show the relationship between principal component analysis and generalized structured component analysis. Let $\mathbf{Z}$ denote a matrix of standardized indicators. The objective can be achieved by minimizing the following least-squares criterion:

$$\phi = SS(\mathbf{Z} - \mathbf{ZWC}) = SS(\mathbf{Z} - \mathbf{\Gamma C}) \tag{2.32}$$

with respect to $\mathbf{W}$ and $\mathbf{C}$, subject to the constraint that components are standardized and uncorrelated to one another (i.e., $\mathbf{W'Z'ZW} = \mathbf{\Gamma'\Gamma} = N\mathbf{I}$) (ten Berge 1993, p. 43). The number of components is usually smaller than that of indicators, thus leading to data reduction. This criterion appears identical to minimizing the sum of squared residuals in the measurement model of generalized structured component analysis. However, in Equation 2.32, all the loadings and weights of $\mathbf{C}$ and $\mathbf{W}$ are free parameters to be estimated.

In other words, no loadings and weights are fixed to certain values (such as zeros), thereby denoting no prespecified structures associated with loadings and weights in **C** and **W**. This makes it possible to solve Equation 2.32 in closed form, which reduces to solving an eigenvalue decomposition problem of the correlation matrix of **Z** (ten Berge 1993, p. 44).

Takane, Kiers, and de Leeuw (1995) proposed constrained component analysis models, where certain loadings of **C** are constrained to fixed values, such as zeros, so as to capture prespecified relationships between components and indicators. Figure 2.4 depicts a constrained component analysis model that Kiers, Takane, and ten Berge (1996) used for a multitrait–multimethod analysis (Campbell and Fiske 1959; also see Browne 1984; Schmitt and Stults 1986; Werts and Linn 1970; Widaman 1985). In this model, six latent variables were specified to influence nine indicators, among which the first three are trait latent variables and the other three are method latent variables. Kiers, Takane, and ten Berge (1996) expressed the model as:

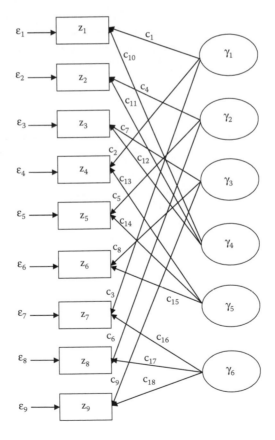

FIGURE 2.4
A path diagram of a constrained component analysis model in the study of Kiers, Takane, and ten Berge (1996).

$$Z = \Gamma C + E, \tag{2.33}$$

where the matrix of loadings (C) was specified as

$$C = \begin{bmatrix} c_1 & 0 & 0 & c_2 & 0 & 0 & c_3 & 0 & 0 \\ 0 & c_4 & 0 & 0 & c_5 & 0 & 0 & c_6 & 0 \\ 0 & 0 & c_7 & 0 & 0 & c_8 & 0 & 0 & c_9 \\ c_{10} & c_{11} & c_{12} & 0 & 0 & 0 & 0 & 0 & 0 \\ 0 & 0 & 0 & c_{13} & c_{14} & c_{15} & 0 & 0 & 0 \\ 0 & 0 & 0 & 0 & 0 & 0 & c_{16} & c_{17} & c_{18} \end{bmatrix}. \tag{2.34}$$

In this example, some loadings were fixed to zeros such that trait and method latent variables underlay each indicator. Another intriguing application of constrained component analysis models can be found in the study of Dillon et al. (2001). The parameters of constrained component analysis models were estimated by minimizing a least-squares criterion, which was essentially of the same form as Equation 2.32. However, the criterion should be minimized iteratively because it involved fixed loadings of C (Takane, Kiers, and de Leeuw 1995), as illustrated in Equation 2.34.

The measurement model of generalized structured component analysis can be viewed as a constrained component analysis model, where certain loadings are fixed to zeros such that a latent variable is assumed for a block of indicators, as shown in Equation 2.2. This in turn indicates that generalized structured component analysis can subsume such a constrained component analysis model as a special case, where all indicators are reflective and no relationships between latent variables are considered. The alternating least-squares algorithm for generalized structured component analysis can be used for fitting constrained component analysis models.

Principal covariates regression (de Jong and Kiers 1992) is a data reduction technique for the analysis of two blocks of indicators. It seeks to extract a series of components from a block of indicators in such a way that they account for the variance of the same block as well as the other block of indicators. Let Z_p denote a matrix consisting of the pth block of indicators. Let C_p denote a matrix of loadings for the pth block of indicators. Let W_p denote a matrix of weights for the pth block of indicators. Let E_p denote a matrix of residuals for the pth block of indicators. Principal covariates regression involves submodels as follows:

$$Z_1 = \Gamma C_1 + E_1$$

$$Z_2 = \Gamma C_2 + E_2$$

$$\Gamma = Z_2 W_2. \tag{2.35}$$

These equations can be combined into a single one:

$$[\mathbf{Z}_1, \mathbf{Z}_2] = \mathbf{Z}_2 \mathbf{W}_2 [\mathbf{C}_1, \mathbf{C}_2] + [\mathbf{E}_1, \mathbf{E}_2]$$

$$\mathbf{Z} = \mathbf{Z}_2 \mathbf{W}_2 \mathbf{C} + \mathbf{E}$$

$$\mathbf{Z} = \boldsymbol{\Gamma} \mathbf{C} + \mathbf{E} \qquad (2.36)$$

where $\mathbf{Z} = [\mathbf{Z}_1, \mathbf{Z}_2]$, $\mathbf{C} = [\mathbf{C}_1, \mathbf{C}_2]$, and $\mathbf{E} = [\mathbf{E}_1, \mathbf{E}_2]$. Figure 2.5 shows a path diagram of a principal covariates regression model. In the diagram, one latent variable is obtained as a weighted composite of three indicators (i.e., $\mathbf{Z}_2 = [z_1, z_2, z_3]$) and it in turn influences all six indicators. In the diagram, straight lines are added to highlight that the latent variable is a component of three indicators only. This model can be expressed into Equation 2.36 as follows:

$$[z_1, z_2, z_3, z_4, z_5, z_6] = [z_4, z_5, z_6] \begin{bmatrix} w_4 \\ w_5 \\ w_6 \end{bmatrix} [c_1, c_2, c_3, c_4, c_5, c_6] + [\varepsilon_1, \varepsilon_2, \varepsilon_3, \varepsilon_4, \varepsilon_5, \varepsilon_6]$$

$$\mathbf{Z} = \mathbf{Z}_2 \mathbf{W}_2 \mathbf{C} + \mathbf{E}$$

$$= \boldsymbol{\Gamma} \mathbf{C} + \mathbf{E}. \qquad (2.37)$$

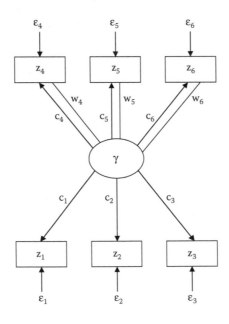

FIGURE 2.5
A path diagram of a principal covariates regression model (the number of dimensions = 1).

To estimate parameters in Equation 2.36, we can minimize the following least-squares criterion:

$$\phi = SS(\mathbf{Z} - \mathbf{\Gamma C}) \tag{2.38}$$

subject to the constraint $\mathbf{\Gamma'\Gamma} = \mathbf{I}$. Again, this is of the same form as Equation 2.32. Computationally, we can solve Equation 2.38 in closed form, which is equivalent to calculating an eigenvalue decomposition of $\mathbf{TZ_1Z_1'T} + \mathbf{Z_2Z_2'}$, where $\mathbf{T} = \mathbf{Z_2}(\mathbf{Z_2'Z_2})^{-1}\mathbf{Z_2'}$ (see de Jong and Kiers 1992).

We can re-express Equation 2.36 as:

$$[\mathbf{Z_1},\mathbf{Z_2}] = [\mathbf{Z_1},\mathbf{Z_2}]\begin{bmatrix}\mathbf{0}\\\mathbf{W_2}\end{bmatrix}[\mathbf{C_1},\mathbf{C_2}] + [\mathbf{E_1},\mathbf{E_2}]$$

$$\mathbf{Z} = \mathbf{ZWC} + \mathbf{E}, \tag{2.39}$$

where

$$\mathbf{W} = \begin{bmatrix}\mathbf{0}\\\mathbf{W_2}\end{bmatrix}.$$

Clearly, this is of the same form as the measurement model of generalized structured component analysis. Thus, generalized structured component analysis can deal with the principal covariates regression model as a special case, although the former typically assumes a single latent variable for each block of indicators, whereas the latter more than one latent variable per block.

A technique related closely to principal covariates regression is redundancy analysis (van den Wollenberg 1977), also called reduced-rank regression (Anderson 1951; Davies and Tso 1982; Izenman 1975) or principal components of instrumental variables (Rao 1964). Redundancy analysis is also a data-reduction technique for analyzing a directional relationship between two blocks of indicators (Lambert, Wildt, and Durand 1988). It aims to extract a series of components from one block of indicators in such a way that they are mutually orthogonal and account for the maximum variance of the other block of indicators. The redundancy analysis model is written as:

$$\mathbf{Z_1} = \mathbf{\Gamma C_1} + \mathbf{E}$$

$$\mathbf{\Gamma} = \mathbf{Z_2W_2}. \tag{2.40}$$

Thus, redundancy analysis can be viewed as a special case of principal covariates regression, in which a set of components are assumed to have no influences on their own indicators, that is, $\mathbf{C_2} = \mathbf{0}$, so that the second equation in Equation 2.35 is removed. To estimate parameters in Equation 2.40, we minimize the following least-squares criterion:

$$\phi = SS(Z_1 - \Gamma C_1) \tag{2.41}$$

subject to the constraint $\Gamma'\Gamma = I$. We can solve Equation 2.41 analytically, which reduces to a singular value decomposition of $(Z_2'Z_2)^{-1/2}Z_2'Z_1$ (e.g., Takane and Shibayama 1991).

Takane and Hwang (2005) developed extended redundancy analysis to generalize redundancy analysis to the analysis of directional relationships between more than two blocks of indicators. In extended redundancy analysis, a single latent variable is typically specified for each block of indicators, although in principle, it is possible to consider more than one latent variable per block. Figure 2.6 shows an example of an extended redundancy analysis model. In this model, a latent variable is specified for each block of two exogenous indicators and the latent variable affects two endogenous indicators.

We can express the extended redundancy analysis model in matrix notation as follows:

$$[z_1, z_2] = [z_3, z_4, z_5, z_6, z_7, z_8] \begin{bmatrix} w_3 & 0 & 0 \\ w_4 & 0 & 0 \\ 0 & w_5 & 0 \\ 0 & w_6 & 0 \\ 0 & 0 & w_7 \\ 0 & 0 & w_8 \end{bmatrix} \begin{bmatrix} c_1 & c_2 \\ c_3 & c_4 \\ c_5 & c_5 \end{bmatrix} + [\varepsilon_1, \varepsilon_2]$$

$$Z_1 = Z_2 W_2 C_1 + E = \Gamma C_1 + E \tag{2.42}$$

where

$$W = \begin{bmatrix} w_3 & 0 & 0 \\ w_4 & 0 & 0 \\ 0 & w_5 & 0 \\ 0 & w_6 & 0 \\ 0 & 0 & w_7 \\ 0 & 0 & w_8 \end{bmatrix}, \quad C_1 = \begin{bmatrix} c_1 & c_2 \\ c_3 & c_4 \\ c_5 & c_5 \end{bmatrix}, \quad \Gamma = Z_2 W_2, \text{ and } E = [\varepsilon_1, \varepsilon_2].$$

This is of essentially the same formulation as Equation 2.40. Thus, extended redundancy analysis can be viewed as a constrained version of redundancy analysis, where certain component weights and/or loadings for exogenous indicators are constrained to fixed values such as zeros in order to account

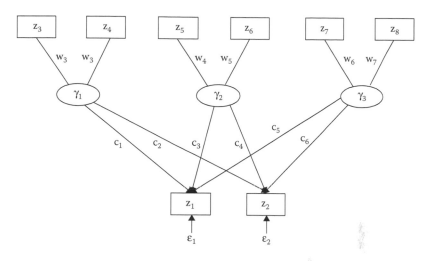

FIGURE 2.6
A path diagram of an extended redundancy analysis model.

for a variety of prespecified relationships among multiple blocks of indicators. Velu (1991) proposed a special case of extended redundancy analysis, where only two blocks of exogenous indicators are considered (also see Reinsel and Velu 1998, p. 73).

To estimate parameters in extended redundancy analysis, we minimize a least-squares criterion, which is of the same form as Equation 2.41. However, owing to the fixed elements of **W** and/or **C**, the criterion is to be minimized in an iterative way. Takane and Hwang (2005) developed an alternating least-squares algorithm to minimize the criterion, which was a simple adaptation of the alternating least-squares algorithm that Kiers and ten Berge (1989) developed for simultaneous components analysis for two or more populations (Millsap and Meredith 1988).

As stated earlier, principal covariates regression subsumes redundancy analysis as a special case. Extended redundancy analysis is a constrained version of redundancy analysis. Thus, naturally, the measurement model of generalized structured component analysis can accommodate both redundancy analysis and extended redundancy analysis models. Specifically, both techniques can be considered a special case of generalized structured component analysis, in which all formative latent variables influence a block of indicators, for example, $C_2 = 0$ in Equation 2.36.

2.5.2 Multivariate Techniques Related to the Structural Model of Generalized Structured Component Analysis

Canonical correlation analysis is used to investigate how low-dimensional representations of two blocks of indicators are related to each other (Horst 1961; Hotelling 1936; Spearman 1913). This is also a data-reduction technique like principal component analysis in that it also aims to obtain a series of weighted composites or components, called canonical variates or variables, from each block of indicators. Nonetheless, the central objective of extracting weighted composites is different from that of principal component analysis. In canonical correlation analysis, weighted composites or components of indicators are obtained in such a way that a component from one block of indicators is correlated maximally with the corresponding component from the other block, while remains uncorrelated with the other components within the same block as well as across the two blocks. The correlation between a pair of components is called a canonical correlation. Thus, canonical correlation analysis focuses on a simpler description of the association among two blocks of indicators, whereas principal component analysis concentrates on a summary of the variability of a single block of indicators. Canonical correlation analysis serves as a general framework that can embrace many statistical techniques for the analysis of two blocks of indicators as special cases, for example, linear regression analysis, discriminant analysis, and correspondence analysis.

There are different ways of formulating the objective of canonical correlation analysis. We will present a data matrix-based formulation because again this can be more useful in understanding the relationship between canonical correlation analysis and generalized structured component analysis. The aim of canonical correlation analysis can be achieved by minimizing the following least-squares criterion:

$$\phi = SS(Z_1 W_1 - Z_2 W_2) \tag{2.43}$$

under the constraints that $W_1' Z_1' Z_1 W_1 = W_2' Z_2' Z_2 W_2 = I$ (e.g., Gifi 1990, p. 220; Young, de Leeuw, and Takane 1976). The maximum number of components extracted is equal to the number of indicators in the smaller block. In Equation 2.43, all the elements of W_1 and W_2 are free parameters to be estimated, indicating that no weights are constrained to fixed values. This makes it possible to solve Equation 2.43 in closed form, which reduces to solving an eigenvalue decomposition problem of

$$(Z_1' Z_1)^{-1/2} Z_1' Z_2 (Z_2' Z_2)^{-1/2} \text{ (ten Berge 1993, p. 53).}$$

It is difficult to directly compare canonical correlation analysis with generalized structured component analysis in that the former focuses on symmetric or nondirectional associations (correlations) between two sets of components, whereas the latter focuses on directional associations (regression coefficients) among multiple components (mostly one component for each block of indicators). Nonetheless, in the case that only one component is extracted from each block of indicators, canonical correlation analysis can be viewed as a special case of generalized structured component analysis. Figure 2.7 shows an example of such a case, where a double-headed arrow indicates a correlation. In the example, there are three indicators for each block.

This case is equivalent to minimizing the following least-squares criterion:

$$\phi = SS \left(\begin{bmatrix} z_1, z_2, z_3, z_4, z_5, z_6 \end{bmatrix} \begin{bmatrix} w_1 & 0 \\ w_2 & 0 \\ w_3 & 0 \\ 0 & w_4 \\ 0 & w_5 \\ 0 & w_6 \end{bmatrix} - \begin{bmatrix} z_1, z_2, z_3, z_4, z_5, z_6 \end{bmatrix} \begin{bmatrix} w_1 & 0 \\ w_2 & 0 \\ w_3 & 0 \\ 0 & w_4 \\ 0 & w_5 \\ 0 & w_6 \end{bmatrix} \begin{bmatrix} 0 & 0 \\ b & 0 \end{bmatrix} \right)$$

$$= SS(\mathbf{ZW} - \mathbf{ZWB})$$

(2.44)

where

$$\mathbf{W} = \begin{bmatrix} w_1 & 0 \\ w_2 & 0 \\ w_3 & 0 \\ 0 & w_4 \\ 0 & w_5 \\ 0 & w_6 \end{bmatrix}$$

and

$$\mathbf{B} = \begin{bmatrix} 0 & b \\ 0 & 0 \end{bmatrix}$$

In Equation 2.44, we assume that the first latent variable affects the second. We will obtain the same path coefficient value when we reverse the relationship

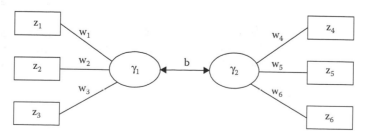

FIGURE 2.7
A path diagram of canonical correlation analysis (the number of dimensions = 1).

$$\text{i.e., } \mathbf{B} = \begin{bmatrix} 0 & b \\ 0 & 0 \end{bmatrix}$$

or assume a bidirectional relationship

$$\text{i.e., } \mathbf{B} = \begin{bmatrix} 0 & b \\ b & 0 \end{bmatrix}$$

between the latent variables. As in simple linear regression, this path coefficient (b) is equivalent to the correlation between the two latent variables, and in turn is equivalent to the largest canonical correlation.

Obviously, the criterion (Equation 2.44) is of the same form as the structural model of generalized structured component analysis. Thus, the one-dimensional solution of canonical correlation analysis can be viewed as a special case of generalized structured component analysis, when two blocks of indicators are all formative (i.e., $\mathbf{C} = \mathbf{0}$).

Another relevant technique is canonical regression analysis (van der Leeden 1990, p. 47). Canonical regression analysis aims to obtain a series of components from each of two blocks of indicators in such a way that one set of components explains the variance of the other set of components as much as possible. This technique assumes

$$\mathbf{Z}_1\mathbf{W}_1 = \mathbf{Z}_2\mathbf{W}_2\mathbf{B}_1 + \mathbf{E}_1$$

$$\mathbf{\Gamma}_1 = \mathbf{\Gamma}_2\mathbf{B}_1 + \mathbf{E}_1 \tag{2.45}$$

where $\mathbf{\Gamma}_1 = \mathbf{Z}_1\mathbf{W}_1$ and $\mathbf{\Gamma}_2 = \mathbf{Z}_2\mathbf{W}_2$. This model is essentially equivalent to the structural model of generalized structured component analysis between two latent variables. Again, in canonical regression analysis, more than one component is typically extracted from each block, whereas in generalized structured component analysis, a single latent variable is considered for each block. Clearly, if only one latent variable is assumed for each block, Equation 2.45 reduces to Equation 2.44.

Glang's (1988) approach can be regarded as a fuller extension of canonical regression analysis, which seeks to maximize the sum of explained variances among latent variables obtained from multiple blocks of indicators. This approach can be used for investigating prespecified directional relationships among latent variables. Technically, it can be viewed as a constrained version of canonical regression analysis, where certain weights in $\mathbf{W}$ and path coefficients in $\mathbf{B}$ are fixed to zeros to accommodate hypothetical relationships among latent variables. The approach aims to minimize the same criterion as Equation 2.44. Thus, when all indicators are formative ($\mathbf{C} = \mathbf{0}$), generalized structured component analysis becomes equivalent to Glang's approach (Glang 2011, personal communication; Hwang 2009; Tenenhaus

2008a). In turn, this indicates that generalized structured component analysis can subsume Glang's approach as a special case.

2.5.3 Partial Least Squares Path Modeling

Partial least squares path modeling (Wold 1966, 1973, 1982; Lohmöller 1989) is a well-known component-based approach to structural equation modeling (Tenenhaus 2008b). It would perhaps be the only technique that is comparable to generalized structured component analysis in scope and capability. Thus, the technical underpinnings of partial least squares path modeling are discussed in detail.

Partial least squares path modeling involves two submodels such as the measurement and structural models. The measurement model can be composed of three different types, depending on how latent variables are related to indicators. One type is the reflective measurement model (or outwards directed model), in which a block of indicators are affected by its latent variable in a way similar to the measurement model of generalized structured component analysis. The reflective measurement model is given as:

$$z_{pj} = \gamma_p c_{pj} + \varepsilon_{pj} \tag{2.46}$$

It is further assumed that the residual ε_{pj} has a zero mean and is uncorrelated with the latent variable γ_p, thereby indicating that $E(z|\gamma) = \gamma_p c_{pj}$. This assumption called the predictor specification condition is not made in generalized structured component analysis.

The second type is the formative measurement model (or inwards directed model), where a latent variable is hypothesized as a linear function of its indicators as follows:

$$\gamma_p = \sum_{j=1}^{J_p} z_{pj} w_{pj} + v_p \tag{2.47}$$

It is also assumed that the residual has a zero mean and is uncorrelated with indicators, so that

$$E(\gamma \mid z) = \sum_{j=1}^{J_p} z_{pj} w_{pj}$$

The last type is the multiple indicator/multiple causes model (Jöreskog and Goldberger 1975; Fornell, Barclay, and Rhee 1988), which is a mixture of the reflective and formative measurement models. This type of the measurement model is adopted when some indicators are formative while others are reflective.

The structural model of partial least squares path modeling includes a series of linear regression models for each endogenous latent variable, as follows:

$$\gamma_p = \sum_{k=1,\,k \neq p}^{P} \gamma_k b_{pk} + \zeta_p. \tag{2.48}$$

This model is virtually identical to the structural model of generalized structured component analysis.

A distinction in model specification between partial least squares path modeling and generalized structured component analysis is that the former has two submodels, whereas the latter involves three submodels. This suggests that at least in model specification, partial least squares path modeling does not define a latent variable as a weighted composite or component of indicators explicitly, whereas generalized structured component analysis does through the specification of the weighted relation model (Equation 2.11). In fact, some researchers assume that even under partial least squares path modeling, the true measurement model is still a factor analytic model, where a latent variable is equivalent to a common factor; they consider that partial least squares path modeling provides a computationally efficient proxy of a latent variable, that is, a component of indicators (e.g., Dijkstra 1981, 2010). Under the assumption of the true factor-analytic measurement model, they showed that the estimates of loadings and path coefficients obtained based on the proxy scores of latent variables from partial least squares path modeling are biased and inconsistent. They also proposed to adjust the estimates to make them consistent (e.g., Croon 2002; Dijkstra 2010; Lu et al. 2011). Consequently, at least at a model-specification level, it is not clear whether partial least squares path modeling can be viewed as a component-based approach to structural equation modeling or simply as an easy-to-implement yet somewhat inferior approach to factor-based structural equation modeling.

Another difference in model specification rests in the number of equations to be used in specifying a structural equation model. Both generalized structured component analysis and partial least squares path modeling entail submodels. However, generalized structured component analysis integrates its submodels into a unified algebraic formulation (i.e., a single equation) as shown in Equation 2.13. On the other hand, partial least squares path modeling does not combine the two submodels into a single equation and instead addresses the two equations separately. This difference in the number of equations for the specification of a structural equation model contributes to characterizing the parameter estimation procedures of the two approaches, as will be discussed subsequently.

In parameter estimation, partial least squares path modeling takes a strategy of estimating a latent variable as a component of indicators. At a parameter estimation level, therefore, it can be clearly viewed as a component-based approach to structural equation modeling. The estimation procedure consists of two main stages. The first stage estimates latent variables as components, which requires the estimation of weights. This stage involves an iterative algorithm for the estimation of the weights. On the other hand, the second stage estimates the remaining parameters (path coefficients and/ or loadings) in the measurement and structural models by means of ordinary linear regression. For example, path coefficients are estimated by regressing endogenous latent variables on their affecting latent variables, whereas loadings are estimated by regressing indicators on their latent variables (refer to Tenenhaus et al. 2005, for more details). The second stage is a noniterative estimation procedure that is based on latent variables obtained from the first stage. Accordingly, the first stage is considered the key estimation procedure in partial least squares path modeling (Hanafi 2007).

We provide here a brief summary of Lohmöller's (1989, p. 29) algorithm for the first stage. This algorithm is best known and implemented into most software programs for partial least squares path modeling, including LVPLS (Lohmöller 1984), PLS Graph (Chin 2001), SmartPLS (Ringle, Wende, and Will 2005), and XLSTAT (Addinsoft 2009). We provide a fuller description of the algorithm in Appendix 2.3.

The algorithm begins by assigning arbitrary initial values to weights and then repeats the following three steps. In the first step, each latent variable is estimated as a weighted composite or component of indicators. In the second step, the so-called inner estimate is calculated for each latent variable. The inner estimate of a latent variable corresponds to a weighted sum of adjacent latent variables that are connected with the latent variable. There exist several ways of calculating the inner estimates (e.g., the centroid, factorial, and path weighting schemes). In the third step, weights are estimated for each block of indicators. There are two ways of estimating the weights: Modes A and B. Mode A is known to be suitable for reflective indicators, whereas Mode B is for formative indicators. The three steps are alternated until no substantial difference occurs between the previous and current estimates of the weights.

Neither of the stages requires a distributional assumption on the data. A nonparametric resampling method such as the jackknife or bootstrap can be used to estimate the standard errors or confidence intervals of parameter estimates (e.g., Chin 2001).

As described above, partial least squares path modeling involves two main "sequential" stages of parameter estimation. In the first stage, weights are estimated in an iterative manner, so that individual latent variable scores are produced. In the second stage, the other parameters are estimated in a least-squares manner, given the latent variable scores. This sequential procedure is carried out because partial least squares path modeling is not equipped with a global optimization function stemming from a unified formulation of the two submodels.

Although there is no single optimization function, we can see that a series of least-squares criteria based on the measurement and structural models is minimized in the second stage. In contrast, it is not clear what criterion is generally being optimized for estimating weights in the first stage. A few attempts have been made to provide a single criterion for the estimation of component weights. For example, Hanafi (2007) provided an association-maximization criterion for the Mode B case. Tenenhaus and Tenenhaus (2011) recently showed that a covariance-maximization criterion could be used for the estimation of weights under Mode B and under a slightly modified version of Mode A. They also showed that the same criterion can accommodate a new mode called Mode ridge. Nevertheless, to our knowledge, no single optimization criterion is yet available for the first stage, which includes both Mode A and Mode B in the Lohmöller's algorithm as special cases.

Conversely, generalized structured component analysis does not involve separate stages of parameter estimation. This approach estimates all parameters simultaneously by minimizing a single least-squares criterion. In this regard, generalized structured component analysis is considered a full-information method, which estimates unknown parameters in all equations simultaneously, while utilizing information available from the entire system of equations. Conversely, partial least squares path modeling is considered a limited or partial information method, which uses a subset of equations at a time for estimating parameters (Tenenhaus 2008b).

It is difficult to evaluate whether the sequential estimation procedure of partial least squares path modeling is inferior (or superior) to the simultaneous estimation procedure of generalized structured component analysis. However, we can point out several differences between the two approaches with respect to parameter estimation. First, the simultaneous estimation procedure of generalized structured component analysis is technically more straightforward and easier to understand, when compared to the sequential procedure of partial least squares path modeling, which has been blamed for its complexity and obscurity (e.g., McDonald 1996; Tenenhaus 2008b). In addition to the ease in understanding the estimation procedure, the alternating least-squares algorithm adopted by generalized structured component analysis has been proven to converge (de Leeuw, Young, and Takane 1976). In contrast, convergence of the partial least squares path modeling algorithm for the estimation of weights has not been fully proven except for the case of dealing with only one or two latent variables (Hanafi 2007; Henseler 2010). Moreover, generalized structured component analysis defines convergence as the decrease in the optimization criterion value beyond a certain threshold, whereas partial least squares path modeling defines convergence as a sort of equilibrium, that is, the point at which no substantial difference occurs between the previous and current estimates of weights.

Second, the availability of measures of overall model fit relies on whether or not some global optimization criterion is present. Because generalized structured component analysis involves a global optimization criterion, it can provide measures of overall model fit, such as FIT and AFIT, which are proportional to the value of the criterion. In contrast, partial least squares path modeling is incapable of providing such a measure of overall model fit. This forces users of this approach to rely on local fit measures for the evaluation of a model. Despite the importance of measures of local fit in evaluating the suitability of models (Bollen 1989, p. 281), they provide little information on how well a model fits to the data as a whole. Moreover, they are of little use for comparisons of a focal model to alternative models. To address this issue, Tenenhaus, Amato, and Esposito Vinzi (2004) introduced a global criterion of goodness of fit (GoF), which is the geometric mean of the average of all squared loadings of indicators (called the average communality) and the average R-squared of endogenous latent variables (called the average R^2), given by:

$$GoF = \sqrt{\text{average communality} \times \text{average } R^2} \tag{2.49}$$

(also see Tenenhaus et al. 2005). Esposito Vinzi, Trinchera, and Amato (2010) further proposed a normalized version of the GoF, the so-called relative GoF (GoF$_{rel}$). The GoF$_{rel}$ is obtained by dividing the average communality and the average R-squared in the GoF by their maximum values (i.e., the average communalities obtained from a principal component analysis and the average R-squared obtained from a canonical correlation analysis, respectively). These measures may be comparable to the FIT in the sense that they also aim to take into account the explained variances of both indicators and latent variables. Nonetheless, it is unclear what the values of the GoF and GoF$_{rel}$ really indicate and how to interpret them. Moreover, they cannot be used when all indicators are formative because the average communality becomes equal to zero. As Tenenhaus et al. (2005) suggested, one may regard these measures as an "operational" alternative to overall model fit. Esposito Vinzi, Trinchera, and Amato (2010) showed how to bootstrap this measure for testing the statistical significance of one or more path coefficients. Henseler and Sarstedt (2013) provided fuller discussions on the two measures and carried out investigations into their properties based on simulated data.

Lastly, it is unclear how to impose certain constraints on model parameters (e.g., equality or zero constraints) in the estimation procedure of partial least squares path modeling (Tenenhaus 2008b). For example, there seems to be no way of constraining all loading estimates to be equal because a set of loadings is updated for each latent variable at a time. Moreover, it seems difficult to constrain the path coefficients relating the same latent variable to different endogenous latent variables, because they are to be estimated based on a separate linear regression analysis for each endogenous latent variable. On the other hand, as will be discussed in Chapter 3, it is straightforward to integrate the imposition of constraints on parameters into the estimation procedure of generalized structured component analysis.

2.5.4 Empirical Comparisons between Generalized Structured Component Analysis and Partial Least Squares Path Modeling

We have discussed various component-based multivariate statistical techniques that are related to generalized structured component analysis. As stated earlier, however, only partial least squares path modeling can be comparable to generalized structured component analysis in terms of data analytic scope and flexibility. We have discussed conceptual similarities and differences between the two approaches in Section 2.5.3. We now focus on comparing their performance empirically based on simulated and real datasets.

2.5.4.1 Simulated Data

To our knowledge, the study of Hwang et al. (2010a) was the first one to investigate the performance of the two component-based approaches to structural equation modeling, as well as of factor-based structural equation modeling, using simulated data. They arrived at three major conclusions. First, generalized structured component analysis may be substituted for partial least squares path modeling. Second, when the model is correctly specified, factor-based structural equation modeling may be used over generalized structured component analysis. Lastly, when the model is incorrectly specified, generalized structured component analysis may be chosen over factor-based structural equation modeling. However, as Hwang et al. (2010b) explained in their web errata, there was a scaling error in the program of generalized structured component analysis used in the simulation study. In the program, indicators were standardized, whereas latent variables were normalized. This scaling inconsistency between indicators and latent variables resulted in less optimal parameter estimates as well as inaccurate overall model fit values. In addition, Henseler (2012) pointed out that the experimental design used by Hwang et al. (2012a), which was adopted mostly from the previous studies (e.g., Paxton et al. 2001), was limited in scope and lacked internal validity. He compared generalized structured component analysis to factor-based structural equation modeling, using a greater variety of experimental designs, and concluded that generalized structured component analysis was found to be inferior to factor-based structural equation modeling in the accuracy of parameter recovery except for the case where the construct measurement was perfectly reliable and valid.

It would be desirable to carry out more thorough investigations into the performance of generalized structured component analysis and factor-based structural equation modeling. Nonetheless, a major concern with such a comparison is that how to compare these approaches in an impartial manner. As we highlighted in Chapter 1, component-based structural equation modeling was developed from conceptually and technically different perspectives, when compared to factor-based structural equation modeling. As a result, it seems to be difficult to find a fair way of generating simulated data for both component- and factor-based approaches to structural equation modeling, as also pointed out by Hwang et al. (2010a). In his study, Henseler (2012) generated simulated data based on the assumption that a factor-based structural equation model was the true model. Obviously, this can yield an unfavorable effect on the performance of component-based approaches to structural equation modeling.

We are not certain what would be an ideal procedure for comparing these conceptually different approaches. At the same time, this is not a main focus of this section. We will focus on comparisons between two component-based approaches to structural equation modeling—partial least squares path modeling and generalized structured component analysis. Unfortunately, it is still not straightforward to generate simulated data in an equally fair way for both the approaches because they involve different model specifications. A promising way may be to construct an implied covariance matrix under the assumption that indicators are not associated with measurement errors, as suggested by McDonald (1996) and echoed by Tenenhaus (2008b). We adopt this procedure for data generation, which is equivalent to Model I in the study of McDonald (1996). As will be shown shortly, the data generation procedure leads the two approaches to yield exactly the same parameter estimates under correct yet simple model specifications. This may be an indication that the data generation procedure is not severely impartial for either of the two approaches.

In this simulation study, we considered two main experimental conditions: model specification and sample size. For model specification conditions, we followed what Henseler (2012) used for his study. He considered under-parameterization and over-parameterization separately. Under-parameterization indicates that one or more parameters are fixed to zeros while their true values are nonzero, whereas over-parameterization denotes that one or more parameters are estimated while their true values are zero (Hu and Bentler 1998, p. 427). He also contemplated four population models (Models A, B, C, and D). In Model A, he used the same model specified by Hwang et al. (2010a) (also see Bollen et al. 2007; Paxton et al. 2000), which involved three positive cross loadings and two positive path coefficients. For the second population model (Model B), he slightly modified Model A to exclude all cross loadings. All other parameters remain unchanged. Models C and D are virtually identical to Models A and B, respectively, except for that the two path coefficients were negative in the same magnitude. For each model, four different conditions of model specification were considered, where cross loadings were either modeled or ignored, and a direct effect of the first latent variable on the third latent variable was either modeled or ignored. This led to 16 different levels of model specification (four models × four parameterizations). The basic model with unstandardized and standardized parameter values is shown in Figure 2.8. For each model, individual-level multivariate normal data were drawn from N(0, Σ), where Σ is the implied population covariance matrix derived from a formulation of factor-based structural equation modeling (i.e., the reticular action model), using the unstandardized parameter values.

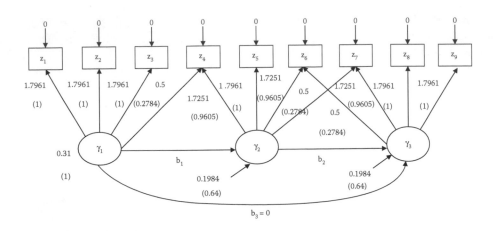

FIGURE 2.8
A path diagram of the basic structural equation model specified for the simulation study. Standardized parameters are given in parentheses.

Additionally, we considered five different levels of sample size (N = 25, 50, 100, 200, and 400). Five hundred samples were generated at each level of the two experimental conditions. Subsequently, all 40,000 samples (5 sample sizes × 16 model specifications × 500 replications) were fitted by generalized structured component analysis and partial least squares path modeling. We used the same initial values per sample for the two approaches. As all indicators were reflective in the models, we used Mode A for partial least squares path modeling in combination with the path weighting scheme that is recommended over the other schemes (Esposito Vinzi, Trinchera, and Amato 2010).

Table 2.7 provides the average mean square errors (MSEs) of parameter estimates obtained from the two component-based approaches. The MSE is the average squared difference between a parameter and its estimate, thereby indicating how far an estimate is, on average, from its parameter, that is, the smaller the MSE, the closer the estimate is to the parameter. Specifically, the MSE is calculated by:

$$\text{MSE}(\hat{\theta}) = E\left[(\hat{\theta}-\theta)^2\right] = E\left(\hat{\theta}-E(\hat{\theta})\right)^2 + \left(E(\hat{\theta})-\hat{\theta}\right)^2, \qquad (2.50)$$

where θ and $\hat{\theta}$ denote a parameter and its estimate, respectively. In Equation 2.50, the first and second terms represent the variance and squared bias of an estimate, respectively. Thus, the MSE entails information on both bias and variability of the estimate (Mood, Graybill, and Boes 1974).

TABLE 2.7

Average Mean Square Errors of Parameter Estimates Obtained from Generalized Structured Component Analysis and Partial Least Squares Path Modeling. Cross Loadings: Yes = Modeled, No = Ignored. Direct Effect (b_3): Yes = Freely Estimated, No = Fixed to Zero

True Model	Cross Loadings	Direct Effect (b_3)	Specification Levels	Parameters	Sample Size	GSCA	PLS-PM
Model A: $b_1 = b_2 =$ 0.6 and cross loadings	Yes	Yes	I (over-parameterized)	Loadings	25	0.05	0.09
					50	0.04	0.09
					100	0.04	0.09
					200	0.04	0.09
					400	0.04	0.09
				Paths	25	1.48	0.15
					50	1.34	0.14
					100	1.05	0.13
					200	1.01	0.13
					400	0.97	0.13
	Yes	No	II (correctly specified)	Loadings	25	0.02	0.09
					50	0.02	0.09
					100	0.02	0.09
					200	0.02	0.09
					400	0.02	0.09
				Paths	25	0.06	0.07
					50	0.05	0.07
					100	0.05	0.06
					200	0.05	0.06
					400	0.05	0.06
	No	Yes	III (under-parameterized)	Loadings	25	0.00	0.00
					50	0.00	0.00
					100	0.00	0.00
					200	0.00	0.00
					400	0.00	0.00
				Paths	25	0.30	0.04
					50	0.27	0.03
					100	0.26	0.02
					200	0.26	0.02
					400	0.25	0.02
	No	No	IV (under-parameterized)	Loadings	25	0.00	0.00
					50	0.00	0.00
					100	0.00	0.00
					200	0.00	0.00
					400	0.00	0.00
				Paths	25	0.05	0.03
					50	0.05	0.03
					100	0.05	0.02
					200	0.05	0.02
					400	0.05	0.02

(Continued)

TABLE 2.7 (*Continued*)

Average Mean Square Errors of Parameter Estimates Obtained from Generalized Structured Component Analysis and Partial Least Squares Path Modeling. Cross Loadings: Yes = Modeled, No = Ignored. Direct Effect (b_3): Yes = Freely Estimated, No = Fixed to Zero

True Model	Cross Loadings	Direct Effect (b_3)	Specification Levels	Parameters	Sample Size	GSCA	PLS-PM
Model B:	Yes	Yes	V	Loadings	25	0.05	0.16
$b_1 = b_2 =$			(over-		50	0.05	0.16
0.6 and no			parameterized)		100	0.05	0.16
cross					200	0.05	0.16
loadings					400	0.05	0.16
				Paths	25	0.08	0.13
					50	0.06	0.11
					100	0.06	0.11
					200	0.06	0.10
					400	0.06	0.10
	Yes	No	VI	Loadings	25	0.05	0.16
			(over-		50	0.05	0.16
			parameterized)		100	0.05	0.16
					200	0.05	0.16
					400	0.05	0.16
				Paths	25	0.05	0.07
					50	0.05	0.07
					100	0.04	0.07
					200	0.04	0.07
					400	0.04	0.07
	No	Yes	VII	Loadings	25	0.00	0.00
			(over-		50	0.00	0.00
			parameterized)		100	0.00	0.00
					200	0.00	0.00
					400	0.00	0.00
				Paths	25	0.03	0.03
					50	0.02	0.02
					100	0.01	0.01
					200	0.00	0.00
					400	0.00	0.00
	No	No	VIII (correctly	Loadings	25	0.00	0.00
			specified)		50	0.00	0.00
					100	0.00	0.00
					200	0.00	0.00
					400	0.00	0.00
				Paths	25	0.02	0.02
					50	0.01	0.01
					100	0.00	0.00
					200	0.00	0.00
					400	0.00	0.00

(Continued)

TABLE 2.7 (*Continued*)

Average Mean Square Errors of Parameter Estimates Obtained from Generalized Structured Component Analysis and Partial Least Squares Path Modeling. Cross Loadings: Yes = Modeled, No = Ignored. Direct Effect (b_3): Yes = Freely Estimated, No = Fixed to Zero

True Model	Cross Loadings	Direct Effect (b_3)	Specification Levels	Parameters	Sample Size	GSCA	PLS-PM
Model C:	Yes	Yes	IX	Loadings	25	0.11	1.28
$b_1 = b_2 =$			(over-		50	0.10	0.94
−0.6 and			parameterized)		100	0.10	1.27
cross					200	0.09	1.03
loadings					400	0.10	1.46
				Paths	25	0.19	0.77
					50	0.17	0.81
					100	0.17	0.85
					200	0.19	0.74
					400	0.15	0.78
	Yes	No	X (correctly	Loadings	25	0.12	1.05
			specified)		50	0.11	1.22
					100	0.10	1.15
					200	0.10	1.36
					400	0.10	1.38
				Paths	25	0.05	0.81
					50	0.04	0.82
					100	0.04	0.69
					200	0.04	0.78
					400	0.04	0.76
	No	Yes	XI	Loadings	25	0.00	0.00
			(under-		50	0.00	0.00
			parameterized)		100	0.00	0.00
					200	0.00	0.00
					400	0.00	0.00
				Paths	25	0.10	0.05
					50	0.06	0.03
					100	0.04	0.03
					200	0.04	0.02
					400	0.04	0.02
	No	No	XII	Loadings	25	0.00	0.00
			(under-		50	0.00	0.00
			parameterized)		100	0.00	0.00
					200	0.00	0.00
					400	0.00	0.00
				Paths	25	0.03	0.03
					50	0.02	0.02
					100	0.01	0.01
					200	0.01	0.01
					400	0.01	0.01

(Continued)

TABLE 2.7 (*Continued*)

Average Mean Square Errors of Parameter Estimates Obtained from Generalized Structured Component Analysis and Partial Least Squares Path Modeling. Cross Loadings: Yes = Modeled, No = Ignored. Direct Effect (b_3): Yes = Freely Estimated, No = Fixed to Zero

True Model	Cross Loadings	Direct Effect (b_3)	Specification Levels	Parameters	Sample Size	GSCA	PLS-PM
Model D: $b_1 = b_2 = -0.6$ and no cross loadings	Yes	Yes	XIII (over-parameterized)	Loadings	25	0.04	0.60
					50	0.04	0.56
					100	0.05	0.53
					200	0.05	0.42
					400	0.05	0.36
				Paths	25	0.09	0.47
					50	0.07	0.39
					100	0.06	0.39
					200	0.06	0.33
					400	0.06	0.28
	Yes	No	XIV (over-parameterized)	Loadings	25	0.05	0.70
					50	0.05	0.68
					100	0.05	0.65
					200	0.04	0.62
					400	0.04	0.65
				Paths	25	0.05	0.46
					50	0.05	0.48
					100	0.04	0.56
					200	0.04	0.57
					400	0.04	0.61
	No	Yes	XV (over-parameterized)	Loadings	25	0.00	0.00
					50	0.00	0.00
					100	0.00	0.00
					200	0.00	0.00
					400	0.00	0.00
				Paths	25	0.03	0.03
					50	0.02	0.02
					100	0.01	0.01
					200	0.00	0.00
					400	0.00	0.00
	No	No	XVI (correctly specified)	Loadings	25	0.00	0.00
					50	0.00	0.00
					100	0.00	0.00
					200	0.00	0.00
					400	0.00	0.00
				Paths	25	0.02	0.02
					50	0.01	0.01
					100	0.00	0.00
					200	0.00	0.00
					400	0.00	0.00

GSCA, Generalized structured component analysis; PLS-PM, Partial least squares path modeling.

We found several intriguing characteristics with respect to parameter recovery. First, under models B and D, which did not include cross loadings and the direct effect (b_3) of the first latent variable (γ_1) on the third latent variable (γ_3), partial least squares path modeling and generalized structured component analysis resulted in identical average MSEs of both loading and path coefficients estimates in correct specification such as levels VIII and XVI. The average MSEs of the loading estimates were virtually equal to zeros; those of the path coefficient estimates were quite small in small samples (e.g., $N \leq 50$) and became equal to zeros in larger samples (e.g., $N > 50$). This suggests that under these conditions, the two approaches produced equally accurate parameter estimates.

Second, under models A and C that involved cross loadings, partial least squares path modeling yielded larger average MSEs of both loading and path coefficient estimates than generalized structured component analysis in correct specifications (see levels II and X). A similar result was also reported by Hwang et al. (2010a).

Third, when a model was over-parameterized (i.e., more than the number of true parameters are estimated), in general, partial least squares path modeling yielded larger average MSEs of both loading and path coefficient estimates than generalized structured component analysis. This tendency was particularly salient when loadings were over-parameterized (see levels V, VI, IX, XIII, and XIV). In fact, when only a path coefficient was over-parameterized within the same model, the differences in the average MSEs of the parameter estimates between the two approaches became negligible (levels VII and XV), except for level I, where the two approaches tended to show differences in parameter recovery.

Fourth, when loadings were under-parameterized (i.e., fewer than the number of true loadings are estimated), generalized structured component analysis showed larger average MSEs of path coefficient estimates than partial least squares path modeling (see levels III, IV, and XI). In this study, under-parameterization of loadings occurred when all cross loadings were ignored.

To summarize, in seven of 16 model specification levels (i.e., levels II, V, VI, IX, X, XIII, and XIV), generalized structured component analysis tended to recover parameters more accurately than partial least squares path modeling. Interestingly, all these cases involved the estimation of cross loadings, regardless of whether modeling cross loadings was correct or incorrect. In other five cases (i.e., levels VII, VIII, XII, XV, and XVI), both the approaches recovered parameters similarly or equally. The five cases did not involve the estimation of cross loadings, regardless of whether disregarding them was correct or incorrect. When ignoring cross loadings was a correct specification, the two approaches performed equally well. Otherwise, they performed similarly. In other three cases (i.e., levels III, IV, and XI), partial least squares path modeling tended to recover path coefficients better than generalized structured component analysis, while recovering loadings similarly to generalized structured component analysis. In these cases, cross loadings were incorrectly ignored, whereas a path coefficient was incorrectly added or correctly ignored. In the remaining case (level I), generalized structured component analysis tended to recover loadings better than partial least squares path modeling, whereas partial least squares path modeling tended to recover path coefficients more accurately than generalized structured component analysis.

Accordingly, the performance of the two approaches seemed to be contingent on whether cross loadings were estimated in the measurement model. When cross loadings were estimated, generalized structured component analysis was generally superior to partial least squares path modeling in recovery of both loadings and path coefficients. This was the case regardless of whether modeling cross loadings was correct or incorrect. In contrast, when no cross loadings were to be estimated, it seems important to check whether ignoring cross loadings is a correct specification. If it was a correct specification, the two approaches performed equally well or similarly. On the other hand, if ignoring cross loadings led to under-parameterization, partial least squares path modeling was generally superior to generalized structured component analysis in recovery of path coefficients. Furthermore, a misspecification in the structural model did not differentiate the performance of the two approaches substantially.

Based on these findings, we may suggest that when researchers assume cross loadings in their measurement model, they should choose generalized structured component analysis over partial least squares path modeling. It shall not be crucial whether they can be ensured that including cross loadings is a correct specification. Partial least squares path modeling is not likely to be well suited to deal with measurement models involving cross loadings. This may be because partial least squares path modeling was originally built on the so-called "basic" design (e.g., Wold 1982; Dijkstra 2010), which refers to the cases in which each block of indicators is associated exclusively with a latent variable, while not reflecting multiple latent variables (i.e., no cross loadings).

When researchers are assured that having no cross loadings is a correct specification, they can utilize either generalized structured component analysis or partial least squares path modeling, because the two approaches performed comparably in such a situation. On the other hand, when researchers are suspicious that their measurement model is under-parameterized by excluding cross loadings, partial least squares path modeling can be an alternative to generalized structured component analysis. In many situations, it would be difficult to decide a priori whether cross loadings should be added to the measurement model. A practice is to compare a model without cross loadings and alternative models with cross loadings. For such comparisons, generalized structured component analysis can be more useful than partial least squares path modeling, because it affords overall model fit measures for model comparison and also performs better for models with cross loadings.

2.5.4.2 Real Data: The ACSI Data

Conventionally, partial least squares path modeling has been adopted for fitting the ACSI model, which is shown in Figure 2.3, owing to its advantages as a component-based approach to structural equation modeling (Anderson and Fornell 2000; Fornell et al. 1996; Johnson et al. 2001). For instance, multivariate normality of indicators is typically violated in customer satisfaction measures (Anderson and Fornell 2000; Johnson et al. 2001). As stated earlier, partial least square path modeling does not rely on such a distributional assumption for parameter estimation. Moreover, in the case of the ACSI, individual-level scores of customer satisfaction are first obtained and subsequently aggregated to provide company-, sector-, or national-level ACSI scores (Fornell et al. 1996). Hence, to calculate the aggregated scores accurately, it is necessary to first obtain the unique individual-level ACSI scores of consumers. This has called for partial least squares path modeling as a procedure to fit the ACSI model. Furthermore, the ultimate outcome variable in the ACSI model is customer loyalty, which, in turn, has been presented as a proxy for profitability (Reichheld and Sasser 1990). Thus, the ACSI model is designed to measure customer satisfaction in such a way that customer satisfaction predicts customer loyalty as accurately as possible. This again entails that partial least squares path modeling is well suited to fit the ASCI model because it is a prediction-oriented method that estimates parameters so that the discrepancies between predictor and outcome variables are minimized in a least-squares sense (Jöreskog and Wold 1982; Fornell and Cha 1994).

For the same reasons described, generalized structured component analysis can be a sensible choice for fitting the ACSI model. We have already demonstrated the use of generalized structured component analysis for the analysis of an ASCI dataset in Section 2.4. We here apply partial least squares path modeling for the analysis of the same ACSI dataset and compare it to generalized structured component analysis. We used SmartPLS (Ringle, Wende, and Will 2005) for the implementation of partial least squares path modeling. Because all indicators are reflective in the ACSI, we chose Mode A for the estimation of latent variables in combination with the path weighting scheme. We used 100 bootstrap samples for the estimation of standard errors. We focused on the output reported from SmartPLS.

Table 2.8 provides several local model fit measures obtained from partial least squares path modeling. For comparison, we reported only the local fit measures which were also calculated by generalized structured component analysis. They included the AVE, composite reliability, and Cronbach's alpha. Table 2.9 reports the R-squared values of endogenous latent variables. The values of these measures were similar to those obtained from generalized structured component analysis, resulting in the same interpretations.

Table 2.8 also presents the estimates of weights and loadings and their standard errors obtained under partial least squares path modeling.

TABLE 2.8

Estimates of Weights and Loadings and Their Standard Errors Obtained from Partial Least Squares Path Modeling for the Original ACSI Model

Latent	Indicator	Weights			Loadings		
		Estimate	SE	$|t|$	Estimate	SE	$|t|$
$AVE = 0.68, a = 0.76, \rho = 0.86$							
CE	z_1	0.45	0.06	8.15	0.87	0.03	27.36
	z_2	0.43	0.06	7.77	0.88	0.04	22.90
	z_3	0.32	0.06	4.98	0.72	0.11	6.63
$AVE = 0.79, a = 0.87, \rho = 0.92$							
PQ	z_4	0.41	0.02	17.10	0.93	0.02	47.67
	z_5	0.40	0.03	15.92	0.93	0.02	51.52
	z_6	0.31	0.02	14.68	0.80	0.07	11.80
$AVE = 0.76, a = 0.70, \rho = 0.86$							
PV	z_7	0.42	0.05	8.26	0.80	0.08	10.18
	z_8	0.71	0.07	10.63	0.93	0.01	62.79
$AVE = 0.85, a = 0.91, \rho = 0.95$							
CS	z_9	0.39	0.01	29.09	0.94	0.02	59.89
	z_{10}	0.34	0.02	20.29	0.92	0.03	36.16
	z_{11}	0.35	0.02	21.32	0.91	0.03	32.02
$AVE = 1.00, a = 1, \rho = 1.00$							
CC	z_{12}	1	0		1	0	
$AVE = 0.88, a = 0.87, \rho = 0.94$							
CL	z_{13}	0.58	0.03	17.77	0.95	0.01	130.94
	z_{14}	0.48	0.03	16.46	0.93	0.02	55.16

AVE, Average variance extracted; CC, Customer complaints; CE, Customer expectations; CL, Customer loyalty; CS, Customer satisfaction; PQ, Perceived quality; PV, Perceived value; SE, Standard errors.

TABLE 2.9

R-Squared Values of Endogenous Latent Variables Obtained from Partial Least Squares Path Modeling for the Original ACSI Model

	R^2
CE	
PQ	0.34
PV	0.53
CS	0.81
CC	0.16
CL	0.39

CC, Customer complaints; CE, Customer expectations; CL, Customer loyalty; CS, Customer satisfaction; PQ, Perceived quality; PV, Perceived value.

TABLE 2.10

Estimates of Path Coefficients and Their Standard Errors Obtained
from Partial Least Squares Path Modeling for the Original ACSI
Model

	Estimate	SE	$|t|$
CE → PQ (b_1)	0.58	0.08	6.96
CE → PV (b_2)	0.12	0.09	1.35
CE → CS (b_3)	0.04	0.06	0.58
PQ → PV (b_4)	0.65	0.09	7.35
PQ → CS (b_5)	0.67	0.08	8.30
PV → CS (b_6)	0.27	0.09	3.01
CS → CC (b_7)	−0.40	0.10	3.98
CS → CL (b_8)	0.58	0.11	5.53
CC → CL (b_9)	−0.10	0.09	1.03

CC, Customer complaints; CE, Customer expectations; CL, Customer loyalty; CS, Customer satisfaction; PQ, Perceived quality; PV, Perceived value;
SE, Standard errors.

SmartPLS did not provide the bootstrapped confidence intervals of parameter estimates. However, there would be no technical difficulty in calculating the confidence intervals in partial least squares path modeling. These
estimates looked similar to those under generalized structured component
analysis, leading to the same interpretations. Nonetheless, it was apparent that the standard errors of the weight estimates were larger than those
obtained from generalized structured component analysis, thus resulting in
consistently smaller t statistic values in absolute value. This indicates that the
weight estimates from partial least squares path modeling were less reliable
than those from generalized structured component analysis. This is also the
case for the standard errors of loading estimates. The difference in magnitude of standard errors between the two approaches was also reported by
Hwang and Takane (2004a) and Hwang et al. (2010a). This may be due to
the full-information estimation feature of generalized structured component
analysis, which uses information on the other parameter estimates simultaneously when estimating the weights.

Table 2.10 exhibits the estimates of path coefficients. The path coefficient
estimates are similar to those obtained via generalized structured component analysis. However, the standard errors of the path coefficient estimates
are consistently larger than those obtained under generalized structured
component analysis. This led to that three path coefficient estimates turned
out to be statistically nonsignificant. That is, *customer expectations* had no sta-
tistically significant effect on *perceived value* (b_2 = 0.12, SE = 0.09, $|t|$ = 1.35)
and *customer satisfaction* (b_3 = 0.04, SE = 0.06, $|t|$ = 0.58). Moreover, *customer
complaints* had no statistically significant effect on *customer loyalty* (b_9 = −0.10,
SE = 0.09, $|t|$ = 1.03). It is difficult to decide which approach provided more

accurate path coefficient estimates because their parameter values are unknown. Nonetheless, the estimates obtained from partial least squares path modeling appear to be less consistent with the relationships hypothesized in the ACSI.

2.6 Summary

Generalized structured component analysis involves the specification of three submodels. Of the submodels, the measurement and structural models are also specified in factor-based structural equation modeling and partial least squares path modeling. Conversely, the weighted relation model is a unique specification to generalized structured component analysis. This submodel clearly exhibits that generalized structured component analysis is a component-based approach to structural equation modeling from the level of model specification. Importantly, generalized structured component analysis integrates the three submodels into a unified algebraic formulation (i.e., a single equation).

The unified formulation, called the generalized structured component analysis model, facilitates the development of a global least-squares optimization criterion for parameter estimation. An alternating least-squares algorithm was developed to minimize the least-squares criterion, although other optimization methods may also be used (Henseler 2012). The algorithm estimates all unknown parameters (weights, path coefficients, and/or loadings) simultaneously by consistently minimizing the criterion. It has been proven to converge in theory and is straightforward to implement in practice. This least-squares estimation does not require the assumption of a distribution underlying the data. Moreover, it can be beneficial in small samples. Once weights are estimated, individual latent variable scores are uniquely calculated based on the weighted relation model.

Owing to the availability of a single optimization procedure, generalized structured component analysis also enables us to provide overall model fit measures. The overall fit measures discussed in Section 2.3 can be of use in summarizing the acceptability of a specified model as a whole and in comparing different models. However, further studies are warranted to investigate the properties of these measures and to recommend some thresholds of their values being indicative of an acceptable fit. In addition to the overall model fit measures, a variety of local fit measures are available to examine the qualities of subconstituents of the model.

Generalized structured component analysis is related to a number of existing techniques that involve the construction of components or weighted composites of indicators mainly for data reduction purposes. A group of such techniques is associated more closely with the measurement model of generalized structured component analysis in the sense that it aims to obtain components to explain indicators. Another group pertains more closely to the structural model in that it focuses on associations between components obtained from different blocks of indicators. In particular, two techniques are more directly comparable to generalized structured component analysis in scope and capability. One is Glang's (1988) approach to maximizing the associations among latent variables from multiple blocks of indicators. This approach can be viewed as a component-based approach to structural equation modeling under the assumption that all indicators are formative. Thus, Glang's approach involves only two submodels of generalized structured component analysis, such as the structural and weighted relation models.

The other technique is partial least squares path modeling, which represents another full-fledged approach to component-based structural equation modeling. We discussed the similarities and differences in model specification and parameter estimation between partial least squares path modeling and generalized structured component analysis. We also compared the two approaches empirically using real and simulated data. Both approaches resulted in quite similar parameter estimates when being applied to customer satisfaction data. However, partial least squares path modeling tended to yield larger standard errors than generalized structured component analysis. This led the approaches to draw divergent conclusions on the statistical significance of certain path coefficient estimates: under partial least squares path modeling, two more path coefficient estimates turned out to be statistically nonsignificant.

In our simulation study, generalized structured component analysis was found to recover parameters more accurately than partial least squares path modeling, when cross loadings were to be estimated. On the other hand, when the correct model involved no cross loadings, both the approaches performed equally or similarly. However, when the exclusion of cross loadings led to an incorrect (under-parameterized) specification, partial least squares path modeling recovered path coefficients better than generalized structured component analysis. Based on this study, it appears hasty to conclude that generalized structured component analysis should be used in lieu of partial least squares path modeling, as made in the study of Hwang et al. (2010a). Nonetheless, given the fact that it would be difficult to judge a priori whether the model is under-parameterized by the deletion of cross loadings, generalized structured component analysis can be an alternative to partial least squares path modeling in many situations where researchers want to compare models with and without cross loadings. Although we considered the same experimental conditions used in the latest, relevant study (Henseler 2012), the range of conditions in our study may still be limited. Thus, it will be necessary to consider a greater variety of experimental conditions for more thorough investigations of the performance of the two component-based approaches. Moreover, it should be noted that our conclusions were given based solely on the accuracy of parameter recovery between the approaches. In practice, however, what would also be important to researchers is the data analytic capabilities of the two approaches. We will discuss various technical extensions of generalized structured component analysis in subsequent chapters.

Appendix 2.1 The Alternating Least-Squares Algorithm for Generalized Structured Component Analysis

We begin by providing arbitrary values for weights, loadings, and path coefficients. We then repeat the following steps until convergence is reached.

Step 1: Update loadings and path coefficients in **A** for fixed weights in **W**. Equation 2.17 can be rewritten as:

$$\phi = SS(\text{vec}(\mathbf{\Psi}) - \text{vec}(\mathbf{\Gamma A}))$$

$$= SS(\text{vec}(\mathbf{\Psi}) - (\mathbf{I} \otimes \mathbf{\Gamma})\,\text{vec}(\mathbf{A})), \qquad (A2.1)$$

where vec(**X**) indicates a supervector formed by stacking all columns of **X** one below another, and $\otimes$ indicates a Kronecker product. Let **a** denote the vector formed by eliminating zero elements from vec(**A**). Let **Φ** denote the matrix formed by eliminating the columns of $\mathbf{I} \otimes \mathbf{\Gamma}$ corresponding to the zero elements in vec(**A**). Then, the least-squares estimate of **a** is obtained by:

$$\hat{\mathbf{a}} = (\mathbf{\Phi'\Phi})^{-1}\mathbf{\Phi'}\text{vec}(\mathbf{\Psi}). \qquad (A2.2)$$

The updated **A** is reconstructed from **â**.

Step 2: Update weights in **W** for fixed loadings and path coefficients in **A**. Let $\mathbf{w}_p$ denote the pth column of unknown component weights in **W**, which is shared by the tth column in **V**, where $t = J + p$ ($p = 1,..., P$). Let $\mathbf{\Lambda} = \mathbf{WA}$. Let $\mathbf{V}_{(-t)}$ denote **V** whose tth column is a vector of zeros. Let $\mathbf{V}_{(t)}$ denote **V** whose columns are all zero vectors except the tth column. Let $\mathbf{\Lambda}_{(-p)}$ denote a product matrix of **W** whose pth column is a vector of zeros and **A** whose pth row is a zero vector. Let $\mathbf{\Lambda}_{(p)}$ denote a product matrix of **W** whose columns are all zero vectors except the pth column and **A** whose rows are all zero vectors except the pth row. Let $\mathbf{m}_{(t)}$ denote a 1 by $J+P$ vector whose elements are all zeros except the tth element being unity. Let $\mathbf{a}_{(p)}$ denote the pth row of **A**. To update $\mathbf{w}_p$, Equation 2.17 can also be rewritten as:

$$\phi = SS(\mathbf{Z}[\mathbf{V} - \mathbf{\Lambda}])$$

$$= SS\Big(\mathbf{Z}\big[(\mathbf{V}_{(t)} + \mathbf{V}_{(-t)}) - (\mathbf{\Lambda}_{(p)} + \mathbf{\Lambda}_{(-p)})\big]\Big)$$

$$= SS\Big(\mathbf{Z}\big[(\mathbf{V}_{(t)} + \mathbf{\Lambda}_{(p)}) - (\mathbf{\Lambda}_{(-p)} - \mathbf{\Lambda}_{(-t)})\big]\Big)$$

$$= SS\Big(\mathbf{Z}\big[(\mathbf{w}_p(\mathbf{m}_{(t)} - \mathbf{a}_{(p)}) - \mathbf{\Delta}\big]\Big)$$

$$= SS\Big(\mathbf{Z}\big[(\mathbf{w}_p\,\boldsymbol{\beta} - \mathbf{\Delta}\big]\Big)$$

$$= SS\Big(vec(\mathbf{Z}\mathbf{w}_p\boldsymbol{\beta}) - vec(\mathbf{Z}\mathbf{\Delta})\Big)$$

$$= SS\Big((\boldsymbol{\beta}' \otimes \mathbf{Z})\mathbf{w}_p - vec(\mathbf{Z}\mathbf{\Delta})\Big), \qquad\qquad (\text{A2.3})$$

where

$$\boldsymbol{\beta} = \mathbf{m}_{(t)} - \mathbf{a}_{(p)},$$

and

$$\mathbf{\Delta} = \mathbf{\Lambda}_{(-p)} - \mathbf{V}_{(-t)}.$$

Thus, Equation A2.3 can generally be expressed as:

$$\phi = \sum_{p=1}^{P} SS((\boldsymbol{\beta}' \otimes \mathbf{Z})\,\mathbf{w}_p - vec(\mathbf{Z}\mathbf{\Delta})) \qquad\qquad (\text{A2.4})$$

Let $\boldsymbol{\theta}_p$ denote the vector formed by eliminating any zero elements from $\mathbf{w}_p$. Let $\boldsymbol{\Xi}$ denote the matrix formed by eliminating the columns of $\boldsymbol{\beta}' \otimes \mathbf{Z}$ corresponding to the zero elements in $\mathbf{w}_p$. Then, the least-squares estimate of $\boldsymbol{\theta}_p$ is obtained by:

$$\hat{\boldsymbol{\theta}}_t = (\boldsymbol{\Xi}'\boldsymbol{\Xi})^{-1}\boldsymbol{\Xi}'vec(\mathbf{Z}\mathbf{\Delta}). \qquad\qquad (\text{A2.5})$$

The updated is recovered from $\hat{\boldsymbol{\theta}}_p$, and subsequently multiplied by

$$\sqrt{\frac{N}{\mathbf{w}_p'\mathbf{Z}'\mathbf{Z}\mathbf{w}_p}}$$

in order to satisfy the standardization constraint imposed on the pth latent variable.

When N is large relative to J, the above algorithm can be made more efficient by the following procedure. Let $\mathbf{M} = \mathbf{Z}'\mathbf{Z} = \mathbf{HH}'$ denote any square root decomposition of $\mathbf{M}$. For example, $\mathbf{H}$ can be a lower triangular matrix obtained by the Cholesky factorization. Then, Equation 2.18 can be rewritten as:

$$\phi = \text{trace}((\mathbf{V} - \mathbf{WA})'\mathbf{M}(\mathbf{V} - \mathbf{WA}))$$

$$= \text{trace}((\mathbf{V} - \mathbf{WA})'\mathbf{HH}'(\mathbf{V} - \mathbf{WA}))$$

$$= \text{SS}(\mathbf{H}'(\mathbf{V} - \mathbf{WA}))$$

$$= \text{SS}(\mathbf{H}'\mathbf{V} - \mathbf{H}'\mathbf{WA}) \tag{A.6}$$

It is computationally more efficient to minimize A2.6 because the size of $\mathbf{H}$ is usually much smaller than that of $\mathbf{Z}$. Moreover, this procedure permits using the covariance or correlation matrix of indicators, from which $\mathbf{M}$ is easily obtained, as discussed in Section 2.2.

A2.1.1 A MATLAB Code of the Alternating Least-Squares Algorithm

We provide a sample MATLAB code of the algorithm below, which is written for the analysis of the ACSI example in Section 2.4. In the code, users need to specify two matrices of weights and path coefficients (i.e., **W0** and **B0**), in which 1 denotes a free parameter to be estimated and 0 is a fixed zero constant.

```
% Z = N by J matrix of standardized indicators
% W = J by P matrix of weights
% C = P by J matrix of loadings
% B = P by P matrix of path coefficients
% A = [C, B];
% V = [I, W];

% load data
Z0 = load('C:\asci.dat');

% specify W and B (1 = free and 0 = fixed)
W0 = [1 0 0 0 0 0
1 0 0 0 0 0
1 0 0 0 0 0
0 1 0 0 0 0
0 1 0 0 0 0
0 1 0 0 0 0
0 0 1 0 0 0
0 0 1 0 0 0
0 0 0 1 0 0
```

```
0 0 0 1 0 0
0 0 0 1 0 0
0 0 0 0 1 0
0 0 0 0 0 1
0 0 0 0 0 1];

B0 = [0 1 1 1 0 0
0 0 1 1 0 0
0 0 0 1 0 0
0 0 0 0 1 1
0 0 0 0 0 1
0 0 0 0 0 0];

N = size(Z0,1);
[J P] = size(W0);
T = J + P;
C0 = W0';
A0 = [C0,B0];
windex = find(W0);
aindex = find(A0);
% standardize data
Z = zscore(Z0)*sqrt(N)/sqrt(N – 1);
% assign random values to W and A
W = W0;
A = A0;
W(windex) = rand(length(windex),1);
A(aindex) = rand(length(aindex),1);
V = [eye(J),W];
Gamma = Z*W;
Psi = Z*V;
vecPsi = reshape(Psi,N*T,1);
it = 0;
f0 = 100000000;
imp = 100000000;
while it < = 100 && imp > 0.00001
        it = it + 1;
        % Step 1: Update A
        Phi = kron(eye(T),Gamma);
        Phi = Phi(:,aindex);
        A(aindex) = (Phi'*Phi)\Phi'*vecPsi;
        % Step 2: Update W
        for p = 1:P
        t = J + p;
        w0 = W0(:,p);
        windex_p = find(w0);
```

```
m = zeros(1,T);
m(t) = 1;
a = A(p,:);
beta = m − a;
H1 = eye(P);
H2 = eye(T);
H1(p,p) = 0;
H2(t,t) = 0;
Delta = W*H1*A − V*H2;
vecZDelta = reshape(Z*Delta, N*T, 1);
XI = kron(beta',Z);
XI = XI(:,windex_p);
theta = (XI'*XI)\XI'*vecZDelta;
zw = Z(:,windex_p)*theta;
theta = sqrt(N)*theta/norm(zw);
W(windex_p,p) = theta;
V(windex_p,t) = theta;
end
Gamma = Z*W;
Psi = Z*V;
dif = Psi-Gamma*A;
f = trace(dif'*dif);
imp = f0 − f;
f0 = f;
vecPsi = reshape(Psi,N*T,1);
end
```

Appendix 2.2 Extensions of the Least-Squares Criterion for Generalized Structured Component Analysis

A2.2.1 Differential Weighting of the Criteria for the Measurement and Structural Models

As stated earlier, Equation 2.17 can be expressed as:

$$\phi = \mathrm{SS}(\mathbf{ZV} − \mathbf{ZWA})$$

$$= \mathrm{SS}(\mathbf{Z} − \mathbf{ZWC}) + \mathrm{SS}(\mathbf{ZW} − \mathbf{ZWB}). \qquad (A2.7)$$

The first term of the criterion is the least-squares criterion for the measurement model, whereas the second is the least-squares criterion for the structural model. In situations where the number of indicators is much greater than that of latent variables, the first term is likely to dominate the value of the criterion. A potential consequence is that resultant latent variable scores can be quite similar to the first principal component scores because, as discussed in Section 2.5.1, the first term appears identical to the criterion for principal component analysis. Although we do not see a problem with this case, we may also consider some mechanism to control for the influence of each term on final solutions. An easy way is to differentially weigh each term in advance and to minimize the following criterion:

$$\phi = \alpha \ SS(\mathbf{Z} - \mathbf{ZWC}) + \beta SS(\mathbf{ZW} - \mathbf{ZWB}), \tag{A2.8}$$

where α and β are scalar weights, subject to $\alpha + \beta = 1$. We should a priori specify the values of α and β based on research objectives or interests. By specifying $\alpha = \beta = 0.5$, we assume that both terms contribute equally to the final solution, indicating that we minimize the original criterion. We may also adjust for the influence of the two terms on the final solution by differently weighing the two terms. For example, we wish to weigh the second term more heavily than the first (i.e., $\beta > \alpha$), so that the structural model has a greater impact on the final solution. Thus, this differential weighting serves as a vehicle for investigating alternative solutions. As stated above, however, their values are situation-dependent and consequently should be determined by the objectives of the analysis. In other words, there is no automated way of selecting the values of the scalar weights.

A2.2.2 Incorporation of Metric Matrices into the Criterion and Robust Estimation

We may consider the integration of metric matrices into Equation 2.17. Two types of metric matrices are available: one is on the row side of and the other on the column side of the matrix $\mathbf{\Psi}$ (Takane and Shibayama 1991). Let $\mathbf{K}$ denote an N by N row-side metric matrix. Let $\mathbf{L}$ denote a T by T column-side metric matrix. Matrices $\mathbf{K}$ and $\mathbf{L}$ are both assumed to be non-negative definite. To estimate $\mathbf{W}$ and $\mathbf{A}$ with these metric matrices incorporated, we minimize

$$\phi = SS(\mathbf{ZV} - \mathbf{ZWA})_{K,L} \tag{A2.9}$$

where

$$SS(\mathbf{X})_{K,L} = \text{trace}(\mathbf{KXLX'})$$

When $\mathbf{K} = \mathbf{I}$ and $\mathbf{L} = \mathbf{I}$, Equation A2.9 reduces to Equation 2.17. When $\mathbf{K} \neq \mathbf{I}$ and/or $\mathbf{L} \neq \mathbf{I}$, Equation A2.9 can still be expressed as Equation 2.17 by a simple transformation (e.g., Rao 1980) as follows. Let $\mathbf{K} = \mathbf{R}_K\mathbf{R}_K'$ and $\mathbf{L} = \mathbf{R}_L\mathbf{R}_L'$

be any square root decompositions of **K** and **L**. Then, Equation A2.9 can be rewritten as

$$\phi = SS(\mathbf{R}_k'\mathbf{Z}\mathbf{V}\mathbf{R}_L - \mathbf{R}_k'\mathbf{Z}\mathbf{W}\mathbf{A}\mathbf{R}_L)$$

$$= SS(\tilde{\mathbf{Z}}\tilde{\mathbf{V}} - \tilde{\mathbf{Z}}\mathbf{W}\tilde{\mathbf{A}}), \qquad (A2.10)$$

where $\tilde{\mathbf{Z}} = \mathbf{R}_K'\mathbf{Z}$, $\tilde{\mathbf{V}} = \mathbf{V}\mathbf{R}_L$, and $\tilde{\mathbf{A}} = \mathbf{A}\mathbf{R}_L$. This criterion is essentially of the same form as Equation 2.17, and can be minimized in a similar way.

Judicious choices on metric matrices may broaden the analytic capacity of generalized structured component analysis. Takane and Hunter (2001) discussed a variety of nonidentity metric matrices, which could lead to more diverse data analyses. For instance, Meredith and Millsap (1985) suggested using the matrix of reliability coefficients or of the inverses of the variances of anti-images (Guttman 1953) as **L**. When the columns of the residual matrix are correlated and/or have markedly different variances after a model is fitted to the data, the variance–covariance matrix among the residuals may be estimated and its inverse be used as **L**. This can be of use in obtaining smaller mean squared errors of parameter estimates by orthonormalizing the residuals in evaluating the overall goodness of fit (Takane and Hunter 2001). In addition, Rao (1964) pointed out that scale invariance might be achieved by specifying certain nonidentity **L** matrices. When the rows of **Z** consist of repeated observations made by the same subject, they are likely to be correlated. In this case, a matrix of serial correlations is estimated and its inverse can be used as **K** (Escoufier 1987) so as to achieve independence of observations. Moreover, when the importance and/or reliability of each row is suspected to be different, a special kind of diagonal matrix may be used for **K** that has the effect of differentially weighting rows of a data matrix. For example, in correspondence analysis, the square root of row totals of a contingency table is used as **K**.

Although **K** and **L** are considered to be fixed matrices above, we may also consider metric matrices that are to be updated iteratively during the optimization procedure. We here provide an example of adopting such an iteratively updated metric matrix for dealing with outliers in generalized structured component analysis.

In general, outlier diagnostics and robust estimation are used for dealing with outliers. In the diagnostic approach, we first attempt to identify potential outliers. This is typically followed by either their removal or some form of adjustment before we fit the data by traditional methods (Belsley, Kuh, and Welsch 1980; Cook and Weisberg 1982). On the other hand, the robust estimation approach first involves fitting a model to a majority of data points and the subsequent detection of outliers that are represented by observations with relatively large residuals (Rousseeuw and Leroy 1987). These two approaches pursued the same goal but proceeded in an opposite direction (Rousseeuw and Leroy 1987). Nonetheless, comparative studies have demonstrated that the diagnostic approach could often fail to detect multiple outliers (Singh 1996; Walczak and Massart 1995).

Thus, we focus here on a robust estimation method for generalized structured component analysis, which is based on an iteratively reweighted least-squares method (Beaton and Tukey 1974). This method is relatively easy to implement and attractive for data reduction techniques (e.g., Griep et al. 1995).

In the robust estimation method, a metric matrix $\mathbf{K}$ is constructed that assigns different weights to observations, depending on the size of outliers. This metric matrix is iteratively re-estimated by the following procedure. Let

$$\mathbf{K} = \mathrm{diag}(k_{11}, k_{22}, \ldots, k_{ll})$$

denote an l by l diagonal matrix of observation weights, where $l = N \times T$. Let $\mathbf{K} = \mathbf{R}_K \mathbf{R}_K'$, then, the least-squares criterion for the robust estimation method can be written as

$$\phi = \mathrm{SS}(\mathrm{vec}(\mathbf{ZV}) - \mathrm{vec}(\mathbf{ZWA}))_{k,I}$$

$$= \mathrm{SS}(\mathbf{R}_k'(\mathrm{vec}(\mathbf{ZV}) - \mathrm{vec}(\mathbf{ZWA}))). \tag{A2.11}$$

In Equation A2.11, $\mathbf{W}$ and $\mathbf{A}$ can be obtained for fixed $\mathbf{K}$ by the same alternating least-squares algorithm described in Appendix 2.1. After $\mathbf{W}$ and $\mathbf{A}$ are updated, a vector of residuals, denoted by $\mathbf{r}$, is calculated by

$$\mathbf{r} = \mathrm{vec}(\mathbf{ZV}) - \mathrm{vec}(\mathbf{ZWA}). \tag{A2.12}$$

The median of the absolute values of residuals, say δ, is then computed as

$$\delta = \mathrm{median}(|r_i|), \text{ for } i = 1, \ldots, l. \tag{A2.13}$$

Using Equations A2.12 and A2.13, the elements of $\mathbf{K}$ are calculated by

$$k_{ii} = \begin{cases} [1 - (r_i / h\delta)^2]^2 & \text{for } |r_i| < h\delta \\ 0 & \text{for } |r_i| \geq h\delta \end{cases}. \tag{A2.14}$$

These steps are repeated until convergence. In Equation A2.14, h represents a variable sensitivity factor that defines a threshold beyond which a weight of zero is assigned to that observation. Either six or nine is usually used for h, but the results are quite similar (Wakeling and Macfie 1992).

Thus, the aforementioned procedure for robust estimation in generalized structured component analysis involves the construction of a nonidentity row-side metric matrix $\mathbf{K}$ in an iterative manner. The metric matrix has the effect of differentially weighting rows or observations of $\boldsymbol{\Psi}$. In Chapter 4, we will show another example of obtaining a row-side metric matrix iteratively, which consists of cluster memberships of individuals, for the capturing of cluster-level heterogeneity in generalized structured component analysis.

Appendix 2.3 A Partial Least Squares Path Modeling Algorithm

We provide a description of Lohmöller's (1989) algorithm for partial least squares path modeling. This algorithm carries out two main stages sequentially. The first stage estimates weights and latent variables iteratively, whereas the second estimates path coefficients and loadings in closed form. The first stage is the most crucial procedure, on which the second stage builds for the estimation of path coefficients and loadings (Hanafi 2007).

Let γ_p and $\mathbf{w}_p$ denote a vector of a latent variable and a vector of weights for a block of indicators for the latent variable, respectively. Let $\mathbf{Z}_p$ denote a matrix of indicators associated with γ_p. Conventionally, both indicators and latent variables are assumed to be standardized in partial least squares path modeling. However, they are to be normalized here, which makes the exposition of equations simpler while producing identical estimates of weights, path coefficients, and loadings. The individual scores of standardized latent variables can always be obtained by multiplying their normalized scores by the square root of sample size.

In the first stage, we repeat the following steps to estimate $\mathbf{w}_p$ and γ_p.

Step 0: Initialize $\mathbf{w}_p$.

Step 1: Compute $\gamma_p = \mathbf{Z}_p \mathbf{w}_p$ and normalize γ_p such that $\gamma_p' \gamma_p = 1$.

Step 2 (internal estimation): Compute a weighed composite of the latent variables connected to γ_p in a given structural model. This weighted composite, denoted by $\mathbf{q}_p$, is called the inner estimate, which takes the general form as follows:

$$\mathbf{q}_p = \sum_{q=1}^{Q} \tau_{qj} \gamma_q ,\qquad (A2.15)$$

where τ_{qp} is a scalar value, called the inner weight, which is assigned to each of Q latent variables (γ_q) connected to γ_p. Three different ways, so-called schemes, are available for the calculation of the inner weight: centroid (Wold 1982), factorial (Lohmöller 1989), and path weighting. In the centroid scheme, τ_{qp}'s are the signs of the correlations between γ_q's and γ_p. In the factorial scheme, τ_{qp}'s are equivalent to the correlations between γ_q's and γ_p. In the path weighting scheme, τ_{qp}'s are the signs the regression coefficients of γ_p on γ_q's when γ_p is the dependent variable, whereas they are the correlations between γ_q's and γ_p when γ_p is a predictor variable. The path weighting scheme is recommended because it takes into account both directions and magnitudes of the relationships between latent variables (Esposito Vinzi, Trinchera, and Amato 2010).

Step 3 (external estimation): Update $\mathbf{w}_p$. For Mode A, $\mathbf{w}_p$ is updated by:

$$\hat{\mathbf{w}}_p = \mathbf{Z}_p'\mathbf{q}_p. \tag{A2.16}$$

For Mode B, $\mathbf{w}_p$ is updated by:

$$\hat{\mathbf{w}}_p = (\mathbf{Z}_p'\mathbf{Z}_p)^{-1}\mathbf{Z}_p'\mathbf{q}_p. \tag{A2.17}$$

Repeat Steps 1–3 until there are no substantial differences in the previous and current weight values for all blocks of indicators.

Once the first stage is complete, we then estimate the path coefficients of latent variables for each endogenous latent variable by means of ordinary least-squares regression. Let $\boldsymbol{\Gamma}_p$ denote a matrix of latent variables that influence γ_p. Let $\mathbf{b}_p$ denote a vector of path coefficients connecting $\boldsymbol{\Gamma}_p$ to γ_p. Then, the least-squares estimate of $\mathbf{b}_p$ is obtained by:

$$\hat{\mathbf{b}}_p = (\boldsymbol{\Gamma}_p'\boldsymbol{\Gamma}_p)^{-1}\boldsymbol{\Gamma}_p'\gamma_p. \tag{A2.18}$$

Moreover, we can estimate the loadings of reflective indicators for their latent variable by means of ordinary least-squares regression. Let $\mathbf{c}_p$ denote a vector of loadings connecting γ_p to $\mathbf{Z}_p$. Then, the least-squares estimate of $\mathbf{c}_p$ is obtained by:

$$\hat{\mathbf{c}}_p = \gamma_p'\mathbf{Z}_p. \tag{A2.19}$$

3

Basic Extensions of Generalized Structured Component Analysis

In this chapter, we discuss several extensions of generalized structured component analysis. They include constrained analyses, modeling of higher-order latent variables, multiple group analyses, the calculation of total and indirect effects, and missing data imputation.

3.1 Constrained Analysis

Parameters can be constrained in generalized structured component analysis. A common practice is to constrain two or more parameters to be related in a particular manner. For example, path coefficient A can be set to be identical to path coefficient B, or path coefficient A is to be equal to the average of path coefficients B and C. This constrained analysis is of use in examining hypothesized relationships between parameters. Another practice is to constrain parameters to fixed values such as zeros or other constants. For example, path coefficients can be omitted by setting them to zeros. In addition, weights or loadings can be fixed to prespecified constants. As will be discussed later, in latent growth curve models (e.g., Meredith and Tisak 1990), loadings are fixed to known (polynomial) basis functions, so that latent variables characterize certain trends of change in repeated measurements. We discuss and exemplify these constrained analyses in this section.

3.1.1 Constraining Parameters for Testing Hypotheses about Them

We can incorporate a range of hypotheses about parameters in the form of linear constraints such as equality constraints. Technically, there are two ways of imposing linear constraints on parameters: the reparametrization method and the null-space method (e.g., Böckenholt and Takane 1994; Takane, Yanai, and Mayekawa 1991).

We describe the two methods in the context of linear regression (also see Takane, Yanai, and Mayekawa 1991). This will also help to understand how to integrate linear constraints into generalized structured component analysis because its estimation procedure is essentially equivalent to solving linear regression problems, as discussed in Chapter 2.

Let us consider a linear regression model

$$y = X\beta + \varepsilon, \tag{3.1}$$

where y is a vector of the dependent variable, X is a matrix of predictor variables, β is a vector consisting of regression coefficients, and ε is a vector of residuals. In the reparametrization method, linear constraints are incorporated into β as follows.

$$\beta = M\beta^*, \tag{3.2}$$

where M is a matrix of linear constraints specified by the reparametrization method and β^* is a vector of regression coefficients reduced by the linear constraints imposed.

On the other hand, in the null-space method, linear constraints are incorporated into β as follows.

$$R'\beta = 0, \tag{3.3}$$

where R is a matrix of linear constraints specified by the null-space method and 0 is a vector of zeros.

Figure 3.1 displays prototypes of unconstrained and constrained linear regression models. Panel (a) of Figure 3.1 shows the unconstrained model, where each of four predictor variables has an effect on the dependent variable. In this model, $\beta = [\beta_1, \beta_2, \beta_3, \beta_4]'$ is a 4 by 1 vector consisting of the regression coefficients of the four predictor variables. Panel (b) exhibits a constrained model in which two equality constraints are imposed on β, such as $\beta_1 = \beta_3$ and $\beta_2 = \beta_4$.

Based on the reparametrization method (Equation 3.2), these equality constraints can be incorporated into β as follows.

$$\begin{bmatrix} \beta_1 \\ \beta_2 \\ \beta_3 \\ \beta_4 \end{bmatrix} = \begin{bmatrix} 1 & 0 \\ 0 & 1 \\ 1 & 0 \\ 0 & 1 \end{bmatrix} \begin{bmatrix} \beta_1^* \\ \beta_2^* \end{bmatrix}$$

$$\beta = M\beta^*. \tag{3.4}$$

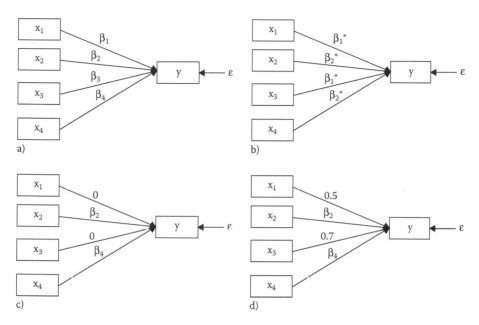

FIGURE 3.1
Prototypes of unconstrained and constrained linear regression models. a) Unconstrained model. b) Constrained model ($\beta_1 = \beta_3$ and $\beta_2 = \beta_4$). c) Constrained model ($\beta_1 = \beta_3 = 0$). d) Constrained model ($\beta_1 = 0.5$ and $\beta_3 = 0.7$).

In Equation 3.4, β_1 and β_3 are set to be equal to β_1^*, and β_2 and β_4 are to be equal to β_2^*. From Equation 3.4, the linear regression model (Equation 3.1) can be re-expressed as

$$y = XM\beta^* + \varepsilon. \tag{3.5}$$

The least-squares estimate of β^* is then obtained by

$$\hat{\beta}^* = (M'X'XM)^{-1}M'X'y. \tag{3.6}$$

Consequently, the estimate of β is given as

$$\hat{\beta} = M(M'X'XM)^{-1}M'X'y. \tag{3.7}$$

This constrained estimation procedure is called the projection method (e.g., Seber 1984, pp. 403–405; also see Takane, Yanai, and Mayekawa 1991).

Based on the null-space method, the same equality constraints can be incorporated into β as follows.

$$\begin{bmatrix} 1 & 0 & -1 & 0 \\ 0 & 1 & 0 & -1 \end{bmatrix} \begin{bmatrix} \beta_1 \\ \beta_2 \\ \beta_3 \\ \beta_4 \end{bmatrix} = \begin{bmatrix} 0 \\ 0 \end{bmatrix}$$

$$\mathbf{R'\beta} = 0. \tag{3.8}$$

This formulation indicates that $\beta_1 - \beta_3 = 0$ or equivalently $\beta_1 = \beta_3$, and $\beta_2 - \beta_4 = 0$ or $\beta_2 = \beta_4$. By combining Equations 3.8 and 3.1, the estimate of $\boldsymbol{\beta}$ is obtained by

$$\hat{\boldsymbol{\beta}} = \left[(\mathbf{X'X})^{-1} - (\mathbf{X'X})^{-1}\mathbf{R}(\mathbf{R'(X'X)}^{-1}\mathbf{R})^{-1}\mathbf{R'(X'X)}^{-1} \right] \mathbf{X'y}. \tag{3.9}$$

The above estimation procedure is called the Lagrangian multiplier method (Searle 1971, pp. 110–123; also see Takane, Yanai, and Mayekawa 1991). Choices between the reparametrization and null-space methods are arbitrary and depend merely on empirical applications (Takane, Yanai, and Mayekawa 1991). Nonetheless, the reparametrization method appears more natural when researchers wish to constrain parameters directly (Böckenholt and Takane 1994).

In generalized structured component analysis, all linear constraints are imposed on parameters via the reparametrization method (Hwang and Takane 2004a). This method is well suited to the estimation procedure of generalized structured component analysis. As an example, we show how to impose linear constraints on loadings and path coefficients. As discussed in Chapter 2, the alternating least-squares algorithm for generalized structured component analysis is composed of two steps. One step updates the estimates of weights included in the $\mathbf{W}$ matrix. The other updates the estimates of loadings and path coefficients in the $\mathbf{A}$ matrix by minimizing the following least-squares criterion (refer to Appendix 2.1).

$$\phi = SS(\mathbf{\Psi} - \mathbf{\Gamma A})$$

$$= SS(\text{vec}(\mathbf{\Psi}) - \text{vec}(\mathbf{\Gamma A}))$$

$$= SS(\text{vec}(\mathbf{\Psi}) - (\mathbf{I} \otimes \mathbf{\Gamma})\text{vec}(\mathbf{A}))$$

$$= SS(\text{vec}(\mathbf{\Psi}) - \mathbf{\Phi a}), \tag{3.10}$$

where $\text{vec}(\mathbf{H})$ indicates a vector formed by stacking all columns of $\mathbf{H}$ one below another, $\otimes$ indicates a Kronecker product, $\mathbf{a}$ is the vector formed by eliminating zero elements from $\text{vec}(\mathbf{A})$, which includes only free loadings and path coefficients to be estimated, and Φ is the matrix formed by eliminating the columns of $\mathbf{I} \otimes \mathbf{\Gamma}$ corresponding to the zero elements in $\text{vec}(\mathbf{A})$.

Minimizing Equation 3.10 with respect to $\mathbf{a}$ is equivalent to solving a linear regression problem, where $\text{vec}(\mathbf{\Psi})$ is considered a vector of the dependent variable and Φ is a matrix of predictor variables. Thus, we can incorporate constraints into $\mathbf{a}$ in the same manner as Equation 3.2, as follows.

$$\mathbf{a} = \mathbf{Ma}^*, \tag{3.11}$$

for some $\mathbf{a}^*$ vector of loadings and path coefficients reduced by linear constraints. By putting Equation 3.11 into Equation 3.10, the least-squares estimate of $\mathbf{a}^*$ is obtained by

$$\hat{\mathbf{a}}^* = (\mathbf{M}'\mathbf{\Phi}'\mathbf{\Phi}\mathbf{M})^{-1}\mathbf{M}'\mathbf{\Phi}'\text{vec}(\mathbf{\Psi}). \tag{3.12}$$

The estimate of $\mathbf{a}$ is subsequently given as

$$\hat{\mathbf{a}} = \mathbf{M}(\mathbf{M}'\mathbf{\Phi}'\mathbf{\Phi}\mathbf{M})^{-1}\mathbf{M}'\mathbf{\Phi}'\text{vec}(\mathbf{\Psi}). \tag{3.13}$$

This constrained estimation step is repeated through iterations until convergence. Linear constraints can be imposed on weights in a similar manner. We demonstrate the imposition of equality constraints on loadings and path coefficients below.

3.1.1.1 Example: The ACSI Model with the Imposition of Equality Constraints

We analyzed the consumer-level ACSI data used in Chapter 2, which included the responses of 774 consumers to the service units within the US sector of public administration. In this example, we assumed equality among loadings as follows: the loadings for z_1 (customer expectations about overall quality) and z_2 (customer expectations about reliability) were identical (i.e., $c_1 = c_2$), the loadings for z_4 (overall quality) and z_5 (reliability) were identical (i.e., $c_4 = c_5$), the loadings for z_9 (overall customer satisfaction, z_{10} (confirmation of expectations), and z_{11} (distance to ideal product or service) were identical (i.e., $c_9 = c_{10} = c_{11}$), and the loadings for z_{13} (repurchase intention) and z_{14} (price tolerance) were identical (i.e., $c_{13} = c_{14}$). Moreover, we assumed that *perceived quality* had identical effects on *perceived value* and *customer satisfaction* (i.e., $b_4 = b_5$).

As described in Chapter 2 (e.g., see Equations 2.26 and 2.28), the ACSI model involved 14 loadings and nine path coefficients. Thus, the $\mathbf{a}$ vector in Equation 3.11 is equivalent to a 23 by 1 vector that stacks the nine path coefficients below the 14 loadings. We incorporated the equality constraints into $\mathbf{a}$ via the reparametrization method, as follows.

$$
\begin{bmatrix} c_1 \\ c_2 \\ c_3 \\ c_4 \\ c_5 \\ c_6 \\ c_7 \\ c_8 \\ c_9 \\ c_{10} \\ c_{11} \\ c_{12} \\ c_{13} \\ c_{14} \\ b_1 \\ b_2 \\ b_3 \\ b_4 \\ b_5 \\ b_6 \\ b_7 \\ b_8 \\ b_9 \end{bmatrix}
=
\begin{bmatrix}
1&0&0&0&0&0&0&0&0&0&0&0&0&0&0&0&0\\
1&0&0&0&0&0&0&0&0&0&0&0&0&0&0&0&0\\
0&1&0&0&0&0&0&0&0&0&0&0&0&0&0&0&0\\
0&0&1&0&0&0&0&0&0&0&0&0&0&0&0&0&0\\
0&0&1&0&0&0&0&0&0&0&0&0&0&0&0&0&0\\
0&0&0&1&0&0&0&0&0&0&0&0&0&0&0&0&0\\
0&0&0&0&1&0&0&0&0&0&0&0&0&0&0&0&0\\
0&0&0&0&0&1&0&0&0&0&0&0&0&0&0&0&0\\
0&0&0&0&0&0&1&0&0&0&0&0&0&0&0&0&0\\
0&0&0&0&0&0&1&0&0&0&0&0&0&0&0&0&0\\
0&0&0&0&0&0&1&0&0&0&0&0&0&0&0&0&0\\
0&0&0&0&0&0&0&1&0&0&0&0&0&0&0&0&0\\
0&0&0&0&0&0&0&0&1&0&0&0&0&0&0&0&0\\
0&0&0&0&0&0&0&0&1&0&0&0&0&0&0&0&0\\
0&0&0&0&0&0&0&0&0&1&0&0&0&0&0&0&0\\
0&0&0&0&0&0&0&0&0&0&1&0&0&0&0&0&0\\
0&0&0&0&0&0&0&0&0&0&0&1&0&0&0&0&0\\
0&0&0&0&0&0&0&0&0&0&0&0&1&0&0&0&0\\
0&0&0&0&0&0&0&0&0&0&0&0&1&0&0&0&0\\
0&0&0&0&0&0&0&0&0&0&0&0&0&1&0&0&0\\
0&0&0&0&0&0&0&0&0&0&0&0&0&0&1&0&0\\
0&0&0&0&0&0&0&0&0&0&0&0&0&0&0&1&0\\
0&0&0&0&0&0&0&0&0&0&0&0&0&0&0&0&1
\end{bmatrix}
\begin{bmatrix} c_1^* \\ c_3 \\ c_2^* \\ c_6 \\ c_7 \\ c_8 \\ c_3^* \\ c_{12} \\ c_4^* \\ b_1 \\ b_2 \\ b_3 \\ b^* \\ b_6 \\ b_7 \\ b_8 \\ b_9 \end{bmatrix}
$$

$$\mathbf{a} = \mathbf{M}\mathbf{a}^*. \tag{3.14}$$

For the ACSI model with the equality constraints imposed, generalized structured component analysis provided that FIT = 0.67 (SE = 0.01, 95% CI = 0.66–0.69), AFIT = 0.67 (SE = 0.01, 95% CI = 0.66–0.69), GFI = 0.99 (SE = 0.00, 95% CI = 0.99–1.00), and SRMR = 0.07 (SE = 0.00, 95% CI = 0.07–0.08). These overall fit values are almost identical to those obtained from the unconstrained ACSI model reported in Chapter 2 (see Section 2.4). In particular, despite the imposition of the constraints, the AFIT value of the constrained model virtually remained the same as that of the unconstrained one. Furthermore, we conducted a *t*-test to compare the FIT values between the unconstrained and constrained models, using 100 bootstrap samples. We found that the *t*-statistic was statistically nonsignificant [$t(99) = 0.53$, $p = 0.60$, 95% CI = −0.00–0.00], indicating that there was no statistically significant difference in FIT between the constrained and unconstrained models. All these suggest that the constrained model be favored to the unconstrained one.

Although they are not reported to conserve space, the weight estimates of the constrained model were quite similar to those of the unconstrained model, thus leading to the same interpretations. Table 3.1 provides the loading estimates obtained from the unconstrained and constrained models. Owing to the imposition of the equality constraints, several loading estimates turned out to be identical. The other loading estimates remained similar to the corresponding estimates from the unconstrained model. The imposition of the constraints on loadings did not change the value of FIT_M (0.80, SE = 0.01, 95% CI = 0.78–0.82) substantially, when compared to that of the unconstrained model (0.80, SE = 0.01, 95% CI = 0.79–0.82). This indicates that the constrained measurement model accounted for essentially the same amount of the total variance of all indicators. Table 3.2 lists the path coefficient estimates obtained from the unconstrained and constrained models. Again, owing to the equality constraint on b_4 and b_5, these path coefficient estimates became identical in the constrained model. The other path coefficient estimates were quite similar to the corresponding estimates from the unconstrained model. The imposition of the equality constraint on the path coefficients led to no substantial change in FIT_S (0.38, SE = 0.02, 95% CI = 0.35 0.41), when compared to that of the unconstrained model (FIT_S = 0.38, SE = 0.01, 95% CI = 0.35–0.40). This indicates that the constrained structural model explained almost the same amount of the total variance of the latent variables. More specifically, the R-squared values of *perceived value* ($R^2 = 0.54$) and *customer satisfaction* ($R^2 = 0.81$) were similar to their counterparts obtained under the unconstrained model. All the loading and path coefficient estimates obtained from the constrained model could be interpreted in the same manner as those from the unconstrained model.

3.1.2 Constraining Parameters to Fixed Values

We now consider another type of constrained analysis, where parameters are set to fixed values in advance. A simple case is to fix parameters to zero. Panel (c) of Figure 3.1 displays such a case in linear regression, where two regression coefficients are set to zero, that is, $\beta_1 = \beta_3 = 0$, indicating that the corresponding predictor variables have no impacts on the dependent variable. The imposition of these zero constraints is equivalent to simply omitting the two predictor variables and their regression coefficients from the model. The remaining regression coefficients are estimated subsequently. As discussed in Appendix 2.1, in generalized structured component analysis, parameters are estimated, while eliminating any zero values in the **W** and **A** matrices and the corresponding columns of their predictor matrices (e.g., **Φ** and **Ξ**). Thus, no additional treatment is necessary for the imposition of zero constraints.

TABLE 3.1

Estimates of Loadings and Their Standard Errors and 95% Confidence Intervals in the Unconstrained and Constrained ACSI Models

Latent	Indicator	Unconstrained				Constrained			
		Estimate	SE	$|t|$	CI	Estimate	SE	$|t|$	CI
CE	z_1	0.86	0.01	66.78	0.83–0.89	0.87	0.01	79.60	0.85–0.89
	z_2	0.88	0.01	76.37	0.84–0.90	0.87	0.01	79.60	0.85–0.90
	z_3	0.73	0.03	27.40	0.68–0.77	0.73	0.03	24.45	0.66–0.78
PQ	z_4	0.94	0.01	133.91	0.93–0.95	0.94	0.01	142.14	0.92–0.95
	z_5	0.94	0.01	126.24	0.92–0.95	0.94	0.01	142.14	0.92–0.95
	z_6	0.79	0.02	34.69	0.74–0.83	0.79	0.02	35.39	0.74–0.83
PV	z_7	0.79	0.02	36.55	0.75–0.82	0.79	0.02	32.75	0.73–0.82
	z_8	0.94	0.01	156.77	0.93–0.95	0.94	0.01	192.37	0.93–0.95
CS	z_9	0.95	0.01	141.10	0.94–0.97	0.92	0.01	139.77	0.91–0.94
	z_{10}	0.91	0.01	73.22	0.88–0.93	0.92	0.01	139.77	0.91–0.94
	z_{11}	0.91	0.01	92.44	0.89–0.93	0.92	0.01	139.77	0.91–0.94
CC	z_{12}	1	0			1	0		
CL	z_{13}	0.96	0.00	238.31	0.95–0.96	0.94	0.00	234.58	0.93–0.95
	z_{14}	0.92	0.01	147.13	0.91–0.93	0.94	0.00	234.58	0.93–0.95

CE, Customer expectations; CI, Confidence intervals; CL, Customer loyalty; CS, Customer satisfaction; PQ, Perceived quality; PV, Perceived value; SE, Standard errors.

TABLE 3.2

Estimates of Path Coefficients and Their Standard Errors and 95% Confidence Intervals in the Unconstrained and Constrained ACSI Models

	Unconstrained				Constrained			
	Estimate	SE	$|t|$	CI	Estimate	SE	$|t|$	CI
CE → PQ	0.58	0.03	19.17	0.52–0.64	0.58	0.03	18.50	0.53–0.65
CE → PV	0.12	0.03	30.71	0.07–0.18	0.12	0.04	3.19	0.05–0.21
CE → CS	0.03	0.02	1.35	−0.02–0.07	0.03	0.02	1.43	−0.01–0.08
PQ → PV	0.65	0.03	21.98	0.60–0.71	0.67	0.02	29.14	0.62–0.71
PQ → CS	0.68	0.03	21.53	0.62–0.73	0.67	0.02	29.14	0.62–0.71
PV → CS	0.26	0.03	8.47	0.21–0.33	0.26	0.03	7.89	0.19–0.33
CS → CC	−0.41	0.04	10.68	−0.48–0.32	−0.40	0.04	10.62	−0.48–0.34
CS → CL	0.59	0.04	19.18	0.53–0.65	0.59	0.04	16.62	0.52–0.66
CC → CL	−0.09	0.04	2.57	−0.16–0.03	−0.09	0.04	2.51	−0.15–0.02

CE, Customer expectations; CI, Confidence intervals; CL, Customer loyalty; CS, Customer satisfaction; PQ, Perceived quality; PV, Perceived value; SE, Standard errors.

Another case is to fix parameters to nonzero constants. Panel (d) of Figure 3.1 displays such a constrained case in linear regression, where two regression coefficients are fixed to nonzero constants, as follows: $\beta_1 = 0.5$ and $\beta_3 = 0.7$. In this case, the linear regression model (Equation 3.1) can be expressed as

$$y = X \begin{bmatrix} 0.5 \\ \beta_2 \\ 0.7 \\ \beta_4 \end{bmatrix} = [x_1, x_3] \begin{bmatrix} 0.5 \\ 0.7 \end{bmatrix} + [x_2, x_4] \begin{bmatrix} \beta_2 \\ \beta_4 \end{bmatrix} + \varepsilon$$

$$y - [x_1, x_3] \begin{bmatrix} 0.5 \\ 0.7 \end{bmatrix} = [x_2, x_4] \begin{bmatrix} \beta_2 \\ \beta_4 \end{bmatrix} + \varepsilon$$

$$y^* = X^2 \beta^2 + \varepsilon \tag{3.15}$$

where $X = [x_1, x_2, x_3, x_4]$, $y^* = y - [x_1, x_3] \begin{bmatrix} 0.5 \\ 0.7 \end{bmatrix}$, $X^2 = [x_2, x_4]$, and $\beta^2 = [\beta_2, \beta_4]'$. The least-squares estimate of β^2 is then obtained by

$$\hat{\beta}^2 = (X^2 {}' X^2)^{-1} X^2 {}' y^*. \tag{3.16}$$

In generalized structured component analysis, nonzero constant constraints can be imposed on parameters in a similar fashion. We again show how to impose nonzero constant constraints on loadings and path coefficients. Under the imposition of nonzero constant constraints, the alternating least-squares algorithm seeks to minimize the following least-squares criterion to update the estimates of loadings and path coefficients.

$$\phi = SS(\boldsymbol{\Psi} - \boldsymbol{\Gamma}\mathbf{A})$$

$$= SS(\text{vec}(\boldsymbol{\Psi}) - \text{vec}(\boldsymbol{\Gamma}\mathbf{A}))$$

$$= SS(\text{vec}(\boldsymbol{\Psi}) - (\mathbf{I} \otimes \boldsymbol{\Gamma})\text{vec}(\mathbf{A}))$$

$$= SS(\text{vec}(\boldsymbol{\Psi}) - \tilde{\boldsymbol{\Phi}}\tilde{\mathbf{a}} - \boldsymbol{\Phi}\mathbf{a}) \tag{3.17}$$

where $\mathbf{a}$ is the vector formed by eliminating any fixed elements from $\text{vec}(\mathbf{A})$, $\tilde{\mathbf{a}}$ is the vector consisting of the nonzero fixed elements in $\text{vec}(\mathbf{A})$, $\boldsymbol{\Phi}$ is the matrix formed by eliminating the columns of $\mathbf{I} \otimes \boldsymbol{\Gamma}$ corresponding to the fixed elements in $\text{vec}(\mathbf{A})$, and $\tilde{\boldsymbol{\Phi}}$ is the matrix consisting of the columns of $\mathbf{I} \otimes \boldsymbol{\Gamma}$ corresponding to the nonzero fixed elements in $\text{vec}(\mathbf{A})$. Then, the least-squares estimate of $\mathbf{a}$ is obtained by

$$\hat{\mathbf{a}} = (\boldsymbol{\Phi}'\boldsymbol{\Phi})^{-1}\boldsymbol{\Phi}'(\text{vec}(\boldsymbol{\Psi}) - \tilde{\boldsymbol{\Phi}}\tilde{\mathbf{a}}). \tag{3.18}$$

In fact, we can use this estimation procedure for the imposition of both zero and nonzero constant constraints. When a parameter is fixed to zero, the corresponding portion of $\tilde{\boldsymbol{\Phi}}\tilde{\mathbf{a}}$ becomes zero, which is equivalent to eliminating the zero element and its corresponding column of $\tilde{\boldsymbol{\Phi}}$. Any constant constraints can be imposed on weights in a similar manner.

Although fixing parameters to constants in advance is technically straightforward, it can result in substantively intriguing and meaningful analyses. Here we consider a constrained model, where loadings are fixed to constants so as to investigate a time-dependent trajectory of change in repeated measures data. This constrained model is called a latent growth curve model (Bollen and Curran 2006; Duncan, Duncan, and Strycker 2006; Meredith and Tisak 1990; Preacher et al. 2008; Willett and Bub 2005). We discuss the latent growth curve model and demonstrate the use of generalized structured component analysis for fitting the model.

3.1.2.1 Latent Growth Curve Models

As stated above, a latent growth curve model can be viewed as a constrained structural equation model where loadings are a priori set to known basis functions (e.g., polynomials) to account for a particular temporal trajectory of observed repeated measures (indicators) on a single characteristic. Moreover, in this model, both mean and variance of a latent variable are to be freely estimated. The mean and variance stand for an average or intra-individual temporal trajectory of repeated assessments and inter-individual differences in the temporal trajectory, respectively. Furthermore, it is typically assumed that observations are taken in the same unit across at least three measurement occasions and are not standardized (Kline 1998, p. 303). The intervals across occasions need not be equal.

A basic latent growth curve model can be regarded as a random-effects growth curve model (e.g., Laird and Ware 1982; Vonesh and Carter 1987) or a two-level hierarchical linear model, where occasions are nested within an individual (Bryk and Raudenbush 1992; Goldstein 1987; Curran 2003). All these techniques can be used for modeling the average trajectory of repeated measures over occasion in addition to inter-individual variations in the trajectory. They are thus distinctive from the traditional fixed-effects growth curve model (Grizzle and Allen 1969; Khatri 1966; Potthoff and Roy 1964; Rao 1965), which focuses only on discerning the average trajectory of change.

Figure 3.2 displays a prototype of a basic latent growth curve model. This prototype model involves four indicators measured repeatedly over four occasions on a single characteristic (z_1, z_2, z_3, and z_4). It also includes two latent variables (γ_1 and γ_2), each of which influences the same four indicators. The mean and variance of each latent variable are denoted by m_p and v_p, respectively ($p = 1, 2$). As shown in the figure, the loadings associated with the first latent variable (γ_1) are all fixed to unity, whereas those associated with the second latent variable (γ_2) are set to 0, 1, 2, and 3 for z_1, z_2, z_3, and z_4, respectively. Thus, the prototype model is equivalent to a constrained measurement model given as

$$z_1 = \gamma_1 + \varepsilon_1$$

$$z_2 = \gamma_1 + \gamma_2 + \varepsilon_2$$

$$z_3 = \gamma_1 + 2\gamma_2 + \varepsilon_3$$

$$z_4 = \gamma_1 + 3\gamma_2 + \varepsilon_4 \tag{3.19}$$

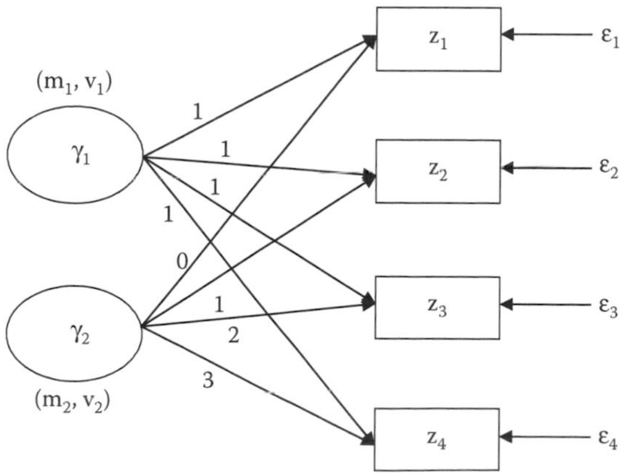

FIGURE 3.2
A prototype latent growth curve model.

or in matrix notation,

$$\begin{bmatrix} z_1 \\ z_2 \\ z_3 \\ z_4 \end{bmatrix} = \begin{bmatrix} 1 & 0 \\ 1 & 1 \\ 1 & 2 \\ 1 & 3 \end{bmatrix} \begin{bmatrix} \gamma_1 \\ \gamma_2 \end{bmatrix} + \begin{bmatrix} \varepsilon_1 \\ \varepsilon_2 \\ \varepsilon_3 \\ \varepsilon_4 \end{bmatrix}$$

$$\mathbf{z} = \mathbf{C}'\gamma + \varepsilon. \tag{3.20}$$

By fixing the loadings this way, the first latent variable is equivalent to the indicator taken when the loading for the second latent variable equals zero, so that in this case it corresponds with the first indicator. Thus, the first latent variable represents the initial status or baseline level of the repeated indicators taken at the first occasion. On the other hand, the second latent variable represents the expected linear rate of change across the four indicators. As such, the means of γ_1 and γ_2 are given specific meanings in the latent curve model. Specifically, they represent the mean score of the first indicator and the mean rate of linear change, respectively. Furthermore, the variance of the first latent variable indicates individual differences around the mean value of the first indicator, whereas that of the second latent variable indicates individual differences in the linear rate of repeated indicators.

Generalized structured component analysis can be used to fit latent growth curve models (e.g., Hwang, Kim, and Tomiuk 2005). We fix loadings to prescribed basis functions, for example, as shown in the prototype model. Conversely, we do not fix weights to constants so that they are to be estimated. This will in turn lead to the calculation of individual latent variable scores. For example, the weighted relation model for the prototype model is given as

$$\gamma_1 = z_1 w_1 + z_2 w_2 + z_3 w_3 + z_4 w_4$$

$$\gamma_2 = z_1 w_5 + z_2 w_6 + z_3 w_7 + z_4 w_8 \tag{3.21}$$

or in matrix notation,

$$\begin{bmatrix} \gamma_1 \\ \gamma_2 \end{bmatrix} = \begin{bmatrix} w_1 & w_2 & w_3 & w_4 \\ w_5 & w_6 & w_7 & w_8 \end{bmatrix} \begin{bmatrix} z_1 \\ z_2 \\ z_3 \\ z_4 \end{bmatrix}$$

$$\gamma = \mathbf{W}'\mathbf{z}. \tag{3.22}$$

We can further specify a structural model if latent variables are influenced by other variables in a given latent growth curve model. In the prototype model, no structured model needs to be specified because all latent variables are exogenous. Moreover, we use unstandardized data to estimate parameters. This indicates that we can calculate latent variable scores by postmultiplying the unstandardized data by weight estimates. We then calculate the means and variances of latent variable scores, and obtain their standard errors via the bootstrap method. In the example that follows, we demonstrate the use of generalized structured component analysis for fitting a latent growth curve model that involves repeated measures on firm-level annual revenue.

3.1.2.1.1 Example: Latent Growth Curve Modeling of the Relationships among Annual Revenue, Customer Loyalty, and Customer Satisfaction

Empirical studies have attempted to show a link between consumer satisfaction and financial performance of the firm (Zahorik and Rust 1992). Although not abundant, the results of these studies have been mixed (Bernhardt, Donthu, and Kennett 2000). On the one hand, a number of researchers have reported a positive association between customer satisfaction and diverse financial outcomes. For instance, Rust and Zahorik (1993) related both customer satisfaction and loyalty to market share and profit, whereas Anderson, Fornell, and Lehmann (1994) showed a positive and significant relationship between customer satisfaction and return-on-investment; and Ittner and Larcker (1998) presented a positive association between customer satisfaction and accounting returns. These findings were generally echoed by Nelson et al. (1992), who reported that a positive relationship held for a number of profitability measures. More recently, Anderson, Fornell, and Mazvancheryl (2004) found a positive and significant association between customer satisfaction and a measure of shareholder values.

On the other hand, several studies have reported conflicting evidence. For instance, Anderson, Fornell, and Rust (1997) suggested that under certain conditions, customer satisfaction could have a negative effect on productivity, and hence on profitability. Also, in various marketing management studies discussed in the study of Bernhardt et al. (2000), mixed results have been reported. For instance, Schneider (1991) concluded that perceived quality and customer satisfaction are not always related to profit, whereas Wiley (1991) as well as Tornow and Wiley (1991) reported negative associations between customer satisfaction (or its antecedents) and financial performance.

Bernhardt et al. (2000) provide an insightful clue as to why conflicting results in studies have emerged. Specifically, they show that in any given time period, a number of extraneous factors could overshadow the effect of customer satisfaction on financial performance and could thus lead to evidence of no significant relationship between satisfaction and economic performance. For instance, a company can spend a large amount of money to improve customer satisfaction in year t. This could lead the company to obtain high customer satisfaction ratings but will also result in lower profits for that year. Similarly, severe weather conditions might prevent highly satisfied customers from purchasing the product of the company for several weeks and thus impact profits. These factors may contribute to a nonsignificant or negative relationship between customer satisfaction and financial measures in a particular time period. In short, Bernhardt et al. (2000) conclude that the true relationship between customer satisfaction and profit may not necessarily be revealed in the short-run or through the analysis of cross-sectional data. Instead, they emphasize the importance of longitudinal data analysis for a more comprehensive investigation of this purported relationship.

It has been argued that when existing customers are satisfied with the products and services they purchase, they often become repeat customers (Oliver 1999). Moreover, most firms find that repeat purchasing on the part of consumers is essential for maintaining a stream of profitability (Oliver 1996). Accordingly, past studies have demonstrated that slight increases in retention rates can have a dramatic effect on company performance (Fornell and Wernerfelt 1987; Reichheld and Sasser 1990). For instance, Reichheld and Sasser (1990) reported that an increase of 5% in customer loyalty can double profitability and much of this profit gain is due to the fact that roughly 70% of all sales come from repeat purchases by loyal customers. Increasing customer loyalty therefore seemingly helps to secure strong financial performance because it apparently ensures long-term revenue growth (see Fornell 1992). In other words, satisfied consumers are more likely to become loyal, and more loyal customers are presumed to result in greater profits. This line of thought points to the mediating role of loyalty in the satisfaction–performance relationship.

We empirically investigated systematic links between customer satisfaction, loyalty, and a financial outcome in a longitudinal manner. Specifically, by using latent growth curve modeling, we aimed to identify an overall temporal trajectory of change in repeated measures of annual revenue and at the same time to evaluate the effect of customer satisfaction on the overall temporal trajectory, which was mediated by customer loyalty.

In the current analysis, five indicators measured across 113 US companies were used: customer satisfaction in 1994 (CS_94), customer loyalty in 1994 (CL_94), as well as three measures of annual revenue from 1994 to 1996 (Rev_94, Rev_95, and Rev_96). Customer satisfaction and loyalty scores for the 113 companies were measured on the basis of the ACSI model (Fornell et al. 1996). All the measures were log-transformed.

The number of time points seemed to be insufficient to capture a complex nonlinear trend of change in annual revenue (MacCallum et al. 1997). Thus, we assumed that in general, there was a linear trend in annual revenues over the three years. As discussed above, we also assumed that customer satisfaction in 1994 had an influence on loyalty in 1994; and, in turn, loyalty in 1994 had effects on the time-dependent trend of change in annual revenues from 1994 to 1996. Customer satisfaction in 1994 was assumed to have only an indirect effect on the temporal trend of change in annual revenue.

Figure 3.3 displays the specified latent growth curve model. In the specified model, all loadings associated with two latent variables, *Initial* and *Linear*, are fixed in advance. That is, the loadings associated with *Initial* are fixed to unity, while those associated with *Linear* are given as 0, 1, and 2 for Rev_94, Rev_95, and Rev_96, respectively. Consequently, *Initial* represents the company-wise scores of annual revenue in 1994, and *Linear* represents the company-wise linear rate of change across the three measures of annual revenue.

Although the linear-trend model was initially specified for the data, we also contemplated two alternative models for model comparison. One model assumed no time-specific change or stability in annual revenue over the three years, where all loadings for a single latent variable called *Initial* were fixed to unity. The other assumed a quadratic trend of change across the three measures of annual revenue. This model involves three latent variables named *Initial*, *Slope*, and *Quadratic*. The loadings associated with *Initial* and *Slope* were constrained in the same manner as those in the linear-trend model. The loadings associated with *Quadratic* were fixed to 0, 1, and 4 for Rev_94, Rev_95, and Rev_96, respectively, so that this latent variable represented the expected acceleration rate of change in annual revenue over the three years. In both alternative models, loyalty had direct effects on all temporal patterns in annual revenue and customer satisfaction had a direct effect on loyalty. The two alternative models are depicted in Figures 3.4 and 3.5.

We applied generalized structured component analysis to fit the linear-trend model as well as the two alternative models. We used 100 bootstrap samples to estimate standard errors and confidence intervals. For the linear-trend model, generalized structured component analysis provided

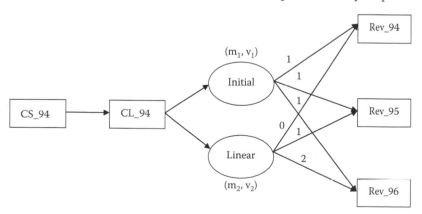

FIGURE 3.3
The linear-trend latent growth curve model specified for the annual revenue data. All residual terms are not displayed.

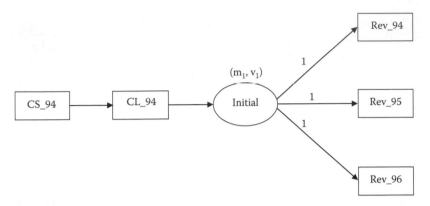

FIGURE 3.4
The stability latent growth curve model specified for the annual revenue data. All residual terms are not displayed.

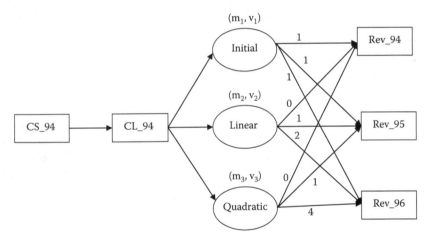

FIGURE 3.5
The quadratic-trend latent growth curve model specified for the annual revenue data. All residual terms are not displayed.

that FIT = 0.95 (SE = 0.00, 95% CI = 0.95–0.95), AFIT = 0.95 (SE = 0.00, 95% CI = 0.95–0.95), GFI = 0.94 (SE = 0.03, 95% CI = 0.87–0.99), and SRMR = 0.10 (SE = 0.03, 95% CI = 0.08–0.22). Estimation of the stability and quadratic-trend models resulted in similar values of the overall model fit measures. This made it difficult to ascertain whether the linear-trend model was most appropriate because the two alternative models also fitted the data similarly. Hence, we further conducted a one-way ANOVA to compare the FIT values among the three models. We found that the F-statistic was statistically significant [F (2,297) = 55.55, p < 0.00]. Subsequent post-hoc comparison procedures using Scheffe's and Fisher's LSD tests showed that there was a statistically significant difference in FIT between the stability and linear-trend models, whereas there was no statistically significant difference between the linear-trend and quadratic-trend models. This suggests that the linear-trend model provided a significantly better fit than the stability model, while it provided essentially the same fit as the quadratic-trend model. Therefore, it was deemed reasonable to choose the linear-trend model over the other alternatives.

In the linear-trend model, the estimated mean of *Initial* was 9.50 (SE = 0.09, 95% CI = 9.33–9.66) and the estimated mean of *Linear* was −0.09 (SE = 0.05, 95% CI = −0.18 to −0.01). Both were statistically significant. This indicates that the average of annual revenue was equal to 9.50 at the first point in time (1994) and that annual revenue seemed to decrease by an average of 0.09 over time. Moreover, the variance estimates of *Initial* (0.95, SE = 0.15, 95% CI = 0.69–1.23) and *Linear* (0.07, SE = 0.01, 95% CI = 0.05–0.10) were statistically significant, suggesting individual differences in both the initial status and linear change of annual revenue.

Loyalty in 1994 had a statistically significant and positive effect on *Initial* (2.24, SE = 0.02, 95% CI = 2.20–2.28). This indicates that the companies with high levels of loyalty showed higher levels of annual revenue in 1994. Loyalty in 1994 also had a statistically significant and negative effect on the linear rate of change (−0.02, SE = 0.01, 95% CI = −0.04 to −0.00). This suggests that the companies with higher levels of loyalty in 1994 showed a decrease in annual revenue at a lower rate than those with lower levels of loyalty. Customer satisfaction in 1994 showed a statistically significant and positive impact on loyalty in 1994 (0.97, SE = 0.00, 95% CI = 0.97–0.98), suggesting that higher levels of customer satisfaction resulted in higher levels of loyalty. Moreover, customer satisfaction in 1994 had statistically significant indirect effects on *Initial* (2.18, SE = 0.02, 95% CI = 2.14–2.23) and *Linear* (−0.02, SE = 0.01, 95% CI = −0.04 to −0.00).

In sum, the analysis of the specified latent growth curve model suggested that there existed a linearly decreasing trend of change in the three yearly measures of annual revenue of firms and customer satisfaction tended to impact the temporal trend of change in a positive way (i.e., the higher the level of customer satisfaction, the lower the rate of decline in annual revenue), which was mediated by customer loyalty.

3.2 Higher-Order Latent Variables

Higher-order latent variables are the ones that have effects on lower-order latent variables and have no influences on indicators (Bollen 1989, pp. 313–314). For instance, the Stanford–Binet intelligence scale (Thorndike, Hagen, and Sattler 1986) was constructed based on the assumption that general ability (g) underlay specific ability constructs such as verbal, visual, and memory.

Under this assumption, general ability may be considered a second-order latent variable, which influences a set of first-order latent variables related to different specific ability constructs (Kline 1998, p. 233). In addition, the inclusion of higher-order latent variables can lead to more parsimonious interpretations.

In generalized structured component analysis, a higher-order latent variable is defined as a weighted composite of its lower-order latent variables. For example, a second-order latent variable is a weighted composite of its first-order latent variables. In other words, lower-order latent variables serve as (latent) indicators for their higher-order latent variables.

For the moment, let us suppose a simple prototype model, which involves one second-order latent variable, displayed in Figure 3.6. In this model, the second-order latent variable, denoted by $\gamma^{(2)}$, affects two first-order latent variables, denoted by $\gamma_1^{(1)}$ and $\gamma_2^{(1)}$, while it is influenced by another first-order latent variable, denoted by $\gamma_3^{(1)}$. Each first-order latent variable underlies two indicators.

For the prototype model, the weighted relation model for the first-order latent variables is given as

$$\gamma_1^{(1)} = z_1w_1 + z_2w_2$$

$$\gamma_2^{(1)} = z_3w_3 + z_4w_4$$

$$\gamma_3^{(1)} = z_5w_5 + z_6w_6 \qquad (3.23)$$

or in matrix notation,

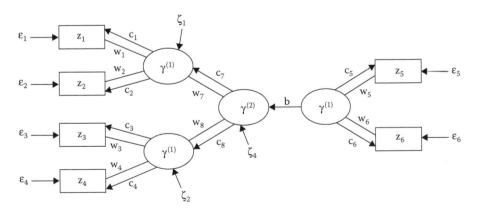

FIGURE 3.6
A prototype structural equation model that involves a second-order latent variable. All residual terms are not displayed.

$$
\begin{bmatrix} \gamma_1^{(1)} \\ \gamma_2^{(1)} \\ \gamma_3^{(1)} \end{bmatrix} = \begin{bmatrix} w_1 & w_2 & 0 & 0 & 0 & 0 \\ 0 & 0 & w_3 & w_4 & 0 & 0 \\ 0 & 0 & 0 & 0 & w_5 & w_6 \end{bmatrix} \begin{bmatrix} z_1 \\ z_2 \\ z_3 \\ z_4 \\ z_5 \\ z_6 \end{bmatrix}
$$

$$
\boldsymbol{\gamma}^{(1)} = \mathbf{W}^{(1)\prime}\mathbf{z}. \tag{3.24}
$$

On the other hand, the second-order latent variable ($\gamma^{(2)}$) is a weighted composite of two first-order latent variables ($\gamma_1^{(1)}$ and $\gamma_2^{(1)}$). Thus, the weighted relation model for the second-order latent variable is given as

$$
\gamma^{(2)} = \gamma_1^{(1)}w_7 + \gamma_2^{(1)}w_8
$$

or in matrix notation,

$$
\gamma^{(2)} = \begin{bmatrix} w_7 & w_8 & 0 \end{bmatrix} \begin{bmatrix} \gamma_1^{(1)} \\ \gamma_2^{(1)} \\ \gamma_3^{(1)} \end{bmatrix}
$$

$$
\boldsymbol{\gamma}^{(2)} = \mathbf{W}^{(2)\prime}\boldsymbol{\gamma}^{(1)} \tag{3.25}
$$

We combine Equations 3.24 and 3.25 into a single equation as follows.

$$
\begin{bmatrix} \boldsymbol{\gamma}^{(1)} \\ \boldsymbol{\gamma}^{(2)} \end{bmatrix} = \begin{bmatrix} \mathbf{W}^{(1)\prime}\mathbf{z} \\ \mathbf{W}^{(2)\prime}\boldsymbol{\gamma}^{(1)} \end{bmatrix} = \begin{bmatrix} \mathbf{I} \\ \mathbf{W}^{(2)\prime} \end{bmatrix} \mathbf{W}^{(1)\prime}\mathbf{z} = \tilde{\mathbf{W}}^{(2)\prime}\tilde{\mathbf{W}}^{(1)\prime}\mathbf{z}
$$

$$
\boldsymbol{\gamma} = \tilde{\mathbf{W}}^{\prime}\mathbf{z} \tag{3.26}
$$

where

$$
\boldsymbol{\gamma} = \begin{bmatrix} \boldsymbol{\gamma}^{(1)} \\ \boldsymbol{\gamma}^{(2)} \end{bmatrix}, \tilde{\mathbf{W}}^{(1)} = \mathbf{W}^{(1)}, \tilde{\mathbf{W}}^{(2)} = [\mathbf{I}, \mathbf{W}^{(2)}], \text{ and } \tilde{\mathbf{W}} = \tilde{\mathbf{W}}^{(1)}\tilde{\mathbf{W}}^{(2)}.
$$

This is the weighted relation model for generalized structured component analysis, which takes into account both first- and second-order latent variables. Equation 3.26 is virtually of the same form as the weighted relation model for generalized structured component analysis described in Chapter 2. In the same way, the weighted relation model that includes all Qth-order latent variables can be generally expressed as

$$\gamma = \prod_{q=1}^{Q} \tilde{W}^{(q)\prime} z = \tilde{W}^{\prime} z \tag{3.27}$$

where

$$\tilde{W}^{\prime} = \prod_{q=1}^{Q} \tilde{W}^{(q)\prime}.$$

The influences of higher-order latent variables on lower-order latent variables can be regarded as loadings, for example, c_7 and c_8 in Figure 3.6, and can be included in the measurement model. In the prototype model, the two first-order latent variables ($\gamma_1^{(1)}$ and $\gamma_2^{(1)}$) can be considered reflective (latent) indicators because they are affected by the second-order latent variable ($\gamma^{(2)}$). If the first-order latent variables are assumed to be formative (i.e., they determine the second-order latent variable), the second-order latent variable has no impact on the first-order latent variables. This indicates that the corresponding loadings are equal to zeros, while the weights for the second-order latent variable remain in the weighted relation model (Equation 3.27). In this way, generalized structured component analysis can accommodate both reflective and formative relationships between higher- and lower-order latent variables. Although the influences of higher-order latent variables on low-order latent variables are considered to be loadings, they can also be viewed as path coefficients connecting higher-order to lower-order "latent variables" and can be specified in the structural model. Both specifications will result in identical solutions. As described in Chapter 2, loadings and path coefficients are combined into a single matrix ($\mathbf{A}$) and are estimated simultaneously within the same step of the alternating least-squares algorithm.

To estimate parameters, we minimize the same least-squares criterion (Equation 2.17), where $\mathbf{W}$ is replaced by $\tilde{\mathbf{W}}$ in Equation 3.27. We can thus utilize the same alternating least-squares algorithm to minimize the criterion. In the step of updating the estimates of weights, the nonfixed elements in each column of $\tilde{\mathbf{W}}$ are updated successively, subject to the standardization constraint imposed on each latent variable.

In partial least squares path modeling, a second-order latent variable is obtained as a weighted composite of all indicators for its first-order latent variables (Lohmöller 1989). The so-called repeated indicator approach seems inconsistent with the fact that a second-order latent variable does not involve indicators. This approach simply results in another first-order latent variable rather than a second-order latent variable. Thus, it is difficult to study how a second-order latent variable is related to its first-order latent variables. Moreover, the approach cannot be used when indicators are reflective for first-order latent variables, whereas the first-order latent variables are formative for second-order latent variables. It may also suffer from an unwieldy number of indicators when handling higher-order latent variables than second-order.

A two-stage sequential approach was also proposed to deal with second-order latent variables in partial least squares path modeling (Agarwal and Karahanna 2000). In the first stage of the approach, the individual scores of first-order latent variables are obtained based on a model with the first-order latent variables only. In the second stage, these latent variable scores are used as indicators for second-order latent variables in a separate model with the second-order latent variables only. This approach is comparable in spirit to how second-order latent variables are handled in generalized structured component analysis (i.e., first-order latent variables serve as indicators for second-order latent variables). Moreover, it can be used when first-order latent variables involve reflective indicators, whereas second-order latent variables involve formative indicators (Diamantopoulos and Winklhofer 2001; Reinartz, Krafft, and Hoyer 2004). Nonetheless, the two-stage approach is unsatisfactory theoretically in that the scores of first-order latent variables are obtained without reference to their second-order latent variables. In addition, none of the two stages considers a model that actually involves second-order latent variables. All models entail first-order latent variables only, which are associated with indicators. Thus, it is difficult to evaluate how well a combined model with both first- and second-order latent variables fits to data.

3.2.1 Example: The Flow Survey Data

The present example comes from the large survey data collected in 1998 on web site environments used in the study of Novak, Hoffman, and Yung (2000). The authors investigated important variables, which played a role in creating attractive web site environments. In particular, they focused on the inter-relationships among various antecedents of *web site flow*, which was defined as "the state occurring during network navigation which is: (1) characterized by a seamless sequence of responses facilitated by machine interactivity, (2) intrinsically enjoyable, (3) accompanied by a loss of self-consciousness, and (4) self-reinforcing" (Hoffman and Novak 1996). The final structural model (called the revised model) of Novak, Hoffman, and Yung (2000) is given in Figure 3.7. In the model, there were three second-order latent variables named *Skill/Control*, *Chall/Arousal*, and *Telepres/Timedis*. Each second-order latent variable was specified to influence two first-order latent variables.

This so-called flow survey data consisted of 45 indicators, which are given in Table 3.3. The number of respondents used in this example was 500. We used 100 bootstrap samples to obtain standard errors and confidence intervals.

Generalized structured component analysis provided that FIT = 0.56 (SE = 0.00, 95% CI = 0.55–0.57), AFIT = 0.56 (SE = 0.00, 95% CI = 0.55–0.57), GFI = 0.98 (SE = 0.00, 95% CI = 0.98–0.98), and SRMR = 0.09 (SE = 0.00, 95%

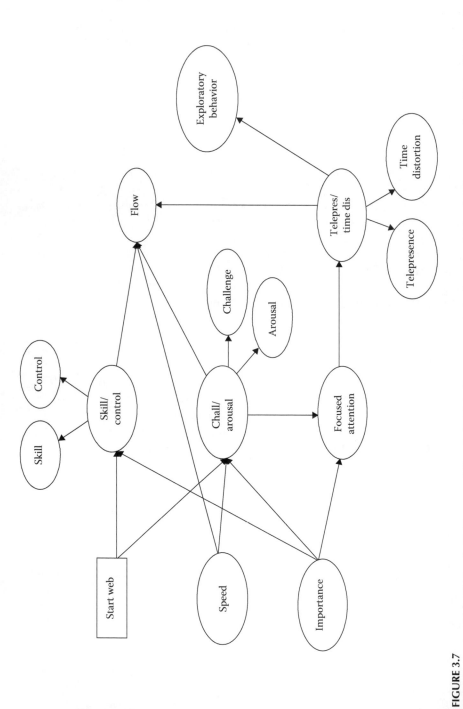

FIGURE 3.7
The structural model of Novak, Hoffman, and Yung's (2000) online customer experience model for the flow survey data. No residual terms are displayed.

TABLE 3.3

45 Indicators Used in Novak, Hoffman, and Yung's (2000) Final Model for the Flow Survey Data ([R] Denotes a Reverse-Coded Indicator)

Startweb: When did you start using the Web? (6 categories)

Speed

sp1: When I use the Web there is very little waiting time between my actions and the computer's response.

sp2: Interacting with the Web is slow and tedious. (R)

sp3: Pages on the websites I visit usually load quickly.

Importance

im1: Important/unimportant.

im2: Irrelevant/relevant. (R)

im3: Means a lot to me/means nothing to me.

im4: Matters to me/doesn't matter.

im5: Of no concern/of concern to me. (R)

Skill

s5: How would you rate your skill at using the Web, compared to other things you do on the computer?

s6: How would you rate your skill at using the Web, compared to the sport or game you are best at?

Control

co1: Controlling/controlled.

co2: Influenced/influential. (R)

co3: Dominant/submissive.

co4: Guided/autonomous. (R)

Challenge

c1: Using the Web challenges me.

c2: Using the Web challenges me to perform to the best of my ability.

c3: Using the Web provides a good test of my skills.

c4: I find that using the Web stretches my capabilities to my limits.

Arousal

a1: Stimulated/relaxed.

a2: Calm/excited. (R)

a3: Frenzied/sluggish.

a4: Unaroused/aroused. (R)

Focused Attention

fa1: Not deeply engrossed/deeply engrossed.

fa2: Absorbed intently/not absorbed intently. (R)

fa3: My attention is not focused/my attention is focused.

fa4: I concentrate fully/I do not concentrate fully. (R)

(Continued)

TABLE 3.3 (*Continued*)

45 Indicators Used in Novak, Hoffman, and Yung's (2000) Final Model for the Flow Survey Data ([R] Denotes a Reverse-Coded Indicator)

Flow

flow1: Do you think you have ever experienced flow on the Web?

flow2: In general, how frequently would you say you have experienced "flow" when you use the Web?

flow3: Most of the time I use the Web I feel that I am in flow.

Telepresence

t2: Using the Web often makes me forget where I am.

t3: After using the Web, I feel like I come back to the "real world" after a journey.

t4: Using the Web creates a new world for me, and this world suddenly disappears when I stop browsing.

t5: When I use the Web, I feel I am in a world created by the websites I visit.

t6: When I use the Web, my body is in the room, but my mind is inside the world created by the websites I visit.

t7: When I use the Web, the world generated by the sites I visit is more real for me than the "real world."

Time Distortion

td1: Time seems to go by very quickly when I use the Web.

td2: When I use the Web, I tend to lose track of time.

Exploratory Behavior

ex1: I enjoy visiting unfamiliar web sites just for the sake of variety.

ex2: I rarely visit web sites I know nothing about. (R)

ex3: Even though there are thousands of different kinds of web sites, I tend to visit the same types of web sites. (R)

ex4: When I hear about a new web site, I'm eager to check it out.

ex5: Surfing the Web to see what's new is a waste of time. (R)

ex6: I like to browse the Web and find out about the latest sites.

ex7: I like to browse shopping sites even if I don't plan to buy anything.

CI = 0.08–0.10). This suggests that the final model accounted for about 56% of the total variance of all variables. Both FIT and AFIT turned out to be statistically significantly different from zero. GFI was close to 1, whereas SRMR was small. Moreover, generalized structured component analysis provided that $FIT_M = 0.62$ (SE = 0.00, 95% CI = 0.61–0.63) and $FIT_S = 0.38$ (SE = 0.01, 95% CI = 0.37–0.40). This indicates that the measurement model accounted for about 62% of the total variance of all indicators, whereas the structural model explained about 38% of the total variance of all latent variables.

Table 3.4 provides the estimates of weights and loadings for the first-order latent variables. All the weight and loading estimates turned out to be statistically significant. Table 3.5 shows the estimates of weights and loadings

TABLE 3.4

Estimates of Weights and Loadings and Their Standard Errors (SE) and 95% Confidence Intervals (CI) for the First-Order Latent Variables Obtained from the Flow Survey Data

First-order Latent	Indicator	Weights				Loadings			
		Estimate	SE	$|t|$	CI	Estimate	SE	$|t|$	CI
	Startweb	1	0			1	0		
Speed	sp1	0.40	0.01	61.70	0.39–0.41	0.84	0.01	64.39	0.81–0.86
	sp2	0.41	0.01	57.51	0.40–0.43	0.83	0.01	64.81	0.80–0.85
	sp3	0.39	0.01	54.76	0.38–0.40	0.83	0.01	82.03	0.81–0.85
Importance	im1	0.23	0.01	23.24	0.21–0.25	0.87	0.01	68.89	0.84–0.89
	im2	0.20	0.01	23.52	0.18–0.21	0.80	0.02	41.62	0.75–0.83
	im3	0.27	0.01	29.42	0.25–0.29	0.93	0.01	149.66	0.92–0.94
	im4	0.23	0.01	20.10	0.21–0.25	0.92	0.01	109.75	0.90–0.94
	im5	0.21	0.01	24.81	0.20–0.23	0.85	0.02	48.89	0.81–0.88
Skill	s5	0.55	0.01	59.02	0.53–0.57	0.85	0.01	86.19	0.83–0.87
	s6	0.60	0.01	58.03	0.58–0.62	0.88	0.01	115.76	0.86–0.89
Control	co1	0.44	0.02	24.44	0.40–0.47	0.72	0.02	33.04	0.67–0.76
	co2	0.26	0.03	8.99	0.20–0.31	0.46	0.05	8.97	0.35–0.55
	co3	0.45	0.02	26.77	0.42–0.49	0.74	0.02	33.16	0.69–0.78
	co4	0.37	0.02	18.59	0.32–0.40	0.63	0.03	21.15	0.56–0.69
Challenge	c1	0.25	0.01	27.39	0.24–0.27	0.65	0.02	29.24	0.62–0.70
	c2	0.34	0.01	36.45	0.32–0.36	0.81	0.01	64.29	0.79–0.83
	c3	0.35	0.01	42.33	0.34–0.37	0.88	0.01	130.20	0.86–0.89
	c4	0.32	0.01	35.34	0.30–0.34	0.80	0.01	56.00	0.77–0.82
Arousal	a1	0.38	0.02	24.32	0.35–0.40	0.68	0.02	28.34	0.62–0.72
	a2	0.34	0.02	19.29	0.30–0.38	0.68	0.03	25.47	0.63–0.73
	a3	0.37	0.02	20.52	0.34–0.40	0.62	0.03	20.70	0.56–0.67
	a4	0.41	0.01	28.27	0.38–0.44	0.69	0.02	31.22	0.65–0.74

(Continued)

TABLE 3.4 (*Continued*)

Estimates of Weights and Loadings and Their Standard Errors (SE) and 95% Confidence Intervals (CI) for the First-Order Latent Variables Obtained from the Flow Survey Data

First-order Latent	Indicator	Weights				Loadings			
		Estimate	SE	|t|	CI	Estimate	SE	|t|	CI
Focused Attention	fa1	0.34	0.02	22.79	0.32–0.38	0.83	0.02	53.16	0.80–0.86
	fa2	0.35	0.01	24.53	0.32–0.37	0.84	0.01	82.46	0.82–0.87
	fa3	0.26	0.01	21.34	0.23–0.28	0.75	0.02	42.55	0.72–0.79
	fa4	0.29	0.01	21.40	0.26–0.32	0.79	.01	580.41	0.75–0.81
Flow	flow1	0.33	0.01	34.10	0.31–0.35	0.89	0.01	116.27	0.87–0.90
	flow2	0.38	0.01	32.68	0.36–0.40	0.96	0.00	326.46	0.96–0.97
	flow3	0.38	0.01	36.66	0.36–0.40	0.92	0.01	142.37	0.91–0.93
Telepresence	t2	0.21	0.01	38.85	0.20–0.22	0.74	0.02	41.04	0.70–0.77
	t3	0.20	0.01	26.64	0.19–0.22	0.76	0.02	47.74	0.73–0.80
	t4	0.22	0.01	35.89	0.20–0.23	0.83	0.01	73.29	0.81–0.85
	t5	0.19	0.01	28.84	0.17–0.20	0.72	0.02	38.95	0.68–0.75
	t6	0.29	0.01	32.97	0.27–0.31	0.87	0.01	113.72	0.86–0.88
	t7	0.18	0.01	26.85	0.16–0.19	0.74	0.02	46.45	0.71–0.77
Time Distortion	td1	0.56	0.01	44.49	0.53–0.58	0.90	0.01	101.98	0.88–0.92
	td2	0.56	0.01	40.76	0.52–0.58	0.90	0.01	114.02	0.88–0.92
Exploratory Behavior	ex1	0.25	0.01	30.51	0.23–0.27	0.79	0.01	55.57	0.77–0.82
	ex2	0.16	0.01	22.01	0.14–0.17	0.62	0.03	23.55	0.56–0.66
	ex3	0.07	0.01	5.70	0.04–0.09	0.26	0.04	6.51	0.17–0.33
	ex4	0.24	0.01	30.09	0.22–0.25	0.75	0.02	46.54	0.71–0.78
	ex5	0.22	0.01	30.66	0.20–0.23	0.71	0.02	36.49	0.67–0.75
	ex6	0.24	0.01	35.00	0.23–0.26	0.83	0.01	76.76	0.81–0.85
	ex7	0.22	0.01	31.37	0.20–0.23	0.72	0.02	36.42	0.68–0.76

CI, Confidence intervals; SE, Standard errors.

TABLE 3.5

Estimates of Weights and Loadings and Their Standard Errors and 95% Confidence Intervals for the Second-Order Latent Variables Obtained from the Flow Survey Data

Second-order Latent	First-order Latent	Weights				Loadings			
		Estimate	SE	$|t|$	CI	Estimate	SE	$|t|$	CI
Skill/Control	Skill	0.71	0.02	38.12	0.68–0.75	0.84	0.01	75.63	0.82–0.87
	Control	0.55	0.02	31.00	0.51–0.58	0.72	0.02	35.03	0.67–0.76
Chall/Arousal	Challenge	0.64	0.02	32.44	0.60–0.68	0.84	0.01	68.68	0.81–0.86
	Arousal	0.58	0.02	34.08	0.55–0.62	0.80	0.02	45.31	0.77–0.83
Telepres/ TimeDis	Telepresence	0.58	0.01	42.58	0.54–0.60	0.87	0.01	84.68	0.85–0.89
	Time Distortion	0.57	0.02	37.70	0.55–0.60	0.87	0.01	106.38	0.85–0.89

CI, Confidence intervals; SE, Standard errors.

for the second-order latent variables. All these weight and loading estimates were also statistically significant.

Table 3.6 exhibits the estimates of path coefficients. All the estimates were statistically significant. In general, the interpretations of the path coefficient estimates are consistent with the relationships hypothesized by Novak et al. (2000). We focus on the interpretations of the path coefficient estimates associated with the three second-order latent variables. *Skill/Control* had a statistically significant and positive effect on *Flow* (0.14, SE = 0.02, 95% CI = 0.10–0.19). *Startweb* had a statistically significant and positive effect on *Skill/Control* (0.23, SE = 0.03, 95% CI = 0.18–0. 28), whereas *Importance* had a statistically significant and positive impact on *Skill/Control* (0.43, SE = 0.02, 95% CI = 0.39–0.48). *Chall/Arousal* showed statistically significant and positive effects on *Focused Attention* (0.42, SE = 0.03, 95% CI = 0.36–0.46) and *Flow* (0.18, SE = 0.03, 95% CI = 0.13–0.23). *Startweb* had a statistically significant and negative effect on *Chall/Arousal* (−0.24, SE = 0.03, 95% CI = −0.29 to −0.19), *Speed* had a statistically significant and positive influence on *Chall/Arousal* (0.13, SE = 0.03, 95% CI = 0.07–0.19), and *Importance* had a statistically significant and positive influence on *Chall/Arousal* (0.36, SE = 0.03, 95% CI = 0.30–0.41). *Telepres/ Timedis* had statistically significant and positive influences on *Flow* (0.48, SE = 0.02, 95% CI = 0.44–0.52) and *Exploratory Behavior* (0.43, SE = 0.03, 95% CI = 0.38–0.48). *Focused Attention* had a statistically significant and positive effect on *Telepres/Timedis* (0.50, SE = 0.02, 95% CI = 0.46–0.54).

TABLE 3.6

Estimates of Path Coefficients and Their Standard Errors and 95% Confidence Intervals Obtained from the Flow Survey Data

	Estimate	SE	$|t|$	CI
Startweb → Skill/Control	0.23	0.03	8.84	0.18–0.28
Startweb → Chall/Arousal	−0.24	0.03	9.01	−0.29 to −0.19
Speed → Flow	0.08	0.03	3.27	0.04–0.13
Speed → Chall/Arousal	0.13	0.03	4.53	0.07–0.19
Importance → Skill/Control	0.43	0.02	19.09	0.39–0.48
Importance → Chall/Arousal	0.36	0.03	12.50	0.30–0.41
Importance → Focused Attention	0.28	0.03	10.51	0.23–0.34
Skill/Control → Flow	0.14	0.02	6.11	0.10–0.19
Chall/Arousal → Focused Attention	0.42	0.03	15.74	0.36–0.46
Chall/Arousal → Flow	0.18	0.03	6.74	0.13–0.23
Telepres/Timedis → Flow	0.48	0.02	22.38	0.44–0.52
Telepres/Timedis → Exploratory Behavior	0.43	0.03	15.41	0.38–0.48
Focused Attention → Telepres/TimeDis	0.50	0.02	22.52	0.46–0.54

CI, Confidence intervals; SE, Standard errors.

3.3 Multiple Group Analysis

A multiple group analysis is conducted to investigate whether model parameters vary across groups. At a basic level, we can also look into whether the model for each group is in the same form (Bollen 1989, p. 356). However, testing invariance in model form is not common in a multiple group analysis. We thus concentrate on cross-group invariance in parameter values. To test this invariance, we can fit the same structural equation model to different groups with the imposition of cross-group equality constraints on parameters. If the model with cross-group equality constraints imposed shows a similar fit to the model without the constraints, we may conclude that the parameters constrained do not differ across groups. If no cross-group constraints are involved in a multiple group analysis, it is equivalent to estimating parameters separately for each group. Although comparing parameter estimates visually across groups can give an idea about similarity of parameters, this unconstrained multiple-group analysis can lead to less objective conclusions.

Generalized structured component analysis can deal with a multiple group analysis with the imposition of cross-group equality constraints. Let us suppose that an identical structural equation model is fitted to G groups simultaneously. The generalized structured component analysis model for the gth group can be expressed as

$$V_g'z_g = A_g'W_g'z_g + e_g \qquad (3.28)$$

$(g = 1,\ldots, G)$. For all G groups, we then have

$$\begin{bmatrix} V_1' & & \\ & \ddots & \\ & & V_G' \end{bmatrix} \begin{bmatrix} z_1 \\ \vdots \\ z_G \end{bmatrix} = \begin{bmatrix} A_1' & & \\ & \ddots & \\ & & A_G' \end{bmatrix} \begin{bmatrix} W_1' & & \\ & \ddots & \\ & & W_G' \end{bmatrix} \begin{bmatrix} z_1 \\ \vdots \\ z_G \end{bmatrix} + \begin{bmatrix} e_1 \\ \vdots \\ e_G \end{bmatrix}$$

$$V'z = A'W'z + e. \qquad (3.29)$$

where

$$V' = \begin{bmatrix} V_1' & & \\ & \ddots & \\ & & V_G' \end{bmatrix}, z = \begin{bmatrix} z_1 \\ \vdots \\ z_G \end{bmatrix}, A' = \begin{bmatrix} A_1' & & \\ & \ddots & \\ & & A_G' \end{bmatrix},$$

$$W' = \begin{bmatrix} W_1' & & \\ & \ddots & \\ & & W_G' \end{bmatrix}, \text{ and } e = \begin{bmatrix} e_1 \\ \vdots \\ e_G \end{bmatrix}.$$

Clearly, Equation 3.29 is essentially of the same form as the generalized structured component analysis model (Equation 2.13) for a single group. This indicates that we can carry out a multiple group analysis in the same fashion as conducting a single group analysis, treating G sets of parameters as a single set of parameters. Consequently, we can minimize the same least-squares criterion to estimate parameters, using the alternating least-squares algorithm described in Chapter 2. Furthermore, any cross-group equality constraints can be imposed on parameters in Equation 3.29 in the same way as in imposing constraints on parameters in a single group, as discussed in Section 3.1.1.

Partial least squares path modeling is incapable of dealing with a multiple-group analysis with the imposition of cross-group constraints (Hwang and Takane 2004). This is because it is not equipped with a single optimization function, which can be used to estimate a whole set of parameters simultaneously.

3.3.1 Example: The Organizational Identification Data

This example is part of the survey data used in the study of Bergami and Bagozzi (2000), which consists of a sample of 305 employees (male = 157 and female = 148) from the electronics division of a large conglomerate in South Korea. From the original data, we used 21 indicators associated with four latent variables, called *organizational prestige*, *organizational identification*, *affective commitment (joy)*, and *affective commitment (love)*. The 21 indicators are presented in Table 3.7. According to Bergami and Bagozzi (2000), *organizational prestige* denotes the perception of a member of the organization that her/his significant others (relatives, friends, etc.) believe that the organization is well-approved. *Organizational identification* is "a form of social identification whereby a person comes to view him- or herself as a member of a particular organization" (Bergami and Bagozzi 2000, p. 557). *Affective commitment (joy)* and *affective commitment (love)* indicate two different types of emotional attachment to the organization: the former is "happiness arising from the organization" (Bergami and Bagozzi 2000, p. 560), whereas the latter is "emotional attraction or affection towards the organization" (Bergami and Bagozzi 2000, p. 560).

Figure 3.8 displays a structural equation model specified for this example. In the model, *organizational prestige* (OP) was assumed to underlie eight indicators (op1–op8), *organizational identification* (OI) was assumed to underlie the six indicators proposed by Mael (1988) (oi1–oi6), and *affective commitment (joy)* (AC_J) and *affective commitment (love)* (AC_L) were assumed to underlie four items (acj1–acj4) and three indicators (acl1–acl3), respectively, among the seven indicators concerning affective commitment developed by Allen and Meyer (1990). Note that the three indicators for *affective commitment (love)* were reverse-coded. According to the findings of Bergami and Bagozzi (2000), it was further assumed that *organizational prestige* had an effect on

TABLE 3.7

21 Indicators Taken from Bergami and Bagozzi's (2000) Organizational
Identification Data

For each of the following items, please indicate how much you agree or disagree. Use the
following 5-point scale: 1: strongly disagree; 2: disagree; 3: neither disagree or agree;
4: agree; 5: strongly agree

Organizational Prestige

op1: My relatives and people close or important to me believe that [Company X] is a
well-known company.

op2: My relatives and people close or important to me believe that [Company X] is a highly
respected company.

op3: My relatives and people close or important to me believe that [Company X] is an
admired company.

op4: My relatives and people close or important to me believe that [Company X] is a
prestigious company.

op5: People in general think that [Company X] is a well-known company.

op6: People in general think that [Company X] is a highly respected company.

op7: People in general think that [Company X] is an admired company.

op8: People in general think that [Company X] is a prestigious company.

Organizational Identification

oi1: When someone criticizes [Company X] it feels like a personal insult.

oi2: I am very interested in what others think about [Company X].

oi3: When I talk about [Company X], I usually say "we" rather than "they".

oi4: [Company X's] successes are my successes.

oi5: When someone praises [Company X] it feels like a personal compliment.

oi6: If a story in the media criticized [Company X], I would feel embarrassed.

Affective Commitment (joy)

acj1: I would be very happy to spend the rest of my career with [Company X].

acj2: I enjoy discussing [Company X] with people outside of it.

acj3: I really feel the problems of [Company X] are my own.

acj4: [Company X] has a great deal of personal meaning for me.

Affective Commitment (love)

acl1: I do not feel like part of a family at [Company X].

acl2: I do not feel emotionally attached to [Company X].

acl3: I do not feel a strong sense of belonging to [Company X].

organizational identification, and *organizational identification* in turn had influences on *affective commitment (joy)* and *affective commitment (love)*.

We fitted the specified model to male and female groups simultaneously. We used 100 bootstrap samples to estimate standard errors and confidence intervals. Generalized structured component analysis provided
that FIT = 0.53 (SE = 0.01, 95% CI = 0.50–0.55), AFIT = 0.53 (SE = 0.01, 95%

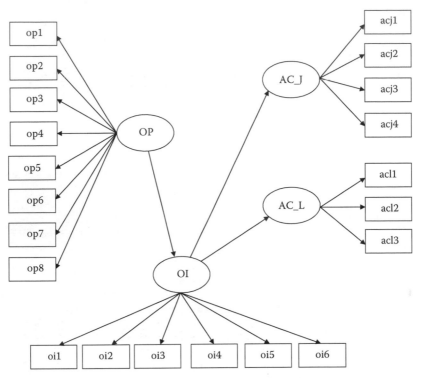

FIGURE 3.8
The structural equation model specified for the organizational identification data. All residual terms are not displayed (OP = Organizational Prestige, OI = Organizational Identification, AC_J = Affective Commitment [joy], and AC_L = Affective Commitment [Love]).

CI = 0.50–0.55), GFI = 0.99 (SE = 0.00, 95% CI = 0.99–0.99), and SRMR = 0.06 (SE = 0.00, 95% CI = 0.06–0.08). This suggests that the model accounted for about 53% of the total variance of all variables in the two groups. Both FIT and AFIT turned out to be statistically significant. GFI was quite close to one, while SRMR was small. In addition, generalized structured component analysis provided that FIT_M = 0.60 (SE = 0.01, 95% CI = 0.57–0.62) and FIT_S = 0.17 (SE = 0.02, 95% CI = 0.13–0.21). This indicates that the measurement model accounted for about 60% of the total variance of all indicators, whereas the structural model explained about 17% of the total variance of all latent variables.

Table 3.8 provides the estimates of weights for the male and female groups. In both groups, the weight estimates turned out to be statistically significant. Table 3.9 presents the estimates of loadings for the two groups. The loading estimates were large and statistically significant in both groups.

Table 3.10 provides the estimates of path coefficients in the groups. In both groups, *organizational prestige* had a statistically significant and positive effect on *organizational identification*. This suggests that if both male and

TABLE 3.8

Estimates of Weights, Standard Errors, and 95% Confidence Intervals Obtained from the Unconstrained Multiple Group Analysis of the Organizational Identification Data

Latent	Indicator	Male				Female			
		Estimate	SE	|t|	CI	Estimate	SE	|t|	CI
OP	op1	0.15	0.01	11.31	0.13–0.18	0.15	0.02	8.98	0.12–0.18
	op2	0.15	0.01	11.57	0.13–0.18	0.17	0.01	11.42	0.14–0.20
	op3	0.17	0.02	8.89	0.14–0.21	0.14	0.01	10.37	0.12–0.17
	op4	0.14	0.02	9.18	0.11–0.17	0.15	0.01	14.96	0.13–0.18
	op5	0.17	0.02	10.60	0.14–0.20	0.16	0.01	12.25	0.14–0.18
	op6	0.17	0.01	11.75	0.14–0.20	0.17	0.01	14.23	0.15–0.20
	op7	0.15	0.02	8.94	0.12–0.18	0.15	0.01	12.48	0.13–0.18
	op8	0.16	0.01	11.53	0.14–0.19	0.15	0.01	13.38	0.12–0.17
OI	oi1	0.20	0.03	6.59	0.15–0.26	0.24	0.03	7.39	0.17–0.31
	oi2	0.19	0.02	8.65	0.16–0.24	0.22	0.03	8.16	0.17–0.27
	oi3	0.17	0.02	7.38	0.12–0.22	0.22	0.02	9.60	0.18–0.27
	oi4	0.30	0.04	8.44	0.23–0.37	0.24	0.03	8.69	0.20–0.30
	oi5	0.20	0.03	6.35	0.15–0.27	0.26	0.03	8.28	0.22–0.34
	oi6	0.18	0.03	6.53	0.15–0.27	0.21	0.03	7.18	0.16–0.27
AC_J	acj1	0.31	0.02	12.73	0.26–0.36	0.29	0.03	9.31	0.23–0.35
	acj2	0.29	0.03	11.52	0.24–0.34	0.41	0.03	14.01	0.35–0.47
	acj3	0.36	0.03	14.14	0.30–0.40	0.36	0.03	11.85	0.30–0.43
	acj4	0.30	0.03	8.85	0.23–0.38	0.30	0.03	9.21	0.22–0.35
AC_L	acl1	0.42	0.03	15.82	0.37–0.48	0.48	0.05	10.49	0.39–0.58
	acl2	0.39	0.03	11.73	0.33–0.45	0.38	0.07	5.73	0.22–0.50
	acl3	0.44	0.04	11.63	0.38–0.52	0.52	0.05	9.92	0.41–0.61

AC_J = Affective commitment (joy); AC_L, Affective commitment (love); CI, Confidence intervals; OI = Organizational identification; OP, Organizational prestige; SE, Standard errors.

TABLE 3.9

Estimates of Loadings, Standard Errors, and 95% Confidence Intervals Obtained from the Unconstrained Multiple Group Analysis of the Organizational Identification Data

Latent	Indicator	Male				Female			
		Estimate	SE	$\lvert t \rvert$	CI	Estimate	SE	$\lvert t \rvert$	CI
OP	op1	0.80	0.05	15.05	0.64–0.87	0.76	0.04	20.60	0.68–0.82
	op2	0.79	0.04	20.37	0.69–0.84	0.85	0.02	39.07	0.80–0.89
	op3	0.73	0.05	13.42	0.61–0.82	0.80	0.03	22.97	0.72–0.85
	op4	0.79	0.05	14.53	0.64–0.86	0.82	0.04	22.49	0.74–0.87
	op5	0.80	0.05	17.72	0.68–0.86	0.81	0.04	19.20	0.71–0.87
	op6	0.84	0.03	29.34	0.78–0.89	0.84	0.04	20.88	0.74–0.90
	op7	0.77	0.04	19.40	0.66–0.83	0.78	0.04	21.73	0.70–0.83
	op8	0.83	0.04	19.94	0.74–0.90	0.77	0.05	14.65	0.62–0.85
OI	oi1	0.83	0.03	26.21	0.75–0.88	0.73	0.04	16.58	0.62–0.80
	oi2	0.79	0.03	24.11	0.73–0.85	0.71	0.05	13.64	0.58–0.80
	oi3	0.64	0.06	11.37	0.53–0.72	0.65	0.05	12.91	0.55–0.72
	oi4	0.88	0.02	43.48	0.84–0.92	0.73	0.06	12.48	0.59–0.82
	oi5	0.85	0.03	25.45	0.77–0.90	0.75	0.03	21.46	0.69–0.82
	oi6	0.76	0.05	16.32	0.66–0.83	0.74	0.06	12.63	0.58–0.81
AC_J	acj1	0.79	0.03	22.61	0.71–0.85	0.69	0.06	12.07	0.58–0.78
	acj2	0.77	0.05	16.29	0.67–0.84	0.83	0.02	34.10	0.78–0.88
	acj3	0.85	0.03	32.37	0.79–0.90	0.76	0.05	16.53	0.65–0.84
	acj4	0.76	0.03	22.18	0.70–0.82	0.62	0.07	8.45	0.42–0.74
AC_L	acl1	0.84	0.04	23.04	0.76–0.89	0.75	0.05	15.01	0.63–0.84
	acl2	0.76	0.05	16.11	0.66–0.83	0.64	0.11	5.87	0.34–0.80
	acl3	0.81	0.04	22.18	0.72–0.88	0.77	0.06	13.23	0.65–0.85

AC_J = Affective commitment (joy); AC_L, Affective commitment (love); CI, Confidence intervals; OI = Organizational identification; OP, Organizational prestige; SE, Standard errors.

TABLE 3.10

Estimates of Path Coefficients and Their Standard Errors and 95% Confidence Intervals from the Unconstrained Multiple Group Analysis of the Organizational Identification Data

	Group	Estimate	SE	$|t|$	CI
OP → OI	Male	0.39	0.09	4.23	0.22–0.55
OI → AC_J		0.71	0.05	13.82	0.60–0.80
OI → AC_L		−0.46	0.09	5.25	−0.63 to −0.31
OP → OI	Female	0.35	0.07	5.27	0.23–0.45
OI → AC_J		0.47	0.04	10.84	0.40–0.56
OI → AC_L		−0.34	0.08	4.03	−0.48 to −0.14

AC_J = Affective commitment (joy); AC_L, Affective commitment (love); CI, Confidence intervals; OI = Organizational identification; OP, Organizational prestige; SE, Standard errors.

female employees were more likely to perceive that their important others believe that their organization was well-regarded, they were more likely to perceive themselves as a member of the organization. This effect appeared similar across the groups. In addition, *organizational identification* had a statistically significant and positive influence on *affective commitment (joy)*, whereas a statistically significant and negative impact on *affective commitment (love)* in both groups. This indicates that in these groups, the employees with a high level of *organizational identification* tended to show a high level of the two types of emotional attachment to the organization. Although these effects of *organizational identification* could be interpreted similarly for both groups, the effects for the male group were larger in size than those for the female group. Consequently, the *R*-squared values of *affective commitment (joy)* and *affective commitment (love)* were greater for the male group, when compared to those for the female group.

Based on the findings of the unconstrained analysis, we further hypothesized that across the two groups, each loading was identical and the effect of *organizational prestige* on *organizational identification* was equal. We conducted a multiple group analysis with the imposition of these cross-group equality constraints on the loadings and path coefficients.

For the constrained model, generalized structured component analysis provided that FIT = 0.53 (SE = 0.01, 95% CI = 0.50–0.56), AFIT = 0.53 (SE = 0.01, 95% CI = 0.50–0.56), GFI = 0.99 (SE = 0.00, 95% CI = 0.99–0.99), and SRMR = 0.06 (SE = 0.00, 95% CI = 0.06–0.09). These overall fit values are almost identical to those obtained for the unconstrained model. In particular, the AFIT value of the constrained model remained virtually the same as that of the unconstrained one. Furthermore, we performed a *t*-test to compare the FIT values between the unconstrained and constrained models based on 100 bootstrap samples. We found that the *t*-statistic was statistically nonsignificant [$t(99) = −0.36$, $p = 0.72$, 95% CI = −0.00–0.00], indicating that there was no statistically significant difference in FIT between the constrained and

TABLE 3.11

Estimates of Loadings, Standard Errors, and 95% Confidence Intervals Obtained from the Constrained Multiple Group Analysis of the Organizational Identification Data

Latent	Indicator	Male				Female			
		Estimate	SE	$|t|$	CI	Estimate	SE	$|t|$	CI
OP	op1	0.78	0.03	23.49	0.68–0.82	0.78	0.03	23.49	0.68–0.82
	op2	0.82	0.02	35.87	0.76–0.86	0.82	0.02	35.87	0.76–0.86
	op3	0.77	0.03	25.45	0.70–0.82	0.77	0.03	25.45	0.70–0.82
	op4	0.80	0.03	23.16	0.74–0.87	0.80	0.03	23.16	0.74–0.87
	op5	0.80	0.03	27.44	0.76–0.86	0.80	0.03	27.44	0.76–0.86
	op6	0.84	0.02	33.88	0.79–0.89	0.84	0.02	33.88	0.79–0.89
	op7	0.78	0.03	29.47	0.72–0.82	0.78	0.03	29.47	0.72–0.82
	op8	0.80	0.03	24.71	0.72–0.86	0.80	0.03	24.71	0.72–0.86
OI	oi1	0.78	0.03	30.21	0.72–0.84	0.78	0.03	30.21	0.72–0.84
	oi2	0.75	0.03	23.96	0.68–0.80	0.75	0.03	23.96	0.68–0.80
	oi3	0.64	0.04	15.73	0.54–0.71	0.64	0.04	15.73	0.54–0.71
	oi4	0.81	0.03	27.80	0.74–0.86	0.81	0.03	27.80	0.74–0.86
	oi5	0.80	0.03	30.15	0.74–0.84	0.80	0.03	30.15	0.74–0.84
	oi6	0.75	0.03	21.49	0.68–0.81	0.75	0.03	21.49	0.68–0.81
AC_J	acj1	0.74	0.03	21.86	0.67–0.80	0.74	0.03	21.86	0.67–0.80
	acj2	0.80	0.03	30.45	0.75–0.85	0.80	0.03	30.45	0.75–0.85
	acj3	0.81	0.03	35.47	0.75–0.85	0.81	0.03	35.47	0.75–0.85
	acj4	0.70	0.04	19.67	0.63–0.78	0.70	0.04	19.67	0.63–0.78
AC_L	acl1	0.79	0.03	25.53	0.71–0.84	0.79	0.03	25.53	0.71–0.84
	acl2	0.71	0.05	13.77	0.59–0.78	0.71	0.05	13.77	0.59–0.78
	acl3	0.79	0.03	26.57	0.73–0.85	0.79	0.03	26.57	0.73–0.85

AC_J = Affective commitment (joy); AC_L, Affective commitment (love); CI, Confidence intervals; OI = Organizational identification; OP, Organizational prestige; SE, Standard errors.

TABLE 3.12

Estimates of Path Coefficients and Their Standard Errors (SE) and 95% Confidence Intervals (CI) from the Constrained Multiple Group Analysis of the Organizational Identification Data

	Group	Estimate	SE	$|t|$	CI
OP → OI	Male	0.37	0.06	6.50	0.21–0.46
OI → AC_J		0.71	0.05	15.55	0.61–0.80
OI → AC_L		−0.46	0.08	5.89	−0.60 to −0.31
OP → OI	Female	0.37	0.06	6.50	0.21–0.46
OI → AC_J		0.47	0.05	9.90	0.40–0.56
OI → AC_L		−0.33	0.07	4.48	−0.46 to −0.14

AC_J = Affective commitment (joy); AC_L, Affective commitment (love); CI, Confidence intervals; OI = Organizational identification; OP, Organizational prestige; SE, Standard errors.

unconstrained models. Thus, it was safe to conclude that the constrained model could be chosen over the unconstrained one.

The weight estimates of the constrained model were found to be quite similar to those of the unconstrained model, although they are not reported here to preserve space. Table 3.11 presents the loading estimates obtained for the constrained model. Owing to the imposition of the cross-group equality constraints, each loading estimate turned out to be identical between the groups. The imposition of the constraints on the loadings did not change the value of FIT_M (0.60, SE = 0.02, 95% CI = 0.57–0.63) substantially, when compared to the value of FIT_M for the unconstrained model. Table 3.12 shows the path coefficient estimates across the two groups obtained for the constrained model. Again, owing to the cross-group equality constraint, the path coefficient estimate relating *organizational identification* to *organizational identification* became identical across the groups. The imposition of the cross-group equality constraint on the path coefficient did not change the value of FIT_S (0.17, SE = 0.02, 95% CI = 0.12–0.22) substantially, indicating that the constrained structural model explained almost the same amount of the total variance of the latent variables in both groups. All the loading and path coefficient estimates obtained for the constrained model could be interpreted in the same manner as those for the unconstrained model.

3.4 Total and Indirect Effects

The effect of one variable on another can be decomposed into direct, indirect, and total effects. A direct effect signifies the impact of one variable on another, which is not transmitted by any other variable. In the

ACSI model depicted in Figure 2.3 in Chapter 2, all nine path coefficients $(b_1, ..., b_9)$ indicate the direct effects of latent variables on other latent variables. For example, the direct effect of *customer satisfaction* on *customer loyalty* is represented by b_8. Likewise, all 14 loadings $(c_1, ..., c_{14})$ denote the direct effects of latent variables on their indicators. For example, the direct effect of *customer expectations* on z_1 (customer expectations about overall quality) is represented by c_1. Thus, in the generalized structured component analysis model, the **C** and **B** matrices are composed of direct effects.

An indirect effect is the one which is mediated by at least one other variable. In the ACSI model, for example, *customer satisfaction* has an indirect effect on *customer loyalty* mediated by *customer complaints*. This indirect effect is calculated by the product of the direct effect of *customer satisfaction* on *customer complaints* (b_7) and the direct effect of *customer complaints* on *customer loyalty* (b_9). The sum of all direct and indirect effects is called a total effect (Alwin and Hauser 1975; Duncan 1975; Finney 1972) or sometimes an effect coefficient (Fox 1980; Lewis-Beck and Mohr 1976). For example, the total effect of *customer satisfaction* on *customer loyalty* is equal to the sum of the direct effect (b_8) and indirect effect $(b_7 \times b_9)$.

Examining total or indirect effects can provide additional information, which testing direct effects only cannot supply (Bollen 1989, pp. 376–377). For instance, in the latent growth curve modeling example discussed in Section 3.1.2.1.1, our main interest was centered in whether customer satisfaction had an impact on the temporal trajectories of change in annual revenue. This information could be obtained only by examining the indirect effect of customer satisfaction on the two latent variables called *Initial* and *Linear*. Moreover, when we investigate the expected change in an endogenous variable that is associated with a unit change in one of its predictor variables, we should look at the total effect of the predictor variable (Pedhazur 1982, p. 603). Furthermore, focusing only on the direct effect of a variable may lead to misleading conclusions. For example, if a latent variable has a small positive direct effect on another latent variable whereas it has a large negative indirect effect, relying solely on the direct effect is likely to disguise or distort the influence of the latent variable. In this case, it may be more sensible to concentrate on the total effect of the latent variable (Bollen 1989, p. 376).

According to the "infinite sum" definition of total effects (e.g., Alwin and Hauser 1975; Graff and Schmidt 1982), Bollen (1987) showed that total and indirect effects could be calculated from direct effects in closed form. In generalized structured component analysis, total and indirect effects can also be calculated based on his procedures (also see Bollen 1989, pp. 377–389). Specifically, let $\mathbf{T}_B$ denote a P by P matrix of total effects of latent variables on other latent variables. Let $\mathbf{ID}_B$ denote a P by P matrix of indirect effects of latent variables on other latent variables. Let $\mathbf{T}_C$ denote a J by P matrix of total effects of latent variables on indicators. Let $\mathbf{ID}_C$ denote a J by P matrix of indirect effects of latent variables on indicators. These matrices of total and indirect effects can be computed as follows.

$$\mathbf{T}_B = \left(\mathbf{I} - \mathbf{B}'\right)^{-1} - \mathbf{I} \tag{3.30}$$

$$\mathbf{ID}_B = \mathbf{T}_B - \mathbf{B}' = \left(\mathbf{I} - \mathbf{B}'\right)^{-1} - \mathbf{I} - \mathbf{B}' \tag{3.31}$$

$$\mathbf{T}_C = \mathbf{C}'\left(\mathbf{I} - \mathbf{B}'\right)^{-1} \tag{3.32}$$

$$\mathbf{ID}_C = \mathbf{T}_C - \mathbf{C}' = \mathbf{C}'\left(\mathbf{I} - \mathbf{B}'\right)^{-1} - \mathbf{C}' \tag{3.33}$$

The definition of total effects is available only when certain conditions are met. A sufficient condition is that the largest eigenvalue of $\mathbf{B}'$ in absolute value must be less than one (Bentler and Freeman 1983). Bollen (1989, p. 381) suggested two practical ways of checking this condition. One is to check whether $\mathbf{B}'$ is a lower triangular matrix. This indicates that the sufficient condition is satisfied under recursive models. The other is to ensure whether the elements of $\mathbf{B}'$ are positive and the sum of the elements in each column is less than one. This heuristic can be used for nonrecursive models. The standard errors and confidence intervals of total and indirect effects can be estimated via the bootstrap method.

The indirect effects calculated by Equations 3.31 and 3.33 comprise all possible indirect effects from one variable to another. We can also calculate specific indirect effects transmitted through a particular mediating variable or path. Again, we apply Bollen's (1987) procedure for the calculation of such specific indirect effects. His procedure begins with the calculation of indirect effects from the original $\mathbf{B}$ or $\mathbf{C}$ matrix. Next, it calls for the calculation of indirect effects using a modified $\mathbf{B}$ or $\mathbf{C}$ matrix, which is obtained based on which mediating variable or path is taken into consideration. Finally, the specific indirect effects transmitted through a mediating variable or path can be calculated by subtracting the indirect effects obtained using the modified $\mathbf{B}$ or $\mathbf{C}$ matrix from the original indirect effects. In the example that follows, we show the calculation of the total and indirect effects of latent variables on other latent variables (i.e., $\mathbf{T}_B$ and $\mathbf{ID}_B$) and of certain specific indirect effects in the ACSI model.

3.4.1 Example: The ACSI Data

We analyzed the consumer-level ACSI data used in Chapter 2 and in Section 3.1.1.1. The $\mathbf{B}$ matrix of path coefficients estimated from the data contains the direct effects of latent variables on other latent variables, as follows.

$$
\mathbf{B'} = \begin{bmatrix}
0 & 0 & 0 & 0 & 0 & 0 \\
0.58 & 0 & 0 & 0 & 0 & 0 \\
0.12 & 0.65 & 0 & 0 & 0 & 0 \\
0.03 & 0.68 & 0.26 & 0 & 0 & 0 \\
0 & 0 & 0 & -0.41 & 0 & 0 \\
0 & 0 & 0 & 0.59 & -0.09 & 0
\end{bmatrix}
\tag{3.34}
$$

The first column of Equation 3.34 consists of the direct effects of *customer expectations,* the second column consists of the direct effects of *perceived quality,* the third column has the direct effect of *perceived value,* the fourth column has the direct effects of *customer satisfaction,* and the fifth column has the direct effect of *customer complaints.*

Then, the total effects and indirect effects of latent variables on other latent variables can be calculated based on Equations 3.30 and 3.31. Table 3.13 provides the total and indirect effects of all latent variables in the ACSI model. All these effects turned out to be statistically significant.

Further, we show the calculation of the specific indirect effects of *customer expectations, perceived quality, perceived value,* and *customer satisfaction* on

TABLE 3.13

Total and Indirect Effects of Latent Variables and Their Standard Errors and 95% Confidence Intervals Obtained from the ASCI Data

	Total Effects				Indirect Effects			
	Estimate	SE	$|t|$	CI	Estimate	SE	$|t|$	CI
CE → PQ	0.58	0.04	15.87	0.51–0.65	0			
CE → PV	0.50	0.04	14.19	0.43–0.57	0.38	0.04	10.84	0.31–0.45
CE → CS	0.56	0.04	15.61	0.48–0.62	0.53	0.03	16.10	0.47–0.59
CE → CC	−0.23	0.03	8.48	−0.30 to −0.19	−0.23	0.03	8.48	−0.30–0.19
CE → CL	0.35	0.03	11.96	0.29–0.41	0.35	0.03	11.96	0.29–0.41
PQ → PV	0.65	0.03	18.87	0.58–0.71	0			
PQ → CS	0.85	0.02	47.89	0.83–0.90	0.17	0.02	7.17	0.13–0.23
PQ → CC	−0.35	0.03	9.98	−0.41 to −0.28	−0.35	0.03	9.98	−0.41 to −0.28
PQ → CL	0.54	0.03	17.70	0.48–0.60	0.54	0.03	17.70	0.48–0.60
PV → CS	0.26	0.03	7.58	0.21–0.35	0			
PV → CC	−0.11	0.02	6.19	−0.14 to −0.08	−0.11	0.02	6.19	−0.14 to −0.08
PV → CL	0.17	0.02	7.22	0.12–0.22	0.17	0.02	7.22	0.12–0.22
CS → CC	−0.41	0.04	10.86	−0.48 to −0.33	0			
CS → CL	0.63	0.03	20.36	0.57–0.69	0.04	0.02	2.42	0.01–0.06
CC → CL	−0.09	0.04	2.56	−0.16 to −0.03	0			

CE, Customer expectations; CI, Confidence intervals; CL, Customer loyalty; CS, Customer satisfaction; PQ, Perceived quality; PV, Perceived value; SE, Standard errors.

TABLE 3.14

Specific Indirect Effects of Latent Variables Transmitted Through
the Path from Customer Complaints to Customer Loyalty, and
Their Standard Errors and 95% Confidence Intervals Obtained
from the ASCI Data

	Estimate	SE	$|t|$	CI
CE → CL	0.02	0.01	2.44	0.01–0.04
PQ → CL	0.03	0.01	2.39	0.01–0.06
PV → CL	0.01	0.00	2.18	0.00–0.02
CS → CL	0.04	0.02	2.42	0.01–0.06

CE, Customer expectations; CI, Confidence intervals; CL, Customer loyalty;
CS, Customer satisfaction; PQ, Perceived quality; PV, Perceived value; SE,
Standard errors.

customer loyalty running through the path from *customer complaints to customer loyalty*, denoted by b_9 in Figure 2.3. In this case, the modified **B** matrix,
denoted by $\mathbf{B}_m$, is obtained by setting the path (b_9) to zero in the original **B**
matrix in Equation 3.34, as follows.

$$\mathbf{B}_m{}' = \begin{bmatrix} 0 & 0 & 0 & 0 & 0 & 0 \\ 0.58 & 0 & 0 & 0 & 0 & 0 \\ 0.12 & 0.65 & 0 & 0 & 0 & 0 \\ 0.03 & 0.68 & 0.26 & 0 & 0 & 0 \\ 0 & 0 & 0 & -0.41 & 0 & 0 \\ 0 & 0 & 0 & 0.59 & 0 & 0 \end{bmatrix} \tag{3.35}$$

We then compute the indirect effects of the latent variables using this modified **B** matrix based on Equation 3.31. Finally, the specific indirect effects
transmitted through the path can be obtained by subtracting the indirect
effects computed based on Equation 3.35 from the original indirect effects
computed based on Equation 3.34. Table 3.14 provides the specific indirect
effects of the four latent variables on *customer loyalty*, which is transmitted
through the single path. All the specific indirect effects turned out to be statistically significant.

3.5 Missing Data

It is not uncommon that observations are missing for reasons such as drop-
out, data coding error, and so on. A simple method for handling missing
observations is to delete any cases containing at least one missing observation (i.e., listwise deletion). However, this method is unsatisfactory if missing

observations are numerous and scattered throughout the dataset, as deletion of the cases may incur substantial loss of information. Another method is to estimate missing observations prior to analysis and then use the estimates in subsequent data analysis. For example, we may use the mean of nonmissing observations (i.e., mean substitution) or any prior knowledge/experience to replace a missing observation by some actual value.

A third method is to iteratively re-estimate values for missing observations (e.g., Gabriel and Zamir 1979; Gifi 1990). We begin by completing the data with some initial estimates for missing observations, obtain model parameter estimates by fitting the model to the complete data, update the estimates of missing observations based on the model parameter estimates, fit the model to the updated data, and so forth. These steps are repeated until no substantial changes occur in the estimates of both missing observations and model parameters. This iterative method can be particularly attractive in generalized structured component analysis for two reasons. First, unlike other imputation techniques such as the expectation–maximization algorithm (Dempster, Laird, and Rubin 1977) or multiple imputation (Rubin 1987; Schafer 1997), the method does not require a distributional assumption for estimation of missing observations. This is well-coupled with generalized structured component analysis that is a distribution-free approach to structural equation modeling. Second, the iterative method can be combined easily into the alternating least-square algorithm for generalized structured component analysis.

To combine the third method into generalized structured component analysis, we treat missing observations as another set of unknown parameters to be estimated, as well as the model parameters (weights, path coefficients, and/or loadings). We minimize the same least-squares criterion for estimating both missing observations and model parameters. However, the alternating least-squares algorithm is slightly extended, so that it consists of two phases. In one phase, the estimates of the model parameters in **W** and **A** are updated given the estimates of missing observations. This model estimation phase is equivalent to repeating the two steps in the original algorithm, given the complete data. In the other phase, the estimates of missing observations are updated with the model parameter estimates fixed. This imputation phase is described in Appendix 3.1. The two phases are alternated until convergence.

3.6 Summary

We discussed various extensions of generalized structured component analysis. These extensions do not require substantial degrees of change in model specification and parameter estimation. That is, imposing constraints on parameters to examine hypotheses about their relationships involves essentially the same least-squares estimation procedure based on the construction of a constraint matrix. Including higher-order latent variables necessitates the specification of additional weight matrices, whereas it does not require a modification in the original estimation procedure of generalized structured component analysis. A multiple group analysis does not call for any further changes in model specification and parameter estimation because it becomes virtually equivalent to a single group analysis. The calculation of total and indirect effects depends only on direct effects that are estimated by the original estimation procedure. The iterative estimation of missing observations is accompanied by an additional estimation step. However, this step is a relatively minor and natural extension of the alternating least-squares algorithm.

Thus, the extensions discussed in this chapter can be considered somewhat straightforward technically. Nevertheless, they contribute greatly to enhancing the data-analytic capability of generalized structured component analysis. Moreover, some of the extensions can serve to render generalized structured component analysis more attractive than partial least squares path modeling, such as multiple-group analyses involving cross-group constraints and higher-order latent variable modeling.

Appendix 3.1 The Imputation Phase of the Alternating Least-Squares Algorithm for Estimating Missing Observations

Let z_j denote an N by 1 vector of observations in a single indicator. Let z_{mj} denote a vector of missing observations in z_j. Let z_{nmj} denote a vector of non-missing observations in z_j, which are considered fixed. The imputation phase of the alternating least-squares algorithm updates the estimate of z_{mj}, given model parameter estimates. Specifically, let $Z_{(-j)}$ denote Z whose jth column is the vector of zeros. Let $Z_{(j)}$ denote an N by J matrix in which the jth column is equal to z_j and the others are all zero vectors. Let $Q = V - WA$. Let $Q_{(-j)}$ denote Q whose jth row is the vector of zeros. Let $q_{(j)}$ denote the jth row of Q. The least-squares criterion for generalized structured component analysis can be re-expressed as

$$\phi = SS(ZQ)$$

$$= SS((Z_{(j)} + Z_{(-j)})Q)$$

$$= SS(z_j q_{(j)} - (-Z_{(-j)}Q_{(-j)}))$$

$$= SS(z_j q_{(j)} - \Pi)$$

$$= SS((q_{(j)}' \otimes I)z_j - vec(\Pi))$$

$$= SS(\Xi z_{mj} + \tilde{\Xi} z_{nmj} - vec(\Pi)), \tag{A3.1}$$

where $\mathbf{\Pi} = -\mathbf{Z}_{(-j)}\mathbf{Q}_{(-j)}$, $\mathbf{\Xi}$ is the matrix formed by eliminating the columns of $\mathbf{q}_{(j)}' \otimes \mathbf{I}$ corresponding to any nonmissing observations in $\mathbf{z}_j$ (i.e., $\mathbf{z}_{nmj}$), and $\tilde{\mathbf{\Xi}}$ is the matrix consisting of the columns of $\mathbf{q}_{(j)}' \otimes \mathbf{I}$ corresponding to the non-missing observations.

The least-squares estimate of $\mathbf{z}_{mj}$ is then obtained by

$$\hat{\mathbf{z}}_{mj} = (\mathbf{\Xi}'\mathbf{\Xi})^{-1}\mathbf{\Xi}'(\mathrm{vec}(\mathbf{\Pi}) - \tilde{\mathbf{\Xi}}\mathbf{z}_{nmj}). \tag{A3.2}$$

The $\mathbf{z}_j$ vector is updated by replacing the missing observations by $\hat{\mathbf{z}}_{mj}$, and then standardized to satisfy $\mathbf{z}_j'\mathbf{z}_j = N$.

4

Fuzzy Clusterwise Generalized Structured Component Analysis

Generalized structured component analysis typically estimates parameters under the implicit assumption that all observations are drawn from a single homogenous population, unless it is known in advance that they belong to different groups. This leads to obtaining a single set of parameter estimates by pooling the data across observations. It is called an aggregate sample analysis. However, in some situations it may be more reasonable to assume that observations come from heterogeneous subgroups in the population, which display distinctive characteristics or response patterns, for example, different behaviors, choices, or preferences. This so-called group- or cluster-level heterogeneity has been considered to be important substantively and studied actively in various disciplines. For example, in developmental psychology, two different longitudinal trends in the development of antisocial behavior, such as life-course persistent and adolescent-limited, were discussed (Moffitt 1993). Moreover, six distinctive pathways in the evolution of adolescent delinquency over age have been identified, including rare offenders, moderate late peakers, high late peakers, decreasers, moderate-level chronics, and high-level chronics (Wiesner and Windle 2004). In marketing, consumer belief structures have been considered to diverge across market segments (Bagozzi 1982). Furthermore, from theoretical perspectives, in the situations where cluster-level heterogeneity is present, an aggregate sample analysis that disregards the heterogeneity is likely to result in biased parameter estimates (e.g., DeSarbo and Cron 1988; Jedidi, Jagpal, and DeSarbo 1997; Muthén 1989).

Cluster-level heterogeneity considered above is *unobserved* in the sense that information on cluster memberships of observations is not available in advance and should be obtained through data analysis. If such cluster information is a priori known or cluster-level heterogeneity is *observed*, a multiple-group analysis can be conducted to investigate heterogeneity in characteristics of observations across groups, as discussed in Chapter 3. In this chapter, we concentrate exclusively on such unobserved cluster-level heterogeneity. We shall not use the term unobserved hereafter.

A simple practice for accommodating cluster-level heterogeneity in generalized structured component analysis is to apply a two-step, sequential procedure. In the first step, some variety of cluster analysis is used to classify groups of observations. In the second step, generalized structured component analysis is applied to each of the clusters. The latter step is equivalent to conducting a multiple-group analysis for the clusters obtained in the first step. Although this two-step procedure is easy to implement, it suffers from a serious drawback in that it classifies observations into groups without taking into account the relationships between indicators and latent variables. In addition, the two steps aim to optimize different objective criteria separately: one for classification of observations into clusters, whereas the other for estimation of the parameters of generalized structured component analysis within each cluster. Thus, a simultaneous use of clustering and generalized structured component analysis, which involves a single optimization criterion, is more desirable.

Hwang, DeSarbo, and Takane (2007) proposed to integrate cluster analysis into generalized structured component analysis in a unified framework so as to take into account cluster-level heterogeneity. The proposed technique is called *fuzzy clusterwise generalized structured component analysis*. As its name suggests, this technique represents an extension of generalized structured component analysis into the framework of fuzzy clustering (e.g., Bezdek 1974a; Dunn 1974; Bezdek et al. 1981; Wedel and Steenkamp 1989). In this chapter, we shall discuss the theoretical underpinnings of the technique as well as its empirical application.

4.1 Fuzzy Clustering

In this section, we provide a brief review of fuzzy clustering to facilitate an understanding of the technical derivation of fuzzy clusterwise generalized structured component analysis. Traditional nonoverlapping clustering techniques, such as k-means or hierarchical clustering algorithms, are based on the assumption of hard/crisp classification (i.e., 1 = member and 0 = nonmember). That is, observations are to belong to only one of the mutually exclusive clusters with clearly separable boundaries. In contrast, fuzzy clustering is an array of clustering algorithms that adopts the fuzzy-set theory (Zadeh 1965), which allows observations to belong partially to a cluster, taking into account the potential ambiguity of cluster boundaries. Specifically, fuzzy clustering is built on two assumptions. First, an observation can be classified into more than one cluster, where the degree of its membership of a cluster lies between 0 and 1. Second, the sum of the memberships of an observation over all clusters must be equal to one (Wedel and Kamakura 1998, pp. 65–66).

Let u_{ki} denote a membership value for the ith observation in the kth group ($i = 1, \cdots, N; k = 1, \cdots, K$), which satisfies the two aforementioned assumptions, that is,

$$0 \leq u_{ki} \leq 1$$

and

$$\sum_{k=1}^{K} u_{ki} = 1$$

Let m denote a prescribed scalar weight, often called the "fuzzifier" (Bezdek 1974a), which influences the degree of fuzziness of memberships. Let y_i and x_i denote vectors of dependent and predictor variables, respectively, measured on the ith observation. Let B_k denote a matrix of regression coefficients of predictor variables in the kth cluster. The main objective of fuzzy clustering is to estimate the fuzzy memberships of observations in a given number of clusters. This objective can be achieved by minimizing the following criterion.

$$\varphi = \sum_{k=1}^{K} \sum_{i=1}^{N} u_{ki}^{m} SS(y_i - B_k x_i), \tag{4.1}$$

with respect to u_{ki} and B_k, subject to

$$\sum_{k=1}^{K} u_{ki} = 1$$

When $x_i = 1$ in Equation 4.1, this is equivalent to solving fuzzy k-means algorithm (Bezdek 1974a; Dunn 1974). On the other hand, when x_i includes other predictor variables, Equation 4.1 becomes the criterion for fuzzy k-regression or fuzzy clusterwise regression (Hathaway and Bezdek 1993; Wedel and Steenkamp 1989). When the number of clusters (K) is equal to 1, Equation 4.1 reduces to the ordinary least-squares criterion for multivariate linear regression because $u_{ki} = 1$ for all observations.

In fuzzy clustering, the value of the fuzzifier m is determined in advance. In principle, we can choose any value of m ranging from 1 to infinity. However, when m approaches 1, all memberships will converge to 0 or 1 as in hard clustering such as k-means. Conversely, when m approaches infinity, all memberships will be equal to $1/K$, leading to no differentiation in cluster memberships of observations (Wedel and Steenkamp 1989). Neither of these values of m is ideal (Arabie et al. 1981). There does not seem to exist a well-grounded way of deciding m in Equation 4.1, although some heuristic procedures were provided (e.g., McBratney and Moore 1985; Okeke and Karnieli 2006; Wedel and Steenkamp 1989). In practice, $m = 2$ is the most popular choice in fuzzy clustering (Bezdek 1981; Gordon 1999; Hruschka 1986; Wedel and Steenkamp 1991).

A number of so-called cluster validity measures (Bezdek 1981; Roubens 1982) can be used for deciding on the number of clusters. Based on his simulation study, Roubens (1982) recommended the fuzziness performance index (FPI) and the normalized classification entropy (NCE) as the most promising cluster validity measures in fuzzy clustering. The FPI and NCE are given by

$$FPI = 1 - (K \times PC - 1)/(K - 1), \tag{4.2}$$

where PC is the partition coefficient (Bezdek 1974b), calculated as

$$PC = \frac{1}{N} \sum_{i=1}^{N} \sum_{k=1}^{K} u_{ki}^2,$$

and

$$NCE = PE/\log K, \tag{4.3}$$

where PE is the partition entropy (Bezdek 1974b), calculated as

$$PE = -\frac{1}{N} \sum_{i=1}^{N} \sum_{k=1}^{K} u_{ki} \log u_{ki}$$

These measures point to the degree of separation between clusters, that is, how well the derived clusters are separated from each other. The smaller the values of FPI and NCE are, the more distinctly separated the clusters are from each other. Thus, the appropriate number of clusters should entail smaller values of FPI and NCE. As can be seen from Equations 4.2 and 4.3, NCE and FPI can be calculated only when more than one cluster is considered, that is, $K > 1$.

In addition, the changes in the value of criterion (Equation 4.1) or other related measures (e.g., the average adjusted R^2 across clusters) may be examined against the different number of clusters in order to determine the number of clusters (Wedel and Steenkamp 1989). The value of Equation 4.1 is to decrease monotonically with the number of clusters. The number of clusters may be chosen as an elbow point in the trajectory of the criterion value varying over the number of clusters, beyond which no substantial decrease in the value occurs.

Besides these heuristics, in practice, nonstatistical criteria for evaluating the usefulness and relevance of clusters (e.g., cluster size, interpretability, etc.) should also be considered for deciding the number of clusters (Arabie and Hubert 1994; Wedel and Kamakura 1998, pp. 4–5). For example, marketing researchers have frequently used six strategic criteria, such as identifiability, substantiality, accessibility, stability, responsiveness, and actionability, to assess the effectiveness of different segments of consumers (e.g., Baker 1988; Frank, Massy, and Wind 1972; Loudon and Della Bitta 1984).

In general, fuzzy clustering offers several advantages over traditional nonoverlapping clustering. For example, fuzzy clustering is likely to provide better fits to a dataset because its partial membership assumption is less restrictive than the binary counterpart. Similarly, fuzzy clustering tends to be more efficient computationally because it is more likely to involve gradual changes in the values of cluster memberships throughout estimation (McBratney and Moore 1985). Moreover, fuzzy clustering was shown to be less afflicted by local optima problems (Heiser and Groenen 1997). Furthermore, the partial memberships of observations can be used to identify the second best cluster that is almost as good as the best cluster, which nonoverlapping clustering methods cannot detect (Everitt, Landau and Leese 2001). Finally, fuzzy clustering is particularly appealing within the context of generalized structured component analysis for two reasons. First, it does not require any distributional assumption in clustering observations. This is well suited to the distribution-free optimization procedure of generalized structured component analysis. Second, the fuzzy clustering algorithm can be integrated coherently into the alternating least-squares algorithm of generalized structured component analysis, as will be discussed subsequently.

4.2 Fuzzy Clusterwise Generalized Structured Component Analysis

Fuzzy clusterwise generalized structured component analysis aims to identify clusters of observations and simultaneously, to estimate the parameters of generalized structured component analysis within each cluster. As stated earlier, this technique builds on *fuzzy clustering*, so that it results in fuzzy or probabilistic cluster memberships of observations. This indicates that *fuzziness* in the technique is associated with cluster memberships only, not with data themselves. Data can be considered fuzzy when they contain nonprecise values stemming from imperfection or inaccuracy in measurement (e.g., Coppi, Gil, and Kiers 2006; Viertl 2011). Although fuzzy clustering regression models have been developed for fuzzy-dependent variables (e.g., D'Urso, Massari, and Santoro 2010; D'Urso and Santoro 2006), no such development has been done yet in generalized structured component analysis. Thus, fuzzy clusterwise generalized structured component analysis can be applied only to nonfuzzy data thus far. In addition, it represents a *clusterwise* method in the sense that generalized structured component analysis is repeatedly applied to each cluster identified. This leads to estimating a separate set of the same parameters in each cluster.

The aim of fuzzy clusterwise generalized structured component analysis can be achieved by minimizing the following criterion.

$$\phi = \sum_{k=1}^{K} \sum_{i=1}^{N} u_{ki}^{m} SS(\mathbf{V}_k' \mathbf{z}_i - \mathbf{A}_k' \mathbf{W}_k' \mathbf{z}_i), \tag{4.4}$$

with respect to u_{ki}, $\mathbf{W}_k$, and $\mathbf{A}_k$, subject to

$$\sum_{k=1}^{K} u_{ki} = 1$$

When $K = 1$, Equation 4.4 becomes equivalent to the least-square criterion (Equation 2.16) for generalized structured component analysis. This indicates that fuzzy clusterwise generalized structured component analysis subsumes generalized structured component analysis as a special case where only one cluster is considered (i.e., an aggregate sample analysis).

Equation 4.4 is comparable to Equation 4.1, in which the sum of squares of all residuals in generalized structured component analysis is replaced by that of residuals in linear regression. A major difference is that generalized structured component analysis allows the specification and testing of more complex path-analytic relationships between indicators and latent variables. In fact, fuzzy clusterwise generalized structured component analysis can include fuzzy clustering as a special case, which focuses on a linear regression analysis for dependent and predictor indicators.

An alternating least-squares algorithm was developed to minimize Equation 4.4. The algorithm repeats two main steps until convergence. One step updates the parameter estimates of generalized structured component analysis, given the estimates of cluster memberships. The other step updates the estimates of cluster memberships, given the parameter estimates. Both steps involve least-squares estimation, so that they do not require any distribution assumption such as multivariate normality of indicators. We provide a description of the algorithm in Appendix 4.1. Although the algorithm is convergent, it is not ensured that the convergence point is the global minimum. To safeguard against potential local minima, we can apply fuzzy k-means (Bezdek 1974a; Dunn 1974) to observations and use the resultant memberships as initial values for u_{ki}. Furthermore, we can run the algorithm many times using different sets of random initial values for parameters, and then choose the solution associated with the smallest value of Equation 4.4.

In fuzzy clusterwise generalized structured component analysis, the FIT measure is calculated as

$$\text{FIT } = 1 - \frac{\displaystyle\sum_{k=1}^{K}\sum_{i=1}^{N} u_{ki}^{m}\text{SS}(\mathbf{V}_k'\mathbf{z}_i - \mathbf{A}_k'\mathbf{W}_k'\mathbf{z}_i)}{\displaystyle\sum_{k=1}^{K}\sum_{i=1}^{N} u_{ki}^{m}\text{SS}(\mathbf{V}_k'\mathbf{z}_i)}. \tag{4.5}$$

As described in Chapter 2, AFIT is calculated based on Equation 4.5, as follows.

$$\text{AFIT} = 1 - (1 - \text{FIT})\frac{d_0}{d_1}, \tag{4.6}$$

where $d_0 = NJ$ is the degrees of freedom for the null model, where J is the number of indicators, and $d_1 = NJ - \delta$ is the degrees of freedom for the model being tested, where δ is the number of free parameters. The free parameters include cluster memberships of observations as well as model parameters, such as weights, path coefficients, and/or loadings, across all clusters. The AFIT cannot be calculated when δ exceeds NJ. In fuzzy clusterwise generalized structured component analysis, δ can exceed NJ if the number of clusters (K) is greater than that of indicators (J), because the number of cluster memberships becomes equivalent to NJ when $K = J$. However, in practice, it is unlikely to encounter such situations because K is usually much smaller than J.

The bootstrap method can be used to estimate the standard errors or confidence intervals of parameter estimates. Once cluster memberships of observations are estimated, observations are assigned to a cluster associated with the highest membership value. Then, a bootstrap sample is constructed by resampling observations in each cluster with replacement.

As in fuzzy clustering, the FPI and NCE can be used for deciding the number of clusters. The values of FIT and AFIT can also be examined against varying numbers of clusters to decide the number of clusters. For example, the number of clusters at which an elbow point of the values of FIT is observed may be chosen as the number of clusters beyond which no substantial increases occur in the values of FIT. Moreover, the number of clusters associated with the largest value of AFIT may be chosen.

The value of the fuzzifier m should also be selected in advance. Similar to fuzzy clustering, there are no theoretically justifiable heuristics that can be used for determining the value of the fuzzifier in fuzzy clusterwise generalized structured component analysis. As discussed in the preceding section, $m = 2$ can be chosen as the default value in fuzzy clusterwise generalized structured component analysis. We propose a regularized version of fuzzy clusterwise generalized structured component analysis in Appendix 4.2. In this version, regularized fuzzy clustering (Miyamoto and Mukaidono 1997) is combined with generalized structured component analysis. A potential advantage of this approach is that the degree of fuzziness of cluster memberships may be determined automatically by using cross validation.

A finite mixture approach to partial least squares path modeling (Hahn et al. 2002) was proposed in an effort to accommodate cluster-level heterogeneity in partial least squares path modeling. This approach is built on the assumption that cluster-level heterogeneity is present only in the structural model. It carries out two steps sequentially. In the first step, the individual scores of latent variables specified in a complete structural equation model are obtained via partial least squares path modeling. In the second step, the measurement model is disregarded and only the structural model is taken into account, where latent variables are now treated as indicators. Under the assumption that each endogenous latent variable is normally distributed, finite mixture linear regression (DeSarbo and Cron 1988) is applied for estimating clusterwise regression coefficients connecting a set of latent variables to each endogenous latent variable, as well as cluster memberships of observations within clusters. Thus, finite mixture partial least squares path modeling is equivalent to a series of finite mixture linear regressions for latent variable scores that are initially estimated by means of partial least squares path modeling.

The premise of this approach that the relationships between indicators and latent variables specified in the measurement model should remain invariant across clusters appears too restrictive and difficult to justify in many situations. In addition, the distributional assumption of multivariate normality of the estimated endogenous latent variables can be arguable and inconsistent with the distribution-free estimation procedure of partial least squares path modeling. Furthermore, the approach can be used only when no latent variables are mediating ones that serve as both dependent and predictor variables in the structural model, because all predictor variables are considered fixed in the second step of the approach. Also importantly, in the second step, using finite mixture covariance structure analysis (Jedidi et al. 1997) can be more sensible than using a series of finite mixture regressions, because finite mixture covariance structure analysis can take into account more complex path-analytic relationships among latent variables, whose scores are a priori calculated via partial least squares path modeling. In fact, it seems less convincing to choose finite mixture partial least squares path modeling over finite mixture covariance structure analysis. Finally, finite mixture modeling generally has limitations of its own. For instance, it often encounters convergence problems such as slow convergence or nonconvergence due to the adoption of the expectation–maximization (Dempster, Laird, and Rubin 1977) or gradient-based estimation procedures (Wedel and Kamakura 1998, p. 88). Moreover, the information criteria for model selection are based on the regularity properties of the likelihood function, which do not typically hold in finite mixture models (McLachlan and Peel 2000). Moreover, because of its reliance on likelihood-based estimation procedures, the correct specification of the probabilistic distribution by which the data are generated is required. However, it is often difficult to decide the correct form of the entire mechanism by which the data are generated.

Fuzzy clusterwise generalized structured component analysis is more general than finite mixture partial least squares path modeling in that it takes into consideration both measurement and structural models in the capturing of cluster-level heterogeneity. Moreover, fuzzy clusterwise generalized structured component analysis does not require a distributional assumption such as multivariate normality.

Another approach to accounting for cluster-level heterogeneity in partial least squares path modeling is a response-based procedure for detecting unit segments in partial least squares path modeling (REBUS-PLS) (Esposito Vinzi et al. 2008). This approach is a generalization of partial least squares topological path modeling (PLS-TPM) (Trinchera, Squillacciotti, and Esposito Vinzi 2006). It carries out several steps sequentially. The approach begins by applying partial least squares path modeling to fit a single model to the entire dataset and computing each observation's measurement and structural model residuals. The measurement model residual is equal to the difference between an indicator and its prediction obtained by regressing the indicator on the corresponding latent variable, whereas the structural model residual is the difference between a latent variable and its prediction obtained by regressing the latent variable on the corresponding predictor latent variables. The approach then applies a hierarchical cluster analysis to the residual scores of all observations in order to decide on the number of clusters as well as cluster memberships of observations. Subsequently, it involves an iterative procedure in which model parameters are estimated via partial least squares path modeling for each cluster of observations, and each observation is re-assigned to a cluster based on the distance between the observation and the model parameters estimated for each cluster. The distance is measured by the so-called closeness measure that is proportional to an observation's measurement and structural model residuals for each cluster. The iterative procedure continues until no substantial changes in cluster memberships occur, for example, fewer than 5% of observations change their cluster memberships between previous and current iterations (Esposito Vinzi, Trinchera, Amato 2010, p. 70).

REBUS-PLS does not require a distributional assumption, which is coherent to the estimation procedure of partial least squares path modeling. In addition, it takes into account both measurement and structural models in uncovering cluster memberships of observations. These can be advantageous over finite mixture partial least squares path modeling. However, REBUS-PLS can be used only when all indicators are reflective, so that each observation's measurement model residual can be calculated (Esposito Vinzi et al. 2008). Although it is relatively straightforward to implement, the approach carries out several steps separately, which involve different optimization criteria. Thus, it is not ensured that solutions obtained from a previous step are most favorable for subsequent steps. Furthermore, it is not yet formally proven that the iterative procedure is convergent, although the procedure appears to converge in empirical applications (Esposito Vinzi, Trinchera, Amato 2010, p. 70). In contrast, fuzzy clusterwise generalized structured component analysis can be employed when indicators are reflective or formative. Also, it is equipped with a single optimization criterion for the estimation of cluster memberships and clusterwise model parameters, which is consistently minimized by an alternating least-squares algorithm.

4.3 Example: Clusterwise Latent Growth Curve Modeling of Alcohol Use among Adolescents

This example is part of a longitudinal survey of substance use among adolescents from two northwestern urban areas in the U.S. (Duncan et al. 1997). The sample consists of 632 adolescents measured on their use of alcohol over four points in time. Alcohol use was assessed by a single self-report item with five response options: (1) life time abstainers, (2) 6-month abstainers, (3) current use of less than four times a month, (4) current use of between 4 and 29 times a month, and (5) current use of 30 times or more a month. Five additional variables were measured once at the initial point in time: parental marital status, family status, socioeconomic status (SES), gender, and age. Parental marital status was coded as follows: 0 = single and 1 = married or living in a committed relationship. Family status was categorized as follows: 0 = step or foster families and 1 = others. SES was calculated as the average of parental annual income and education level. Parental annual income was assessed based on a 16-point scale ranging from "6,000 dollars and below" to "50,000 dollars or more." Education levels range from "Grade level 6 or less" to "Graduate level." Male and female were coded as 0 and 1, respectively.

4.3.1 The Single-Cluster Latent Growth Curve Model

A latent growth curve model was specified for identifying an average or intra-individual change in alcohol use over time as well as interindividual differences in the temporal change; and simultaneously, for examining the effects of the five time-invariant variables on the temporal change. From descriptive statistics, it was found that the mean levels of alcohol use increased monotonically over time ($t_1 = 2.23$, $t_2 = 2.46$, $t_3 = 2.65$, and $t_4 = 2.94$). This suggests that there existed a linear trend of change in the consumption of alcohol over the four assessments. Moreover, alcohol use was merely measured over four points in time. This relatively small number of measurements may be insufficient to reveal a complex nonlinear temporal pattern in the repeated measures of alcohol use (MacCallum et al. 1997). In fact, previous studies with the same data also supported such a linear trend of change in alcohol use over the four measurement intervals (Duncan et al. 1997; Hwang and Takane 2004b). Accordingly, a linear-trend latent growth curve model was specified for the data.

Figure 4.1 displays the specified single-cluster linear-trend model. As depicted in Figure 4.1, the loadings were a priori fixed for two latent variables labeled *Intercept* and *Linear*. Every loading relating *Intercept* to the repeated measures of alcohol use was fixed to 1. On the other hand, the loadings relating *Linear* to the repeated measures were set to 0, 1, 2, and 3. Moreover, the specified model included the effects of the five time-invariant variables on the two latent variables.

At first, no cluster-level heterogeneity was taken into account for the latent growth curve model, that is, $K = 1$. That is, fuzzy clusterwise generalized structured component analysis was used to fit the aggregate sample, single-cluster latent growth curve model to the data, as discussed in Chapter 3. This aggregate analysis provided that FIT = 0.96 (SE = 0.00, 95% CI = 0.96–0.96) and AFIT = 0.96 (SE = 0.00, 95% CI = 0.96–0.96).

The mean estimates of *Intercept* and *Linear* were 2.12 (SE = 0.05, 95% CI = 2.02–2.21) and 0.27 (SE = 0.01, 95% CI = 0.25–0.29), respectively. This is indicative of significant levels of alcohol use at the initial status as well as significant linear growth in alcohol use over the four time points. These results appear consistent with previous studies, which reported significant levels of both initial status and growth rate of alcohol consumption (Duncan et al. 1997;

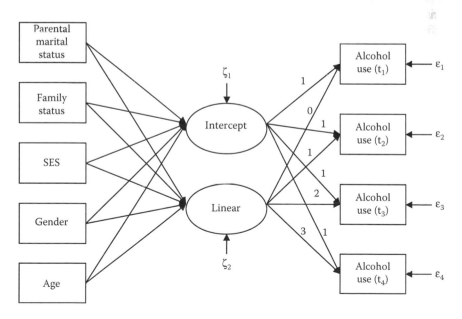

FIGURE 4.1
The latent growth curve model specified for the alcohol use data.

Hwang and Takane 2004b). The variance estimates of *Intercept* and *Linear* were 0.60 (SE = 0.03, 95% CI = 0.53–0.65) and 0.04 (SE = 0.00, 95% CI = 0.03–0.05), respectively, indicating that substantial individual differences existed in both initial status and growth rate of alcohol use. This is also consistent with what is reported in the study of Duncan et al. (1997). The correlation estimate between the two latent variables was equal to −0.15 (SE = 0.04, 95% CI = −0.24 to −0.08). This significant correlation indicates that the level of alcohol use at the initial time point appeared to be negatively related to the growth rate of alcohol use. In other words, adolescents showing high levels of alcohol use at the initial status were likely to show low growth rates of alcohol use over the four assessments. This significant and negative correlation is also reported in the study of Duncan et al. (1997).

Table 4.1 provides the path coefficient estimates of the five time-invariant variables. The effect of parental marital status on *Intercept* was equal to −0.22 (SE = 0.10, 95% CI = −0.40 to −0.04). It points to higher initial levels of alcohol use by adolescents living with single parents compared to those living with both parents. The effect of parental marital status on *Linear* was 0.02 (SE = 0.03, 95% CI = −0.03–0.08), indicating that adolescents living with non-single parents seemed to increase their use of substance at a higher rate. Yet, this estimate was statistically nonsignificant. The effects of family status on *Intercept* and *Linear* were 0.22 (SE = 0.06, 95% CI = 0.08–0.32) and 0.05 (SE = 0.02, 95% CI = 0.02–0.08), respectively. These estimates suggest that adolescents living with other families rather than step or foster families showed higher levels of alcohol use at the initial status, and also their alcohol use increased at a higher rate over time. The effects of SES on *Intercept* and *Linear* were equal to −0.06 (SE = 0.06, 95% CI = −0.16–0.06) and −0.02 (SE = 0.01, 95% CI = −0.05–0.00), respectively. It suggests that socially and economically more disadvantaged adolescents displayed higher levels of alcohol use, and also showed a higher rate of increase in alcohol use than those less disadvantaged. Yet, both effects were statistically nonsignificant. The effect of gender

TABLE 4.1

Path Coefficient Estimates and Their Standard Errors and 95% Confidence Intervals in the Parenthesis of the Single-Cluster Latent Growth Curve Model Specified for the Alcohol Use Data

	Intercept			Linear		
	Estimate	**SE**	**95% CI**	**Estimate**	**SE**	**95% CI**
Parental marital status	−0.22	0.10	−0.40 to −0.04	0.02	0.03	−0.03–0.08
Family status	0.22	0.06	0.08–0.32	0.05	0.02	0.02–0.08
SES	−0.06	0.06	−0.16–0.06	−0.02	0.01	−0.05–0.00
Gender	0.03	0.06	−0.08–0.14	0.04	0.02	−0.00–0.07
Age	0.14	0.01	0.13–0.16	0.01	0.00	0.01–0.02

CI, Confidence intervals; SE, Standard errors; SES, Socioeconomic status.

on *Intercept* was 0.03 (SE = 0.06, 95% CI = −0.08–0.14). It suggests that female adolescents seemed to show a higher level of alcohol use at the initial status. However, this effect was statistically nonsignificant. On the other hand, the effect of gender on *Linear* was 0.04 (SE = 0.02, 95% CI = −0.00–0.07). This suggests that female adolescents tended to increase use of alcohol at a higher rate over the four time points. The effect of age on *Intercept* was equal to 0.14 (SE = 0.01, 95% CI = 0.13–0.16), indicating that older adolescents showed higher levels of alcohol use at the initial status. On the other hand, the effect of age on *Linear* was 0.01 (SE = 0.00, 95% CI = 0.01–0.02), indicating that older adolescents were likely to increase use of alcohol at a higher rate compared to younger adolescents.

4.3.2 Multiple-Cluster Latent Growth Curve Models

To take into account cluster-level heterogeneity, fuzzy clusterwise generalized structured component analysis was applied to fit the same linear-trend model to these data, varying the number of clusters. The problem of nonconvergence did not occur.

Figure 4.2 shows the values of FIT, AFIT, FPI, and NCE for different multiple-cluster latent growth curve models. It was shown that the values of FIT increased gradually beyond $K = 2$, suggesting that no substantial changes in FIT were obtained by having more than two clusters. This is consistent with a value of AFIT, which was maximized when $K = 2$. Furthermore, the minimum values of both FPI and NCE were obtained at $K = 2$. Thus, $K = 2$ was adopted for further analyses.

For the two-cluster latent growth curve model, the mean estimates of *Intercept* and *Linear* were 0.77 (SE = 0.02, 95% CI = 0.73–0.82) and 0.18

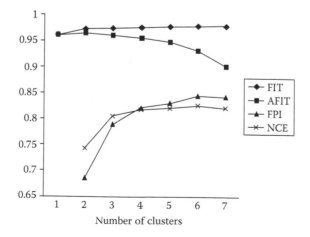

FIGURE 4.2
The values of FIT, AFIT, fuzziness performance index, and normalized classification entropy for different multiple-cluster latent curve models for the alcohol use data.

(SE = 0.01, 95% CI = 0.17–0.19), respectively, in cluster 1. On the other hand, the mean estimates of *Intercept* and *Linear* were 1.38 (SE = 0.02, 95% CI = 1.33–1.41) and 0.08 (SE = 0.01, 95% CI = 0.07–0.09), respectively, in cluster 2. Both clusters indicated significant levels of alcohol use at the initial status and also significant linear growth in alcohol use over the four time points. The two clusters involved substantively different temporal patterns of intra-individual change: The observations in cluster 1 were likely to show relatively low levels of alcohol use at the initial assessment while their use of alcohol tended to increase at a higher rate over the four time points. On the other hand, the observations in cluster 2 involved a relatively high level of alcohol use at the first assessment, but a lower rate of increase in alcohol use over the four time points. Thus, cluster 1 may be characterized as "a low initial level of alcohol use accompanied by a high growth rate" and cluster 2 as "a high initial level of alcohol use accompanied by a low growth rate." Li, Duncan, and Hops (2001) have also classified 179 U.S. adolescents living in a similar geographical area into two clusters, based on the temporal patterns of their alcohol use from grade 6 to grade 12 (i.e., eight time points). In their study, one cluster showed a low level of alcohol use at grade 6 and showed a monotonic increase until grade 12. The other cluster showed a high level of alcohol use at grade 6, no substantial increase until grade 8, and a linear increase during later grades. Although it is not feasible to directly compare our results to theirs for various reasons, for example, different sample sizes, the different number of time points, and so on, both analyses appear to convey similar information concerning the temporal patterns of alcohol use within two clusters of adolescents.

In cluster 1, the variance estimates of *Intercept* and *Linear* were 0.11 (SE = 0.01, 95% CI = 0.10–0.13) and 0.03 (SE = 0.00, 95% CI = 0.02–0.03), respectively. In cluster 2, the variance estimates of *Intercept* and *Linear* were 1.0 (SE = 0.04, 95% CI = 0.92–1.09) and 0.01 (SE = 0.00, 95% CI = 0.01–0.01), respectively. This indicates that both clusters involved substantial interindividual differences in the initial status and the growth rate of alcohol use. The correlation between the two latent variables was 0.51 (SE = 0.03, 95% CI = 0.46–0.56) in cluster 1 indicating that the levels of alcohol use at the initial status were significantly and positively correlated with the growth rate of alcohol use. On the other hand, the correlation between the latent variables was −0.04 (SE = 0.06, 95% CI = −0.15–0.06) in cluster 2. This suggests that the levels of alcohol use at the initial status were negatively associated with the growth rate of alcohol use. Yet, this latter correlation turned out statistically nonsignificant.

Table 4.2 provides the path coefficients of the five exogenous indicators estimated for each cluster. In order to focus on distinct characteristics of the path coefficient estimates across the two clusters, statistically significant estimates are considered for interpretational purposes here.

In cluster 1, family status had a significant and positive effect on *Intercept* (0.15, SE = 0.06, 95% CI = 0.03–0.28), indicating that adolescents living with other families rather than step or foster families displayed higher levels of

TABLE 4.2

Path Coefficient Estimates, and Their Standard Errors and 95% Confidence Intervals in the Parenthesis of the Two-Cluster Latent Growth Curve Model Specified for the Alcohol Use Data

		Intercept			Linear		
		Estimate	SE	95% CI	Estimate	SE	95% CI
Cluster 1	Parental marital status	−0.15	0.11	−0.34–0.11	0.09	0.04	−0.00–0.19
	Family status	0.15	0.06	0.03–0.28	0.05	0.03	−0.01–0.10
	SES	−0.09	0.07	−0.22–0.03	−0.09	0.03	−0.15 to −0.03
	Gender	−0.01	0.06	−0.13–0.11	0.04	0.02	0.00–0.10
	Age	0.09	0.01	0.08–0.11	0.02	0.00	0.01–0.02
Cluster 2	Parental marital status	−0.25	0.11	−0.47 to −0.02	−0.03	0.04	−0.11–0.04
	Family status	0.30	0.10	0.11–0.47	0.05	0.02	0.02–0.10
	SES	0.02	0.06	−0.08–0.13	0.02	0.02	−0.02–0.06
	Gender	0.03	0.07	−0.10–0.19	0.03	0.02	−0.01–0.08
	Age	0.19	0.01	0.17–0.21	0.00	0.00	−0.00–0.01

CI, Confidence intervals; SE, Standard errors.

alcohol use at the initial status. Also, SES had a significant and negative impact on *Linear* (−0.09, SE = 0.03, 95% CI = −0.15 to −0.03) suggesting that socially and economically more disadvantaged adolescents were likely to show a higher rate of increase in alcohol use than those less disadvantaged. Gender had a significant and positive influence on *Linear* (0.04, SE = 0.02, 95% CI = 0.00–0.10), indicating that female adolescents were likely to show a higher rate of increase in alcohol use than male adolescents. Moreover, in cluster 1, age exhibited significant and positive effects on both *Intercept* (0.09, SE = 0.01, 95% CI = 0.08–0.11) and *Linear* (0.02, SE = 0.00, 95% CI = 0.01–0.02). It suggests that older adolescents displayed higher levels of alcohol use at the initial status; and at the same time, tended to increase use of alcohol at a higher rate compared to younger adolescents.

In cluster 2, parental marital status had a significant and negative impact on *Intercept* (−0.25, SE = 0.11, 95% CI = −0.47 to −0.02). This indicates higher initial levels of alcohol use by adolescents living with single parents compared to those living with both parents. In this cluster, family status showed significant and positive impacts on *Intercept* (0.30, SE = 0.10, 95% CI = 0.11–0.47) and *Linear* (0.05, SE = 0.02, 95% CI = 0.02–0.10). It suggests that adolescents living with other families rather than step or foster families showed higher levels of alcohol use at the first time point and also that their alcohol use increased at a higher rate over time. Also, age had a significant and positive effect on *Intercept* (0.19, SE = 0.01, 95% CI = 0.17–0.21) indicating that older adolescents exhibited higher levels of alcohol use at the initial status than younger adolescents in cluster 2.

Thus, in the first cluster characterized by a low initial level of alcohol use accompanied by a high growth rate, family status and age had statistically significant influences on the initial level of alcohol use, while SES, gender, and age had statistically significant effects on the growth rate of alcohol use over time. On the other hand, in the second cluster involving a high initial level of alcohol use accompanied by a low growth rate, parental marital status, family status, and age showed statistically significant impacts on the initial level of alcohol use, while only family status displayed a statistically significant effect on the linear increase of alcohol use over time. When classifying respondents based on a cut-off point of 0.5, the sizes of clusters 1 and 2 arrived at 293 (46%) and 339 (54%), respectively.

4.4 Summary

We discussed a disaggregate extension of generalized structured component analysis to account for cluster-level heterogeneity. This extension, called fuzzy clusterwise generalized structured component analysis, aims to classify observations into heterogeneous clusters and estimate the parameters of generalized structured component analysis in each of the clusters simultaneously. Technically, it combines generalized structured component analysis with fuzzy clustering in a unified manner. It is equipped with a single optimization criterion for parameter estimation, which is minimized by an alternating least-squares algorithm.

We also demonstrated the use of fuzzy clusterwise generalized structured component analysis in the context of a latent growth curve modeling of the repeated measures of alcohol use among adolescents. This technique enabled the determination that $K = 2$ represented the reasonable number of clusters based on model selection heuristics. In particular, the two-cluster latent growth curve model seemed to reveal substantively distinct temporal trajectories on alcohol use within the two clusters of adolescents. That is, one cluster showed a lower level of alcohol use at the initial status but a higher rate of increase in alcohol use over time, whereas the other cluster displayed a higher level of alcohol use at the initial status but a lower rate of growth in alcohol use. Moreover, the influences of the time-invariant variables on the two latent variables of temporal change were shown to be different across the two clusters. Importantly, all these insights could not be gained by fitting the single-cluster model. Therefore, fuzzy clusterwise generalized structured component analysis appears to be helpful in studying qualitatively different longitudinal processes on alcohol use and different consequences of its antecedents across two relatively heterogeneous subgroups of adolescents.

Appendix 4.1 The Alternating Least-Squares Algorithm for Fuzzy Clusterwise Generalized Structured Component Analysis

Let $\mathbf{Z} = [\mathbf{z}_i, \cdots, \mathbf{z}_N]'$ denote an N by J matrix of indicators for N observations. Let $\mathbf{U}_k = \text{diag}[u_{k1}, \cdots, u_{kN}]$ denote a diagonal matrix consisting of fuzzy membership values of observations in cluster k ($k = 1, \ldots, K$). We seek to minimize the following criterion.

$$\phi = \sum_{k=1}^{K} \sum_{i=1}^{N} u_{ki}^m SS(\mathbf{V}_k'\mathbf{z}_i - \mathbf{A}_k'\mathbf{W}_k'\mathbf{z}_i)$$

$$= \sum_{k=1}^{K} SS(\mathbf{Z}\mathbf{V}_k - \mathbf{Z}\mathbf{W}_k\mathbf{A}_k)_{\mathbf{U}_k^m} \qquad (A4.1)$$

with respect to u_{ki}, $\mathbf{W}_k$, and $\mathbf{A}_k$, subject to

$$\sum_{k=1}^{K} u_{ki} = 1$$

and

$$\sum_{i=1}^{N} \gamma_{ip}^2 = N,$$

where

$$SS(\mathbf{M})_{\mathbf{H}} = \text{trace}(\mathbf{M}'\mathbf{H}\mathbf{M}).$$

To minimize this criterion, the following two steps are alternated until convergence.

Step 1: The parameters of generalized structured component analysis ($\mathbf{W}_k$ and $\mathbf{A}_k$) in each cluster are updated for fixed $\mathbf{U}_k$. This step is equivalent to minimizing

$$\phi = \sum_{k=1}^{K} SS((\mathbf{U}_k^m)^{1/2}(\mathbf{Z}\mathbf{V}_k - \mathbf{Z}\mathbf{W}_k\mathbf{A}_k))$$

$$= \sum_{k=1}^{K} SS(\mathbf{Z}_k\mathbf{V}_k - \mathbf{Z}_k\mathbf{W}_k\mathbf{A}_k), \qquad (A4.2)$$

subject to

$$\sum_{i=1}^{N} \gamma_{ip}^{2} = N,$$

where

$$\mathbf{Z}_{k} = \left(\mathbf{U}_{k}^{m}\right)^{1/2} \mathbf{Z}.$$

This is equivalent to minimizing the original least-squares criterion of generalized structured component analysis for each of K clusters. Thus, the same alternating least-squares algorithm for generalized structured component analysis can be used to update the parameters $\mathbf{W}_k$ and $\mathbf{A}_k$ within each cluster (refer to Appendix 2.1).

Step 2: The membership parameter u_{ki} is updated for fixed $\mathbf{W}_k$ and $\mathbf{A}_k$. Let

$$d_{ki} = \text{SS}(\mathbf{V}_k' \mathbf{z}_i - \mathbf{A}_k' \mathbf{W}_k' \mathbf{z}_i).$$

Minimizing Equation A4.1 with respect to u_{ki} under the constraint

$$\sum_{k=1}^{K} u_{ki} = 1$$

is equivalent to minimizing

$$\phi^{*} = \sum_{k=1}^{K} \sum_{i=1}^{N} u_{ki}^{m} d_{ki} + \lambda \left(\sum_{k=1}^{K} u_{ki} - 1\right), \qquad (A4.3)$$

where λ is a Lagrangian multiplier. The estimate of u_{ki} is given by

$$\hat{u}_{ki} = \frac{1}{\sum_{c=1}^{K} \left(\dfrac{d_{ki}}{d_{ci}}\right)^{1/(m-1)}}. \qquad (A4.4)$$

Equation A4.4 can be derived as follows. By solving

$$\frac{\partial \phi^{*}}{\partial u_{ki}} = m u_{ki}^{m-1} d_{ki} - \lambda = 0$$

for u_{ki}, we have

$$\hat{u}_{ki} = \left(\frac{\lambda}{m d_{ki}} \right)^{1/(m-1)}.$$

(A4.5)

By using

$$\frac{\partial \phi^*}{\partial \lambda} = \sum_{k=1}^{K} u_{ki} - 1 = 0$$

and Equation A4.5, we have

$$\hat{\lambda} = \left(\left(\sum_{k=1}^{K} 1 / (m d_{ki}) \right)^{1/(m-1)} \right)^{1}{}^{m}.$$

(A4.6)

Then, Equation A4.4 is obtained by inserting Equation A4.6 in Equation A4.5 (see also Wedel and Steenkamp 1989).

Appendix 4.2 Regularized Fuzzy Clusterwise Generalized Structured Component Analysis

From the standpoint of fuzzy clustering that permits an observation to belong partially to multiple clusters, that is, $u_{ki} \in [1, 0]$, hard clustering, where an observation is to belong exclusively to one cluster, may be viewed as an ill-posed problem that involves only two extreme membership values, that is, $u_{ki} \in \{1, 0\}$ (Li and Mukaidono 1995). In this regard, Miyamoto and Mukaidono (1997) proposed *regularized fuzzy clustering* or *fuzzy clustering regularized by entropy*, where fuzzy k-means was considered a regularized version of k-means, which aims to convert the ill-posed (binary) solution of hard clustering to a well-posed (probabilistic) solution. They used the negative sum of entropies of membership values, that is,

$$\sum_{k=1}^{K} \sum_{i=1}^{N} u_{ki} \log(u_{ki})$$

(Bezdek 1981), as a penalty term because the entropy term is minimized when $u_{ki} = 1/K$, and thus tends to deviate membership values from the extreme values (Miyamoto 1998).

As another approach to capturing cluster-level heterogeneity, we can combine regularized fuzzy clustering with generalized structured component analysis in a unified framework. This technique can be called *regularized fuzzy clusterwise generalized structured component analysis*. We seek to minimize the following criterion to estimate parameters.

$$
\begin{aligned}
\phi &= \sum_{k=1}^{K} \sum_{i=1}^{N} u_{ki} SS(\mathbf{V}_k' \mathbf{z}_i - \mathbf{A}_k' \mathbf{W}_k' \mathbf{z}_i) - m \sum_{k=1}^{K} \sum_{i=1}^{N} u_{ki} \log u_{ki} \\
&= \sum_{k=1}^{K} SS(\mathbf{Z}\mathbf{V}_k - \mathbf{Z}\mathbf{W}_k\mathbf{A}_k)_{\mathbf{U}_k} - m \sum_{k=1}^{K} \sum_{i=1}^{N} u_{ki} \log u_{ki},
\end{aligned} \tag{A4.7}
$$

with respect to u_{ki}, $\mathbf{W}_k$, and $\mathbf{A}_k$, subject to

$$
\sum_{k=1}^{K} u_{ki} = 1
$$

and

$$
\sum_{i=1}^{N} \gamma_{ip}^2 = N.
$$

In the above criterion, m is a positive tuning parameter that controls for the degree of fuzziness of memberships, that is, the larger the m the fuzzier memberships by imposing a greater weight on the entropy term. Thus, this tuning parameter plays the same role as the fuzzifier in standard fuzzy clustering.

To minimize Equation A4.7, we can develop an alternating least-squares algorithm. In the algorithm, two steps are repeated as follows.

Step 1: For fixed $\mathbf{U}_k$, the parameters of generalized structured component analysis ($\mathbf{W}_k$ and $\mathbf{A}_k$) in each cluster are updated. The first step is equivalent to minimizing

$$
\begin{aligned}
\phi &= \sum_{k=1}^{K} SS(\mathbf{U}_k^{1/2}(\mathbf{Z}\mathbf{V}_k - \mathbf{Z}\mathbf{W}_k\mathbf{A}_k)) \\
&= \sum_{k=1}^{K} SS(\mathbf{Z}_k\mathbf{V}_k - \mathbf{Z}_k\mathbf{W}_k\mathbf{A}_k),
\end{aligned} \tag{A4.8}
$$

where

$$Z_k = U_m^{1/2}Z.$$

This is identical to the first step of the alternating least-squares algorithm for fuzzy clusterwise generalized structured component analysis.

Step 2: For fixed W_k and A_k, u_{ki} is updated by

$$\hat{u}_{ki} = \frac{\exp\left(-\dfrac{d_{ki}}{m}\right)}{\displaystyle\sum_{c=1}^{K}\exp\left(-\dfrac{d_{ci}}{m}\right)}. \tag{A4.9}$$

Equation A4.9 can be derived as follows. Minimizing Equation A4.7 under the constraint

$$\sum_{k=1}^{K} u_{ki} = 1$$

is equivalent to minimizing

$$\phi^{*} = \sum_{k=1}^{K}\sum_{i=1}^{N} u_{ki}SS(V_k'z_i - A_k'W_k'z_i) - m\sum_{k=1}^{K}\sum_{i=1}^{N} u_{ki}\log u_{ki} + \lambda(\sum_{k=1}^{K} u_{ki} - 1)$$

$$= \sum_{k=1}^{K}\sum_{i=1}^{N} u_{ki}d_{ki} - m\sum_{k=1}^{K}\sum_{i=1}^{N} u_{ki}\log u_{ki} + \lambda(\sum_{k=1}^{K} u_{ki} - 1). \tag{A4.10}$$

Solving $\dfrac{\partial\phi^{*}}{\partial u_{ki}} = 0$ yields

$$u_{ki} = \exp\left(-\left(\frac{d_{ki}}{m} + \frac{\lambda}{m} + 1\right)\right). \tag{A4.11}$$

Using $\dfrac{\partial\phi^{*}}{\partial\lambda} = \sum_{k=1}^{K} u_{ki} - 1 = 0$ and Equation A4.11 yields

$$\exp\left(-\left(\frac{\lambda}{m} + 1\right)\right) = \frac{1}{\displaystyle\sum_{c=1}^{K}\exp\left(-\dfrac{d_{ci}}{m}\right)}. \tag{A4.12}$$

Then, Equation A4.9 is obtained by combining Equations A4.11 with A4.12 (also see Hwang et al. 2010a; Suk and Hwang 2014).

A major advantage of using the regularized version over fuzzy cluster-wise generalized structured component analysis is that it can open up the possibility of determining the degree of fuzziness in cluster memberships in an automatic manner, because the tuning parameter can be chosen by cross validation (e.g., Hastie, Tibshirani, and Friedman 2009, p. 214) as in regularized linear regression (e.g., Hoerl and Kennard 1970; Le Cessie and Van Houwelingen 1992; Lee and Silvapulle 1988). More specifically, in the cross validation method, the entire dataset is divided into subsets. One of the subsets is set as a test sample, whereas the remaining subsets are used as a calibration sample, from which parameters are estimated under a given value of m. The resultant parameter estimates are applied to the test sample in order to calculate a goodness of fit measure such as FIT. This procedure is repeated as many times as the number of the subsets, changing the test and calibration samples systematically. The goodness of fit measure is averaged over the test samples. The value of m yielding the smallest average value of the goodness of fit measure can be chosen as the final one.

5

Nonlinear Generalized Structured Component Analysis

5.1 Introduction

So far, we have assumed that the variables we analyze by generalized structured component analysis are all quantitative, that is, they are measured on relatively continuous scales and are linearly related to their latent variables. However, not all variables satisfy these conditions. Some variables are purely qualitative in the sense that no *a priori* quantifications are given. Some other variables are relatively continuous, but are nonlinearly related to their latent variables. In this chapter, we discuss extensions of generalized structured component analysis, called "nonlinear" generalized structured component analysis (Hwang and Takane 2010), that allow optimal quantifications (scaling, transformations) of the observed variables. The word "nonlinear" here refers to the nonlinearity of the data transformations, and not to the nonlinearity in the structural model.

Qualitative data have traditionally been analyzed by a technique called (multiple) correspondence analysis (Benzécri 1973; Greenacre 1984; Lebart, Morineau, and Warwick 1984) or dual scaling (Nishisato 1980). This technique, like principal component analysis of quantitative variables, attempts to explain the observed relationships among categorical variables parsimoniously by postulating a small number of underlying components. It provides a graphical display of the relationships among the observed categories in a form easily assimilable to human eyes. However, the technique typically extracts orthogonal components, for which no structural models may be specified. Thus, there is a good reason to develop a technique for structural equation models for qualitative data.

5.1.1 A Variety of Qualitative Data

Below are some more detailed examples of variables requiring quantifications or requantifications. Suppose that we are interested in finding out how religions affect attitude toward abortion. Respondent's religion may be recorded in a multiple choice format with categories: (1) Christian, (2) Muslim, (3) Buddhist, and (4) Others. Here, numbers are merely used as category labels, and have no numerical meaning. In psychology, such categories are called nominal (unordered) categories, which must be quantified in such a way that the resultant values are best predictive of attitude toward abortion.

Attitude toward abortion, on the other hand, may be measured on a rating scale with: (1) Strongly agree, (2) Agree, (3) Neutral, (4) Disagree, and (5) Strongly disagree, as response categories. Contrary to the religion variable above, these categories have a prescribed order according to some attribute (e.g., favorableness toward abortion), and the numbers tentatively given to these categories reflect this order. One may be tempted to use them as quantified values of the categories in subsequent analyses. Although this often gives a good approximation, there is no assurance that these scores are linearly related to the quantities they are supposed to measure. For example, suppose that one intends to measure the degree of "conservatism" of respondents via their attitude toward abortion. It may be that the latter is merely monotonically related to the former. That is, attitude toward abortion is only an ordinal measure of conservatism. In such cases, *a priori* given (observed) scores must be requantified in such a way that they are more linearly related to the quantities they are intended to measure, while maintaining the prescribed order of the categories.

You may recall that the organizational identification data (Bergami and Bagozzi 2000) used in Chapter 3 had 21 observed variables, which were all measured on 5-point rating scales. (See Table 3.7 for a detailed description of the variables.) There we applied generalized structured component analysis using the *a priori* given scores of 1–5. Later in this chapter (Section 5.3.3), we examine the adequacy of this procedure by optimally requantifying the categories in these variables using nonlinear generalized structured component analysis.

As another example, consider a situation in which a requantification is even more critical. Prisoners are classified in terms of how satisfied they are with the sentences they have received: (1) Completely satisfied, (2) Somewhat dissatisfied, and (3) Very dissatisfied. Suppose that we would like to predict the risk of a second offence based on how satisfied they are with their sentences. One may be tempted to believe that the more dissatisfied the prisoners are, the higher the probability of reoffending, and so the categories above are properly ordered by the *a priori* given scores. It turns out (Hayashi 1993) that the prisoners who are most satisfied with their sentences in fact have the highest chance of committing a second crime, followed by those who are least satisfied. Those who are in-between have the lowest chance of committing a second offence. This indicates that the *a priori* assigned scores are not only nonlinearly but also nonmonotonically related to the chance of a second crime, suggesting again a need for requantification. In this case, observed categories should be treated nominally (i.e., as unordered categories) in order to deliberately ignore the prescribed order.

Variables measured on a relatively continuous scale are usually treated as quantitative variables for which no transformations are necessary. In some cases, however, they may be treated as ordinal variables because it is anticipated that they are nonlinearly related to underlying latent variables and/or to other observed variables. Suppose that we are predicting the mortality rate of lower intestinal cancer from the daily intake of meat products. These two variables are typically nonlinearly related (Segi 1979). The log transformation is often applied to the mortality rate, which may approximately linearize the relationship between the two variables. Often, however, the exact form of the relationship between the two variables is unknown, except that they are monotonically related to each other. In such cases, it would make more sense to transform one variable monotonically to render their relationship as linear as possible. The exact form of the monotonic transformation is left unspecified. The specific monotonic transformation is determined in such a way that the prescribed model "best" fits the transformed data. Examples 1 and 2 in Sections 5.3.1 and 5.3.2 further illustrate this situation.

5.1.2 The Optimal Scaling Idea

Qualitative data are quantified by assigning numerical values to the qualitative data. This is done in such a way that the association between the model and the transformed data is maximized while strictly maintaining the measurement characteristics of the observed data. This is called optimal scaling or optimal data transformation. This implies that if a procedure is available to fit a model to quantitative data, it can also be used for the analysis of qualitative data because the qualitative data become quantitative through optimal scaling. This idea fits seamlessly into the generalized structured component analysis situation, in which there is already a procedure to fit a model to quantitative data. All we have to do is to link an appropriate optimal scaling procedure to the conventional generalized structured component analysis for quantitative data.

An appropriate optimal scaling procedure must preserve measurement characteristics of the observed data. Young (1981) described six types of measurement as combinations of three scale levels (nominal, ordinal, and numerical) and two levels of measurement process (discrete and continuous). Each type involves its own measurement characteristics and restrictions. Of the three scale levels, the numerical level corresponds with quantitative data, for which no optimal scaling is necessary. Of the remaining four types, the continuous nominal type is rarely used. So we mainly focus on the remaining three types: discrete nominal, discrete ordinal, and continuous ordinal. Since we consider only one type of nominal data, the discrete nominal type is simply referred to as nominal hereafter. In nominal data, optimally scaled data corresponding to observations that fall in the same category are restricted to take an identical value. Ordinal variables are required to maintain their order after the transformation. Discrete ordinal data should keep tied observations tied, whereas continuous ordinal data allow tied observations to become untied. The latter is called the primary, and the former the secondary, approach to ties in Kruskal's (1964a, b) least-squares monotonic transformations.

In nonlinear generalized structured component analysis, optimal scaling of qualitative data and estimation of model parameters (i.e., those in conventional generalized structured component analysis) are carried out in a unified way, that is, by minimizing a single least-squares criterion (see Equation 5.1). This is done by alternately minimizing the criterion with respect to optimal data transformations and model estimation, leading naturally to an alternating least-squares algorithm that we have been using consistently in generalized structured component analysis.

5.2 Nonlinear Generalized Structured Component Analysis

In accordance with the optimal scaling idea described above, data are considered qualitative in nonlinear generalized structured component analysis, and the data matrix $\mathbf{Z}$ in Equation 2.17 is replaced by the optimally scaled data matrix $\mathbf{S}$. Let $\mathbf{z}_j$ and $\mathbf{s}_j$ denote N by 1 vectors of the jth observed qualitative variable and its optimally scaled counterpart, respectively $(j = 1, \cdots, J)$. Then,

$$\phi = \text{SS}(\mathbf{SV} - \mathbf{SWA}) = \text{SS}(\mathbf{\Psi}^* - \mathbf{\Gamma}^*\mathbf{A}), \tag{5.1}$$

is minimized with respect to $\mathbf{W}$, $\mathbf{A}$, and $\mathbf{S}$, subject to the restrictions that $\text{diag}(\mathbf{\Gamma}^{*\prime}\mathbf{\Gamma}^*) = \mathbf{I}$, $\mathbf{s}_j'\mathbf{s}_j = 1$, and $\mathbf{s}_j = \omega(\mathbf{z}_j)$, where $\mathbf{\Psi}^* = \mathbf{SV}$, $\mathbf{\Gamma}^* = \mathbf{SW}$, and ω refers to a transformation of the original variable to the optimally scaled counterpart subject to the measurement characteristics of the data. This criterion is minimized by alternating two phases. One phase is identical to the alternating least-squares estimation procedure for updating model parameters ($\mathbf{W}$ and $\mathbf{A}$) in the conventional generalized structured component analysis for quantitative data. This phase remains the same as before; readers are referred to Section 2.2 to review any details. The other is the optimal scaling phase, in which qualitative data are transformed to quantitative data $\mathbf{S}$ in such a way that they agree maximally with their model predictions, while preserving the measurement characteristics of the data. In practice, variables in the original data matrix $\mathbf{Z}$ may have a mix of different measurement characteristics; for example, some variables are nominal, others are ordinal, and yet others are numerical. Thus, each variable in $\mathbf{Z}$ is transformed by $\mathbf{s}_j = \omega(\mathbf{z}_j)$, where ω depends on the measurement characteristics of the variable $\mathbf{z}_j$. These two phases are alternately applied and repeated until convergence.

Given $\mathbf{W}$ and $\mathbf{A}$, the optimal scaling phase updates each variable s_j sequentially for $j = 1, \cdots, J$. If $\mathbf{z}_j$ is numerical, the optimal scaling phase can be skipped entirely for the variable. The optimal scaling phase consists of two steps. In the first step, the model prediction $\hat{\mathbf{s}}_j$ corresponding to $\mathbf{s}_j$ is calculated from Equation 5.1. In the second step, the model prediction is transformed into the optimally scaled data $\mathbf{s}_j$ in such a way that it minimizes Equation 5.1 while respecting its measurement characteristics. Specifically, these two steps are described as follows.

Step 1: The first step is to update the model prediction $\hat{\mathbf{s}}_j$ corresponding to $\mathbf{s}_j$ for fixed $\mathbf{W}$ and $\mathbf{A}$. Let $\mathbf{S}_{(-j)}$ denote $\mathbf{S}$ whose jth column is replaced by the vector of zeros. Let $\mathbf{S}_{(j)}$ denote an N by J matrix in which the jth column is equal to s_j and the others are all zero vectors. Let $\mathbf{\Sigma} = \mathbf{V} - \mathbf{WA}$, and let $\mathbf{\Sigma}_{(-j)}$ denote $\mathbf{\Sigma}$ whose jth row is replaced by the vector of zeros. Let $\mathbf{\sigma}_{(j)}$ denote the jth row of $\mathbf{\Sigma}$. Then, Equation 5.1 can be re-expressed as:

$$\phi = \mathrm{SS}(\mathbf{S}\mathbf{\Sigma})$$

$$= \mathrm{SS}((\mathbf{S}_{(j)} + \mathbf{S}_{(-j)})\mathbf{\Sigma})$$

$$= \mathrm{SS}(\mathbf{s}_j\mathbf{\sigma}_{(j)} - (-\mathbf{S}_{(-j)}\mathbf{\Sigma}_{(-j)}))$$

$$= \mathrm{SS}(\mathbf{s}_j\mathbf{\sigma}_{(j)} - \mathbf{K}_{(-j)}), \tag{5.2}$$

where $\mathbf{K}_{(-j)} = -\mathbf{S}_{(-j)}\mathbf{\Sigma}_{(-j)}$. The model prediction $\hat{\mathbf{s}}_j$ is calculated by:

$$\hat{\mathbf{s}}_j = \mathbf{K}_{(-j)}\mathbf{\sigma}'_{(j)}(\mathbf{\sigma}_{(j)}\mathbf{\sigma}'_{(j)})^{-1}. \tag{5.3}$$

Step 2: In the second step, the optimally transformed data s_j is obtained such that it is as close to $\hat{\mathbf{s}}_j$ as possible in the least-squares sense while satisfying its measurement restrictions. That is, s_j is updated by minimizing the normalized differences between s_j and $\hat{\mathbf{s}}_j$. This amounts to regressing $\hat{\mathbf{s}}_j$ onto the space of $\mathbf{z}_j$, which represents the measurement restriction for variable j. The least-squares estimate of s_j may generally be expressed as

$$\mathbf{s}_j = \mathbf{T}_j(\mathbf{T}'_j\mathbf{T}_j)^{-1}\mathbf{T}'_j\hat{\mathbf{s}}_j = \mathbf{T}_j\mathbf{q}_j \tag{5.4}$$

where $\mathbf{q}_j = (\mathbf{T}'_j\mathbf{T}_j)^{-1}\mathbf{T}'_j\hat{\mathbf{s}}_j$ is a vector of optimal scaling weights. In Equation 5.4, $\mathbf{T}_j$ is constructed to satisfy the measurement restriction imposed on the variable. For example, for a nominal variable, $\mathbf{T}_j$ is given as a dummy-coded indicator matrix, which shows the category choices of respondents (cf. Gifi 1990). In this case, once $\mathbf{T}_j$ is constructed, it remains fixed throughout the iterations. This means that the data transformation for a nominal variable can be carried out by simply replacing the original variable by its appropriate dummy-coded indicator matrix. For an ordinal variable, $\mathbf{T}_j$ is iteratively updated by Kruskal's (1964b) least-squares monotonic transformation so that the elements of $\mathbf{q}_j$ satisfy the order restriction. (To be more exact, $\mathbf{T}_j$ is not explicitly constructed in Kruskal's algorithm, but only algorithmically represented. The appendix to this chapter gives details of the algorithm.) As noted above, there are two approaches to tied observations in an ordinal variable. In the primary approach, tied observations can be untied in transformed data, whereas in the secondary approach, they remain tied. The updated $\mathbf{s}_j$ is subsequently normalized to satisfy $\mathbf{s}_j'\mathbf{s}_j = 1$.

The alternating least-squares algorithm proposed above repeats the two major phases (estimation of model parameters and optimal scaling) until convergence. Although the algorithm is monotonically convergent like all other alternating least-squares algorithms, there is no guarantee that it converges to a global minimum. This is similar to conventional, linear generalized structured component analysis, but the problem could be more aggravated in nonlinear structured component analysis due to a larger number of parameters to be estimated. To avoid a suboptimal solution, this algorithm may be repeatedly applied to the data, with varying starting values (e.g., 10–20 different sets of starting values). The smallest value of Equation 5.1 may then be regarded as the global minimum. The parameter estimates associated with the smallest criterion value are chosen as the final solutions. As will be shown in the next section, given many different sets of random starts, the algorithm seems to converge to the same optimum with high frequency.

As in the linear version, nonlinear generalized structured component analysis adopts the bootstrap method (Efron 1982) to assess the reliability of parameter estimates and examine their significance without recourse to the assumption of multivariate normality of the observed variables.

5.3 Examples

5.3.1 Kempler's Size Judgement Data

The first example concerns the size judgement data used in Kempler's (1971) study (see also Takane, Young, and de Leeuw 1980). This study was originally motivated by a psychological phenomenon in which younger children tend to see taller objects larger than flat objects. To investigate this phenomenon, four age groups consisting of 16–25 children in grades 1, 3, 5, and 7 were asked to judge the size of 100 rectangles as either "large" or "small." The 100 different rectangles were created by factorial combinations of 10 height levels and 10 width levels each varying from 10 to 14.5 inches in one half-inch intervals. For each age group, the number of children who judged a rectangle as large was counted, and used as a measure of perceived largeness of the rectangle. Consequently, these data include two nominal variables representing 10 different levels of height and width (these variables are treated as nominal because the psychological values of these variables are unknown and to be quantified); and four ordinal variables indicating the number of counts for the four age groups. The sample size was equal to the number of rectangles judged (i.e., $N = 100$).

Following Takane, Young, and de Leeuw (1980), we specified the weighted addictive model for the data, in which two exogenous variables (height and width) were hypothesized to influence four endogenous variables (four age groups' perceived largeness of rectangles). It would not be suitable to apply linear generalized structured component analysis, since the data consisted of all qualitative variables (either nominal or ordinal). We thus applied nonlinear generalized structured component analysis to fit the model to the data. In the optimal scaling phase, the two original nominal variables were converted to two 100 (rectangles) by 10 (categories) dummy-coded indicator matrices, which were subsequently used as a new set of exogenous variables (i.e., 20 dummy-coded indicators in total) in the model. We applied the continuous-ordinal transformation (Kruskal's primary monotonic transformation) to the four original endogenous variables.

The specified model is depicted in Figure 5.1. In the figure, as shown in Equation 5.4, each optimally scaled nominal variable (HEIGHT or WIDTH) is expressed as a weighted composite of the corresponding 10 dummy-coded indicators (i.e., z_1–z_{10} for HEIGHT and z_{11}–z_{20} for WIDTH). Thus, the optimal scaling weights for each nominal variable can be considered equivalent to component weights, and the optimally scaled nominal variables are comparable to latent variables in the context of generalized structured component analysis, as given in Equation 2.12. In addition, these optimally scaled nominal variables are hypothesized to affect four endogenous, continuous-ordinal variables (s_1–s_4). All residual terms for the endogenous variables were removed to make the figure concise. It may be worthwhile noting that this weighted additive model can be viewed as a structural equation model, where latent variables are formed as weighted composites of exogenous observed variables, and they are supposed to influence endogenous observed variables. More precisely, this model is called extended redundancy analysis, a special case of generalized structured component analysis, as described in Section 2.5.1.

We repeated the alternating least-squares algorithm with 20 sets of random starting values for parameter estimates. The algorithm was found to converge to the same optimum for all repetitions. Nonlinear generalized structured component analysis provided FIT = 0.96 and AFIT = 0.96, indicating that the specified model accounted for about 96% of the total variance in the variables. Table 5.1 provides parameter estimates and 95% bootstrap confidence intervals calculated from 200 bootstrap samples. For each bootstrap we used a different set of random starting values for parameter estimates.

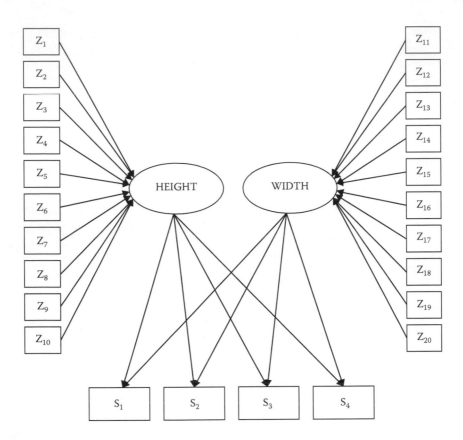

FIGURE 5.1
The specified model for Kempler's data.

Figure 5.2 displays the original observed values of the four endogenous variables on the *x*-axis and their monotonically transformed values on the *y*-axis. The connected line segments indicate the least-squares monotonic transformations, while the dots wobbling around the transformations depict pairs of observed data and the corresponding model predictions.

As shown in Table 5.1, the optimal scaling weight estimates for HEIGHT turned out to be all statistically significant except the weight for z_5 representing the category for 12 inches high. Moreover, these weight estimates tended to get larger with height. In addition, the optimal scaling weight estimates for WIDTH were all statistically significant except for z_{15} related to the category for 12 inches wide. Again, the weight estimates for WIDTH appeared larger when the level of width increased. These indicate that both HEIGHT and WIDTH are at least ordinal, although they were treated as nominal. Furthermore, HEIGHT and WIDTH had statistically significant and positive effects on the four optimally scaled endogenous variables (s_1–s_4). The impact of HEIGHT on perceived largeness of rectangles tended to decrease with age,

TABLE 5.1

Parameter Estimates and 95% Bootstrap Confidence Intervals
Obtained from the Proposed Nonlinear Generalized Structured
Component Analysis for Kempler's Data

			Estimate	95% CI
Weights	HEIGHT	z_1	−1.43	−1.66, −1.27
		z_2	−1.39	−1.67, −1.16
		z_3	−0.91	−1.12, −0.70
		z_4	−0.57	−0.74, −0.41
		z_5	−0.20	−0.46, 0.03
		z_6	0.33	0.52, 0.17
		z_7	0.58	0.79, 0.37
		z_8	1.08	1.34, 0.83
		z_9	1.02	1.24, 0.80
		z_{10}	1.48	1.72, 1.28
	WIDTH	z_{11}	−1.56	−1.84, −1.27
		z_{12}	−1.27	−1.47, −1.01
		z_{13}	−0.96	−1.27, −0.73
		z_{14}	−0.53	−0.73, −0.32
		z_{15}	−0.10	−0.46, 0.14
		z_{16}	0.33	0.64, 0.03
		z_{17}	0.48	0.73, 0.21
		z_{18}	0.99	1.29, 0.74
		z_{19}	1.32	1.71, 1.09
		z_{20}	1.30	1.56, 1.07
Path coefficients	HEIGHT → s_1		0.87	0.80, 0.97
	HEIGHT → s_2		0.77	0.70, 0.88
	HEIGHT → s_3		0.79	0.71, 0.91
	HEIGHT → s_4		0.73	0.65, 0.82
	WIDTH → s_1		0.44	0.35, 0.55
	WIDTH → s_2		0.60	0.52, 0.70
	WIDTH → s_3		0.58	0.49, 0.68
	WIDTH → s_4		0.67	0.60, 0.79

CI, Confidence interval.

whereas that of WIDTH was likely to increase with age. This suggests that
younger children tend to focus more heavily on the height than the width
of rectangles in their size judgments. This tendency appears to diminish as
they become older. These findings validate Kempler's (1971) hypothesis with
respect to children's perception on the size of rectangles.

5.3.2 The Basic Health Indicator Data

The second example data set was analyzed to compare the performance of the
linear and nonlinear versions of generalized structured component analysis.

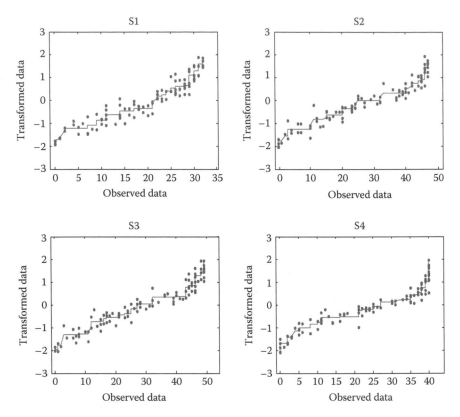

FIGURE 5.2
The monotonic transformations of the four endogenous variables (s_1–s_4) in Kempler's data.

This example pertains to part of the so-called Basic Health Indicator data collected by the World Health Organization in 1999. From the entire database, five observed variables were selected in this application: (1) real gross domestic product (GDP) per capita adjusted for purchasing power parity in 1985 US dollars, (2) the average number of years of education given to females aged 25 years and above (female education; FEDU), (3) the percentage of children immunized against measles in 1997 (Measles), (4) infant mortality rate (IMR), defined as the number of deaths per 1,000 live births between birth and 1 year of age in 1998, and (5) maternal mortality ratio (MMR), defined as the number of maternal deaths per 100,000 live births in 1990. The sample size was 50, indicating the total number of countries for which these variables were measured.

For illustrative purposes, we specified a model for these data, in which a latent variable was defined as a weighted composite of the first three observed variables (GDP, FEDU, and Measles), which, in turn, were assumed to influence the three observed variables from which the latent variable was defined. This latent variable was named "Social and Economic Growth (SEG)." In

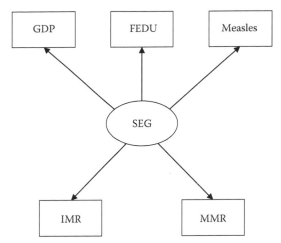

FIGURE 5.3
The specified model for the Basic Health Indicator data.

addition, the latent variable was assumed to affect two endogenous observed variables (IMR and MMR). The specified model is depicted in Figure 5.3. Only component loadings and path coefficients are displayed to make the figure concise. (This figure may give a false impression that the five observed variables have exactly the same status in the model. Note, however, that only the first three variables are used to define the latent variable, which, in turn, are assumed to affect all five variables.) This model is a special case of generalized structured component analysis, called principal covariate regression (de Jong and Kiers 1992), which is briefly discussed in Section 2.5.1.

At first, the linear version of generalized structured component analysis was applied to fit the model to the data, which provided FIT = 0.79 and AFIT = 0.78, indicating that the specified model accounted for about 79% of the total variance in the data. Table 5.2 (the third and fourth columns) gives parameter estimates and 95% bootstrap confidence intervals based on 200 bootstrap samples. The weight estimates of the three observed variables for the latent variable (i.e., GDP, FEDU, and Measles) were similar and statistically significant, indicating that they played significant roles in defining the latent variable. Moreover, all loading estimates for these variables were large and statistically significant. This suggests that the latent variable was well defined and explains a large portion of the variances of the observed variables. However, the loading estimate for Measles was smaller than those for GDP and FEDU. Furthermore, the latent variable SEG had statistically significant and negative effects on IMR and MMR. This indicates that a high level of SEG is likely to decrease both infant and maternal mortality rates.

Next, nonlinear generalized structured component analysis was applied to fit the same model to the data, assuming that all five observed variables were

TABLE 5.2

Parameter Estimates and 95% Bootstrap Confidence Intervals Obtained from Linear and Nonlinear Generalized Structured Component Analysis (GSCA) for the Basic Health Indicator Data

		Linear GSCA		Nonlinear GSCA	
		Estimate	95% CI	Estimate	95% CI
Weights	Gross domestic product (GDP)	0.43	0.39, 0.49	0.35	0.33, 0.40
	FEDU	0.46	0.42, 0.52	0.35	0.33, 0.37
	Measles	0.33	0.19, 0.37	0.34	0.21, 0.35
Loadings	GDP	0.87	0.80, 0.93	0.98	0.95, 1.00
	FEDU	0.93	0.89, 0.96	0.94	0.90, 1.00
	Measles	0.62	0.35, 0.77	0.93	0.86, 1.00
Path coefficients	SEG → infant mortality rate	−0.85	−0.91, −0.80	−0.89	−1.00, −0.81
	SEG → maternal mortality ratio	−0.74	−0.82, −0.67	−0.87	−1.00, −0.76

CI, Confidence interval.

continuous-ordinal. We applied Kruskal's (1964b) primary monotonic transformation to the five variables. We repeated the alternating least-squares algorithm with 20 sets of random starting values for parameter estimation. The algorithm always converged to the same optimum. Figure 5.4 displays the original values of the five variables on the x-axis and their monotonically transformed values on the y-axis. Again, the connected line segments indicate the best monotonic transformations, and the dots scattered around the transformations indicate model predictions plotted against the corresponding observed data. Table 5.2 (the last two columns) presents parameter estimates and 95% bootstrap confidence intervals obtained from nonlinear generalized structured component analysis.

Estimation by nonlinear generalized structured component analysis resulted in a much higher level of model fit (FIT = 0.94 and AFIT = 0.94), compared to its linear counterpart. The interpretations of the parameter estimates are basically the same as those obtained by the linear analysis. It is worth noting, however, that all loading estimates for the transformed variables were larger than those for the original (nontransformed) ones. In particular, the loading for Measles was much larger, indicating that it explained a greater portion of the variance of the latent variable. In addition, the estimates of path coefficients became larger than those from the linear analysis. Thus, nonlinear generalized structured component analysis was able to enhance the associations among the variables substantially by taking into account the nonlinear relationships among them. This suggests that nonlinear

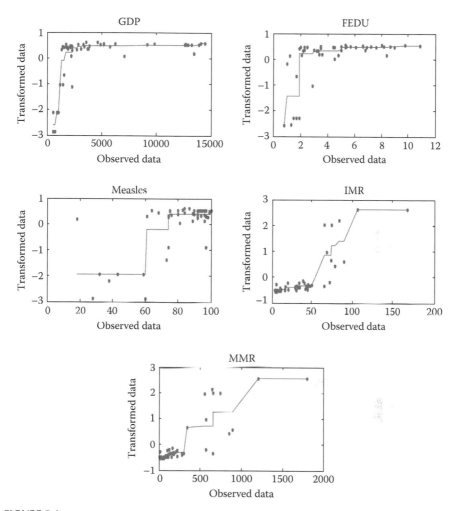

FIGURE 5.4
The monotonic transformations of the five variables in the Basic Health Indicator data.

generalized structured component analysis is beneficial in analyzing the interrelationships among the health-indicator variables, due to its capability to accommodate potential nonlinear relationships among variables.

5.3.3 The Organizational Identification Data Revisited

In Section 3.4, we analyzed Bergami and Bagozzi's (2000) organizational identification data by the conventional (linear) generalized structured component analysis, assuming that the data are quantitative. However, the 21 observed variables were all measured on 5-point rating scales, which may also be regarded as qualitative. The reader is referred to Table 3.7 for full description of the variables. It is of interest to see if any substantial differences would occur by treating them as qualitative (discrete-ordinal). The data were reanalyzed by fitting exactly the same structural model as given in Section 3.4 by nonlinear generalized structured component analysis in combination with Kruskal's secondary approach to ties. Recall that this model had four latent variables (OP, OI, AC_J, and AC_L) with eight, six, four, and three indicators, respectively, and it was assumed that OP affects OI, which in turn affects both AC_J and AC_L (see Figure 3.7). This is the first full-fledged example of nonlinear generalized structured component analysis. There were two groups of subjects, males and females, so a multiple-group analysis was applied with weights and loadings constrained to be equal across the groups, while path coefficients might differ.

TABLE 5.3

Estimates of Loadings, Standard Errors, $|t|$, and 95% Confidence Intervals from Nonlinear Constrained Multiple Group Analysis of the Organizational Identification Data

Latent	Indicator	Both Male and Female			
		Estimate	SE	$|t|$	95% CI
OP	op1	0.79	0.03	23.85	0.73, 0.86
	op2	0.84	0.03	27.74	0.78, 0.90
	op3	0.88	0.04	19.81	0.79, 0.96
	op4	0.90	0.04	20.14	0.81, 0.98
	op5	0.79	0.03	21.53	0.72, 0.86
	op6	0.91	0.03	26.03	0.84, 0.98
	op7	0.89	0.04	20.17	0.80, 0.97
	op8	0.83	0.04	22.46	0.75, 0.90
OI	oi1	0.82	0.03	27.94	0.76, 0.87
	oi2	0.77	0.03	25.32	0.71, 0.83
	oi3	0.69	0.04	16.09	0.60, 0.77
	oi4	0.83	0.03	28.54	0.77, 0.89
	oi5	0.81	0.03	30.27	0.76, 0.86
	oi6	0.77	0.04	20.86	0.70, 0.85
AC_J	acj1	0.73	0.03	26.28	0.68, 0.79
	acj2	0.81	0.02	34.09	0.76, 0.85
	acj3	0.81	0.02	40.14	0.77, 0.85
	acj4	0.71	0.03	24.68	0.65, 0.76
AC_L	acl1	0.82	0.03	23.18	0.75, 0.89
	acl2	0.82	0.05	16.25	0.72, 0.92
	acl3	0.79	0.03	26.19	0.73, 0.85

AC_J, Affective commitment (joy); AC_L, Affective commitment (love); CI, Confidence interval; OI, Organizational identification; SE, Standard errors; OP, Organizational prestige.

TABLE 5.4

Estimates of Path Coefficients, Standard Errors, $|t|$, and 95% Confidence Intervals from Nonlinear Constrained Multiple Group Analysis of Organizational Identification Data

Path Coefficient	Group	Estimate	SE	$|t|$	95% CI
OP → OI	Male	0.37	0.06	5.29	0.23, 0.51
OI → AC_J		0.72	0.04	16.65	0.63, 0.81
OI → AC_L		−0.31	0.10	3.15	−0.51, −0.12
OP → OI	Female	0.37	0.06	5.29	0.23, 0.51
OI → AC_J		0.41	0.05	8.13	0.31, 0.51
OI → AC_L		−0.35	0.07	4.41	−0.51, −0.20

AC_J, Affective commitment (joy); AC_L, Affective commitment (love); CI, Confidence interval; OI, Organizational identification; SE, Standard errors; OP, Organizational prestige.

The FIT from the nonlinear analysis was 0.60, which was an improvement from 0.53 obtained in the linear analysis. The estimates of loadings and path coefficients from the nonlinear analysis reported in Tables 5.3 and 5.4 are all similar to those obtained from the linear analysis reported in Tables 3.11 and 3.12, respectively. This indicates that the linear analysis was indeed appropriate. Note that the standard errors of the estimates tend to be slightly larger in the nonlinear analysis, indicating that the estimates are somewhat less reliable than those obtained by the linear analysis. This is due to the fact that a much larger number of parameters were estimated in the nonlinear analysis. This indicates that the nonlinear analysis can even be harmful by producing less reliable estimates when the linear analysis is appropriate. The nonlinear analysis is beneficial only when nonlinearity is substantial, as in the second example.

Figure 5.5 depicts the least-squares monotonic transformations for two of the variables in the two groups. The transformations are all nearly linear, again indicating that the initial assignment of values of 1–5 for each variable is more or less justified.

5.4 Summary

In this chapter, generalized structured component analysis was extended to the analysis of qualitative variables. This nonlinear extension of generalized component analysis integrates the model estimation in the original linear version and the extra data transformation into a unified procedure. In particular, it adopts the optimal scaling approach to data transformation that preserves measurement characteristics of the observed qualitative

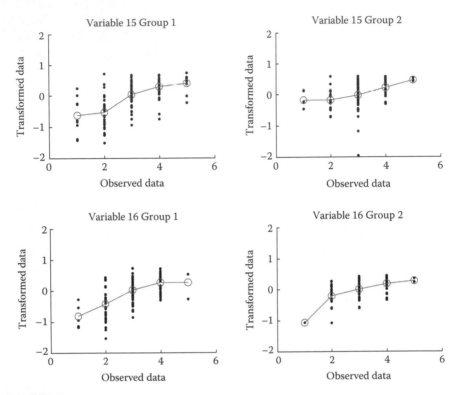

FIGURE 5.5
Kruskal's least-squares monotonic transformation (the secondary approach) obtained for the organizational identification data (Variable 15 and Variable 16 for Group 1: male and Group 2: female).

variables. As demonstrated by empirical applications, nonlinear generalized structured component analysis could effectively be applied for the specification and analysis of path-analytic relationships among qualitative variables, taking into account their nonlinear associations. In contrast, the partial least squares path modeling has no place for optimal scaling since it has no global optimization criterion, although discrete nominal variables can perhaps be incorporated by coding them into dummy variables. The factor-based structural equation models can accommodate discrete nominal variables, if they are exogenous variables. Certain factor-based structural equation modeling software (e.g., Mplus) can handle ordered categorical variables, if the number of categories is not too excessive, by discretizing continuous variables at certain thresholds. However, no structural equation modeling approaches other than generalized structured component analysis can handle relatively continuous ordinal data, such as those given in Example 2.

Despite its significant implications, nonlinear generalized structured component analysis is not yet implemented in a software program, so that it would be difficult for researchers and practitioners to employ the method in their substantive research. It would thus be desirable to integrate this new development into a software program in the near future.

Appendix 5.1 Algorithms for Kruskal's (1964a, b) Least-Squares Monotonic Transformations

In this appendix, we illustrate algorithms for Kruskal's (1964a, b) least-squares monotonic transformations using concrete examples. We first discuss a situation in which there are no tied observations in the original variable, for which there is no need to distinguish between the primary and secondary approaches. This algorithm, however, serves as the basic algorithm for both approaches, which are distinguished by different preprocessing steps required for tied observations.

Table 5.5 shows the steps involved in the basic algorithm. The first column displays the observed data sorted in ascending order of magnitude (the elements of z_j). The second column shows the corresponding model predictions (the elements of $\hat{s}_j$). We would like to derive s_j whose elements are weakly monotonically related to the elements of z_j. The weak monotonicity means that if $z_i > z_k$, then $s_i \geq s_k$, where z_i and s_i are the ith element of z_j and s_j, respectively.

In Step 1, we duplicate as many elements of $\hat{s}_j$ from the top until we encounter the first violation of monotonicity. The third element of $\hat{s}_j$ is 2, which is smaller than the second element of value 4, and which therefore violates monotonicity. To restore monotonicity, we take the average of the two [i.e., $(2 + 4)/2 = 3$], and tentatively assign this value to both the second and the third elements of s_j. Taking the average incurs the smallest possible change in s_j with regard to $\hat{s}_j$. This averaging operation creates no new violations of monotonicity in s_j because 3 is larger than 1 (the first element). We then continue to copy the elements of $\hat{s}_j$ from where it was interrupted (the fourth

TABLE 5.5

Steps to Follow in the Least-Squares Monotonic Transformation with No Ties

Data	Model	Step 1	Step 2	Step 3	Step 4	Step 5	Loss
1	1	1	1	1	1	1	0
2	4	4	3	3	3	3	1
3	2	2	3	3	3	3	1
4	6		6	5.5	5.5	5	1
5	5		5	5.5	4.75	5	0
6	4			4	4.75	5	1
7	7					7	0

element) until another violation is found. The fifth element violates monotonicity (Step 2), and so we again take the average of these two elements to remove the violation (Step 3). This averaging generates no new violations in s_j because 5.5 is larger than 3. So we resume copying the elements of $\hat{s}_j$ from the sixth element, but we immediatcly find that this element violates monotonicity. We take the average of these two elements in violation, and assign this average to both the fifth and the sixth elements of s_j (Step 4). However, this operation creates a new violation because 4.75 is smaller than 5.5, the fourth element of s_j. So we need to take the average of the three elements in violation to obtain the value of 5, which is assigned to the corresponding three elements of s_j (Step 5). We then copy the seventh element of $\hat{s}_j$ into s_j. Since this element satisfies the monotonicity, the algorithm is completed. The (unnormalized) loss incurred by the transformation is equal to the sum of squares of the differences between the corresponding elements of $\hat{s}_j$ and s_j, which in this case is equal to 4. This algorithm minimizes $(\hat{s}_j - s_j)'(\hat{s}_j - s_j)$ under the order restriction implied by z_j. Although an explicit construction of the T_j matrix in Equation 5.4 is unnecessary, it would look like:

$$
T_j = \begin{bmatrix} 1 & 0 & 0 & 0 \\ 0 & 1 & 0 & 0 \\ 0 & 1 & 0 & 0 \\ 0 & 0 & 1 & 0 \\ 0 & 0 & 1 & 0 \\ 0 & 0 & 1 & 0 \\ 0 & 0 & 0 & 1 \end{bmatrix},
$$

which would produce an identical result to the above.

TABLE 5.6

Steps to Follow in the Least-Squares Monotonic Transformation When There Are Ties

						Secondary Approach	
			Primary Approach				
Data	Model	Step 1	Step 2	Step 3	Loss	Model	Loss
1	1	1	1	1	0	1	0
2	2	2	2	2	0	4	0
2	4	4	4	4	0	4	4
2	6 }	5.5	5.5	5	1	4	4
3	5 }	5.5 }	4.75 }	5	0	5	0
4	4	4 }	4.75 }	5	1	5.5	2.25
4	7			7	0	5.5	2.25

As alluded to earlier, there are two approaches to tied observations. The primary approach allows unequal transformed values corresponding to tied observations, whereas the second approach keeps tied observations tied in transformed data. Suppose that the second, third, and fourth observations in z_j are tied, and also the sixth and seventh observations are tied, as shown in the first column of Table 5.6. In the primary approach, we reorder model predictions corresponding to the tied observations in the order of their own magnitude. Thus, we reorder 4, 2, and 6 corresponding to the first set of tied observations into 2, 4, and 6. The model predictions of 4 and 7 corresponding to the second set of ties remain the same because their order is consistent with their magnitude. Once this pre-reordering step is carried out (the second column of Table 5.6), we simply apply the algorithm illustrated above. In the present case, the fourth and fifth model predictions with the values of 6 and 5 are in the first violation of monotonicity. So we take the average of the two and tentatively assign the value 5.5 to the corresponding elements of s_j (Step 1). This averaging creates no new violations of monotonicity. However, we immediately find that the sixth element of s_j of value 4 (shown at the bottom of Step 1) is smaller than this average. We take the average of this element and the one above it to resolve this violation and find 4.75, which is also smaller than the fourth element (Step 2). Consequently, we are obliged to take the average of the fourth, fifth, and sixth elements (Step 3). We then copy the seventh element of s_j into s_j without invoking any new violation. The loss incurred by the transformation is the sum of squared differences between $\hat{s}_j$ (the second column) and the final s_j (Step 3), which is equal to 2 in the present case. The T_j matrix that would give equivalent results is given by:

$$T_j = \begin{bmatrix} 1 & 0 & 0 & 0 & 0 \\ 0 & 1 & 0 & 0 & 0 \\ 0 & 0 & 1 & 0 & 0 \\ 0 & 0 & 0 & 1 & 0 \\ 0 & 0 & 0 & 1 & 0 \\ 0 & 0 & 0 & 1 & 0 \\ 0 & 0 & 0 & 0 & 1 \end{bmatrix}$$

In the secondary approach, we take the average of the initial model predictions (4, 2, and 6) corresponding to tied observations before we apply the basic algorithm described earlier. We thus have 4 for the second, third, and fourth model predictions (corresponding to the first set of tied observations), and 5.5 for the last two model predictions (corresponding to the second set of tied observations). It happens that this new set of model predictions already satisfies the order restriction implied by the observed data. So there is no need to further apply the basic algorithm in this case. The loss incurred by the transformation is calculated by the sum of squared differences between the initial model predictions (before averages are taken for tied observations) given in the second column of Table 5.5 and the final s_j given in the seventh column of Table 5.6, which is equal to 12.5 in the present case. The matrix T_j that gives equivalent results is given by:

$$T_j = \begin{bmatrix} 1 & 0 & 0 & 0 \\ 0 & 1 & 0 & 0 \\ 0 & 1 & 0 & 0 \\ 0 & 1 & 0 & 0 \\ 0 & 0 & 1 & 0 \\ 0 & 0 & 0 & 1 \\ 0 & 0 & 0 & 1 \end{bmatrix}$$

6

Generalized Structured Component Analysis with Latent Interactions

We have thus far focused on examining structural equation models that include only *main effects* of variables, sometimes called *simple effects* (Baron and Kenny 1986) or *first-order effects* (Marsh, Wen, and Hau 2004). In this chapter, we consider *interaction* or *moderation effects* to address whether the form and/or magnitude of the relationship between variables is influenced by another variable often called a moderator (Baron and Kenny 1986; Lazarsfeld 1955). The importance of investigating interaction effects has been well recognized and studied in a wide range of scientific fields (e.g., Bagozzi, Baumgartner, and Yi 1992; Busemeyer and Jones 1983; Kenny and Judd 1984; Sharma, Durand, and Gur-Arie 1981; Schumacker and Marcoulides 1998).

In general, a moderator can be either continuous or categorical (Baron and Kenny 1986). In the context of structural equation modeling, a moderator can be further assumed to be either observed or latent. For example, gender, ethnicity, or occupation can serve as a categorical and observed moderator, whereas a latent class variable eliciting cluster-level heterogeneity can be treated as a categorical and latent moderator (e.g., Lazarsfeld and Henry 1968; Magidson and Vermunt 2004; McLachlan and Peel 2000; Wedel and Kamakura 2000). To examine the effect of an observed moderator, a multiple group analysis can be carried out to investigate whether the effect of a variable on another variable varies across different levels or groups of the moderator (Baron and Kenny 1986; Ridgon, Schumacker, and Worthe 1998). In Chapter 3, we discussed how to conduct a multiple group analysis in generalized structured component analysis. On the other hand, fuzzy clusterwise generalized structured component analysis can be applied to identify different clusters/classes of a latent class variable, which are unknown a priori, and simultaneously to examine whether this variable affects the relationship between variables. We discussed fuzzy clusterwise generalized structured component analysis in Chapter 4.

In addition, a single continuous variable, for example, "level of reward" (Baron and Kenny 1986), can be considered to be a continuous and observed moderator. The effect of this moderator can be examined by adding an interaction term in a fashion similar to that used in multiple linear regression (Aiken and West 1991; Baron and Kenny 1986). The interaction term is formed as the product of the moderator and a predictor variable. Once the product term is constructed, generalized structured component analysis can be applied to examine the effect of the moderator without any technical alteration.

In the present chapter, we focus on a *continuous latent moderator*, which involves a block of indicators. As will be discussed in the empirical example in Section 6.4, the level of materialism can be regarded as a continuous latent moderator, which affects the relationship between two latent variables such as stress and depression. The effect of such a moderator is called a *latent inter-action effect* (Marsh et al. 2004). This latent interaction effect can be modeled and tested only within the context of structural equation modeling, because a moderator is a latent variable associated with indicators. We discuss how to examine a latent interaction effect in generalized structured component analysis. In testing the effect of a continuous latent moderator on the relationship between variables, the variables can be either observed or latent. For generality, however, we assume that they are all latent variables.

A practical approach to testing a latent interaction effect in generalized structured component analysis can be a so-called product-indicator approach (e.g., Algina and Moulder 2001; Chin, Marcolin, and Newsted 1996). In this approach, some multiplicative terms of indicators for a moderator and a latent variable are obtained and subsequently used as indicators for a latent interaction term. As will also be discussed in Section 6.3, this approach can be modified to use the residuals of product terms, which are orthogonal to original indicators (Little, Bovaird, and Widaman 2006). The product-indicator approach can be easily implemented into generalized structured component analysis with no further technical refinement. In particular, it is well-coupled with the distribution-free estimation procedure of generalized structured component analysis, which does not require normality of product indicators. This is beneficial because the normality assumption of product indicators is unlikely to hold.

In spite of the ease with which it can be implemented, the product-indicator approach has drawbacks. First, it is unclear which and how many indicators should be chosen to form product indicators (e.g., Jaccard and Wan 1995; Jöreskog and Yang 1996; Marsh et al. 2004). A solution can be to use all possible product terms of indicators (Henseler and Chin 2010). However, this is likely to yield an unwieldy number of product indicators particularly when contemplating higher-way latent interaction terms than two-way (e.g., three- or four-way latent interaction terms). Second, the product-indicator approach is not applicable when indicators for moderators and/or latent variables are formative or equivalently, an interaction between formative latent variables is considered (Chin, Marcolin, and Newsted 2003; Henseler and Fassott 2010).

An alternative approach can be a two-stage procedure (e.g., Chin et al. 2003; Henseler and Fassott 2010; Yang Jonsson 1998). This approach carries out two steps sequentially. In the first stage, individual latent variable scores are estimated via generalized structured component analysis, using a simpler model that excludes latent interaction effects from the original one. Then, a latent interaction term between latent variables is obtained as the elementwise product of the latent variable scores. In the second stage, the latent interaction effect is estimated by fitting the original *structural* model. This stage is equivalent to fitting a path-analytic model, treating the latent variables as well as latent interaction terms obtained in the first stage as indicators. A more detailed description of the two-stage approach is provided in Section 6.3. The two-stage approach defines a latent interaction term as the product of interacting latent variables (moderating and predicting latent variables). This way of modeling is consistent with that adopted for modeling the effect of an observed continuous moderator in multiple linear regression (e.g., Cohen and Cohen 1983). Consequently, the two-stage approach does not need to form any new (product) indicators for latent interaction terms. This can be of particular use in testing higher-way latent interaction effects, for which a clear rule lacks to obtain a suitable set of product indicators. Nevertheless, in the two-stage approach, there is no assurance that the solution obtained in the first stage is most appropriate for the second stage because the first stage is applied with no reference to the second stage; for example, the individual scores of all latent variables are obtained without consideration of latent interaction terms. Each stage addresses a different optimization criterion.

Hwang, Ho, and Lee (2010) extended generalized structured component analysis to examine latent interaction effects. In their approach, a latent interaction term is defined as the product of interacting latent variables, whose individual scores are uniquely obtained as the weighted composites of indicators. As a result, it does not require the construction of product indicators for latent interaction terms, and can be of help in examining higher-way latent interaction effects. This is comparable to the two-stage approach aforementioned. Unlike the two-stage approach, however, their approach is a single-stage or simultaneous procedure that estimates all model parameters including those for latent interaction effects, while considering both measurement and structural models at the same time. It is equipped with a single least-squares optimization criterion for parameter estimation.

This chapter is devoted to providing technical accounts and an empirical application of Hwang et al.'s (2010) approach. In the chapter, latent variables that influence other latent variables denote main-effect latent variables, and a moderator indicates a continuous latent variable, unless stated otherwise.

6.1 Generalized Structured Component Analysis with Latent Interactions

6.1.1 Model Specification

We begin by describing model specification in Hwang et al.'s (2010) approach. In their approach, two different structural models were originally considered to accommodate all possible latent interaction terms. One model takes into account only exogenous latent interaction terms that affect latent variables. The other includes endogenous latent interaction terms, which are affected by other latent variables and/or latent interaction terms. However, it is not necessary to distinguish these different structural models for specifying latent interaction terms. We here provide a single formulation that takes into account both structural models.

Although it was not clearly described by Hwang et al. (2010), in their approach, a moderator is assumed to influence the relationship between latent variables in a linear manner. That is, as the moderator changes, the effect of a latent variable on another latent variable changes at a constant rate. The assumption of such linear-function moderation is typically made, although different types of moderation such as step- or quadratic-function moderation can also be assumed (Baron and Kenny 1986; Edwards 1995). Making this assumption implies that a product term of interacting latent variables can be used to examine a latent interaction effect (Baron and Kenny 1986; Cleary and Kessler 1982; Cohen and Cohen 1983).

Let γ^* denote a P_1 by 1 vector consisting of latent variables. Let γ^{**} denote a P_2 by 1 vector consisting of latent interaction terms that are obtained as products of interacting latent variables. Then, $\gamma = [\gamma^*; \gamma^{**}]$ denotes a P by 1 vector stacking all latent interaction terms below latent variables ($P = P_1 + P_2$). In Hwang et al.'s (2010) approach, the measurement model is generally given as

$$\mathbf{z} = \begin{bmatrix} \mathbf{C}', \mathbf{0} \end{bmatrix} \begin{bmatrix} \gamma^* \\ \gamma^{**} \end{bmatrix} + \boldsymbol{\varepsilon} = \dot{\mathbf{C}}'\gamma + \boldsymbol{\varepsilon}, \tag{6.1}$$

where $\mathbf{0}$ is a J by P_2 matrix of zeros. The structural model is generally expressed as follows.

$$\gamma = \mathbf{B}'\gamma + \boldsymbol{\zeta}. \tag{6.2}$$

The weighted relation model is given as

$$\gamma = \begin{bmatrix} \mathbf{W}' & \mathbf{0} \\ \mathbf{0} & \mathbf{I}_{P_2} \end{bmatrix} \begin{bmatrix} \mathbf{z} \\ \gamma^{**} \end{bmatrix} = \dot{\mathbf{W}}'\dot{\mathbf{z}}, \tag{6.3}$$

where $\mathbf{I}_{P_2}$ is an identity matrix of order P_2,

$$\dot{\mathbf{z}} = \begin{bmatrix} \mathbf{z} \\ \gamma^{**} \end{bmatrix}$$

denotes a $J + P_2$ by 1 vector consisting of indicators and latent interaction terms.

These submodels are then combined into a single equation, as follows.

$$\begin{bmatrix} \mathbf{z} \\ \boldsymbol{\gamma} \end{bmatrix} = \begin{bmatrix} \dot{\mathbf{C}}' \\ \mathbf{B}' \end{bmatrix} \boldsymbol{\gamma} + \mathbf{e}$$

$$\begin{bmatrix} \mathbf{I}^* \\ \dot{\mathbf{W}}' \end{bmatrix} \begin{bmatrix} \mathbf{z} \\ \boldsymbol{\gamma}^{**} \end{bmatrix} = \begin{bmatrix} \dot{\mathbf{C}}' \\ \mathbf{B}' \end{bmatrix} \dot{\mathbf{W}}' \begin{bmatrix} \mathbf{z} \\ \boldsymbol{\gamma}^{**} \end{bmatrix} + \mathbf{e}$$

$$\dot{\mathbf{V}}\dot{\mathbf{z}} = \dot{\mathbf{A}}'\dot{\mathbf{W}}'\dot{\mathbf{z}} + \mathbf{e}, \tag{6.4}$$

where $\mathbf{I}^* = [\mathbf{I}_J, \mathbf{0}]$,

$\mathbf{I}_J$ is an identity matrix of order J, and $\dot{\mathbf{V}}' = \begin{bmatrix} \mathbf{I}^* \\ \dot{\mathbf{W}}' \end{bmatrix}$.

It is easy to see that Equation 6.4 is essentially of the same form as the generalized structured component analysis model. The only difference is that Equation 6.4 includes both latent variables ($\boldsymbol{\gamma}^*$) and latent interaction terms ($\boldsymbol{\gamma}^{**}$). When no latent interaction terms are involved (i.e., $\boldsymbol{\gamma}^{**} = \mathbf{0}$), Equation 6.4 reduces to the generalized structured component analysis model.

To illustrate the above model specification, let us contemplate a prototype model depicted in Figure 6.1. This model includes two exogenous latent variables (γ_1 and γ_2) and a latent interaction between them (γ_{12}). All the three variables are assumed to influence an endogenous latent variable (γ_3). Thus, the latent interaction term is exogenous. Each latent variable is associated with two reflective indicators. The latent interaction term is determined as the product of the exogenous latent variables, that is, $\gamma_{12} = \gamma_1 \gamma_2$, and thus is not associated with any indicators. Note that although it is not displayed in

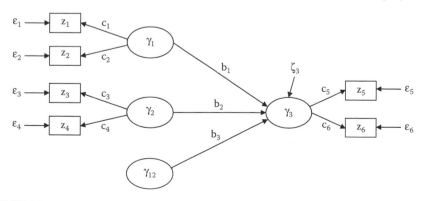

FIGURE 6.1
A prototype model with an exogenous two-way latent interaction term. All weights assigned to indicators are not displayed.

Figure 6.1 to make the figure concise, a weight is assigned to each indicator for γ_1, γ_2, and γ_3.

The measurement model for the prototype model is given in matrix notation, as follows.

$$
\begin{bmatrix} z_1 \\ z_2 \\ z_3 \\ z_4 \\ z_5 \\ z_6 \end{bmatrix} = \begin{bmatrix} c_1 & 0 & 0 & 0 \\ c_2 & 0 & 0 & 0 \\ 0 & c_3 & 0 & 0 \\ 0 & c_4 & 0 & 0 \\ 0 & 0 & c_5 & 0 \\ 0 & 0 & c_6 & 0 \end{bmatrix} \begin{bmatrix} \gamma_1 \\ \gamma_2 \\ \gamma_3 \\ \gamma_{12} \end{bmatrix} + \begin{bmatrix} \varepsilon_1 \\ \varepsilon_2 \\ \varepsilon_3 \\ \varepsilon_4 \\ \varepsilon_5 \\ \varepsilon_6 \end{bmatrix}
$$

$$
\mathbf{z} = \begin{bmatrix} \mathbf{C'}, 0 \end{bmatrix} \begin{bmatrix} \boldsymbol{\gamma}^* \\ \boldsymbol{\gamma}^{**} \end{bmatrix} + \boldsymbol{\varepsilon} = \dot{\mathbf{C}}' \boldsymbol{\gamma} + \boldsymbol{\varepsilon}, \tag{6.5}
$$

where $\gamma^* = [\gamma_1, \gamma_2, \gamma_3]'$ and $\gamma^{**} = \gamma_{12}$.

The structural model for the prototype model is given as follows.

$$
\gamma_3 = \gamma_1 b_1 + \gamma_2 b_2 + \gamma_{12} b_3 + \zeta_3. \tag{6.6}
$$

Equivalently, in matrix notation,

$$
\begin{bmatrix} \gamma_1 \\ \gamma_2 \\ \gamma_3 \\ \gamma_{12} \end{bmatrix} = \begin{bmatrix} 0 & 0 & 0 & 0 \\ 0 & 0 & 0 & 0 \\ b_1 & b_2 & 0 & b_3 \\ 0 & 0 & 0 & 0 \end{bmatrix} \begin{bmatrix} \gamma_1 \\ \gamma_2 \\ \gamma_3 \\ \gamma_{12} \end{bmatrix} + \begin{bmatrix} \gamma_1 \\ \gamma_2 \\ \zeta_3 \\ \gamma_{12} \end{bmatrix}
$$

$$
\boldsymbol{\gamma} = \mathbf{B}' \boldsymbol{\gamma} + \boldsymbol{\zeta}. \tag{6.7}
$$

The weighted relation model for the prototype model is given in matrix notation, as follows.

$$
\begin{bmatrix} \gamma_1 \\ \gamma_2 \\ \gamma_3 \\ \gamma_{12} \end{bmatrix} = \begin{bmatrix} w_1 & w_2 & 0 & 0 & 0 & 0 & 0 \\ 0 & 0 & w_3 & w_4 & 0 & 0 & 0 \\ 0 & 0 & 0 & 0 & w_5 & w_6 & 0 \\ 0 & 0 & 0 & 0 & 0 & 0 & 1 \end{bmatrix} \begin{bmatrix} z_1 \\ z_2 \\ z_3 \\ z_4 \\ z_5 \\ z_6 \\ \gamma_{12} \end{bmatrix}
$$

$$\gamma = \begin{bmatrix} \mathbf{W}' & \mathbf{0} \\ \mathbf{0} & 1 \end{bmatrix} \dot{\mathbf{z}} = \dot{\mathbf{W}}'\dot{\mathbf{z}}. \qquad (6.8)$$

The three submodels above are combined into Equation 6.4, where $J = 6$, $P_1 = 3$, and $P_2 = 1$.

The prototype model depicted in Figure 6.1 involves only an exogenous two-way interaction term. It is also straightforward to incorporate endogenous latent interaction terms as well as higher-way latent interaction terms into Equation 6.4. As an example, let us consider another prototype model, displayed in Figure 6.2. This model has 10 indicators for five latent variables (two indicators per latent variable). The three latent variables (γ_1, γ_2, and γ_3) are hypothesized to influence an endogenous latent variable (γ_5). Moreover, the model specifies three two-way latent interactions (γ_{12}, γ_{13}, and γ_{23}) and a three-way latent interaction (γ_{123}) as products of their corresponding latent variables, that is, $\gamma_{12} = \gamma_1 \gamma_2$, $\gamma_{13} = \gamma_1 \gamma_3$, $\gamma_{23} = \gamma_2 \gamma_3$, and $\gamma_{123} = \gamma_1\gamma_2\gamma_3$. All of them are assumed to affect the endogenous latent variable. Furthermore, a latent

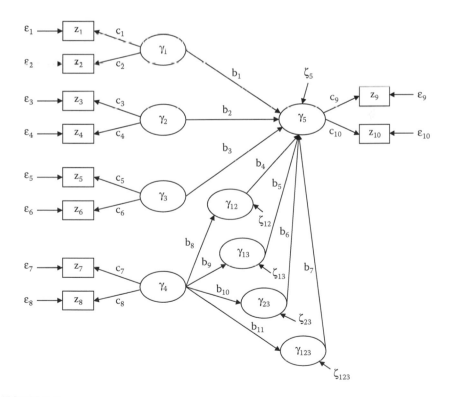

FIGURE 6.2
A prototype model with endogenous two-way and three-way latent interaction terms. All weights assigned to indicators are not displayed.

variable (γ_4) is specified to influence all the latent interaction terms. Thus, the latent interaction terms are considered endogenous. Again, a weight is assigned to each indicator for the five latent variables, although it is not displayed in Figure 6.2.

The measurement model for this prototype model is given in matrix notation, as follows.

$$
\begin{bmatrix} z_1 \\ z_2 \\ z_3 \\ z_4 \\ z_5 \\ z_6 \\ z_7 \\ z_8 \\ z_9 \\ z_{10} \end{bmatrix} =
\begin{bmatrix}
c_1 & 0 & 0 & 0 & 0 & 0 & 0 & 0 & 0 \\
c_2 & 0 & 0 & 0 & 0 & 0 & 0 & 0 & 0 \\
0 & c_3 & 0 & 0 & 0 & 0 & 0 & 0 & 0 \\
0 & c_4 & 0 & 0 & 0 & 0 & 0 & 0 & 0 \\
0 & 0 & c_5 & 0 & 0 & 0 & 0 & 0 & 0 \\
0 & 0 & c_6 & 0 & 0 & 0 & 0 & 0 & 0 \\
0 & 0 & 0 & c_7 & 0 & 0 & 0 & 0 & 0 \\
0 & 0 & 0 & c_8 & 0 & 0 & 0 & 0 & 0 \\
0 & 0 & 0 & 0 & c_9 & 0 & 0 & 0 & 0 \\
0 & 0 & 0 & 0 & c_{10} & 0 & 0 & 0 & 0
\end{bmatrix}
\begin{bmatrix} \gamma_1 \\ \gamma_2 \\ \gamma_3 \\ \gamma_4 \\ \gamma_5 \\ \gamma_{12} \\ \gamma_{13} \\ \gamma_{23} \\ \gamma_{123} \end{bmatrix} +
\begin{bmatrix} \varepsilon_1 \\ \varepsilon_2 \\ \varepsilon_3 \\ \varepsilon_4 \\ \varepsilon_5 \\ \varepsilon_6 \\ \varepsilon_7 \\ \varepsilon_8 \\ \varepsilon_9 \\ \varepsilon_{10} \end{bmatrix}
$$

$$
\mathbf{z} = \dot{\mathbf{C}}' \begin{bmatrix} \boldsymbol{\gamma}^* \\ \boldsymbol{\gamma}^{**} \end{bmatrix} + \boldsymbol{\varepsilon} = \dot{\mathbf{C}}' \boldsymbol{\gamma} + \boldsymbol{\varepsilon}, \tag{6.9}
$$

where $\boldsymbol{\gamma}^* = [\gamma_1, \gamma_2, \gamma_3, \gamma_4, \gamma_5]'$ and $\boldsymbol{\gamma}^{**} = [\gamma_{12}, \gamma_{13}, \gamma_{23}, \gamma_{123}]'$.

The structural model for the prototype model is given in matrix notation, as follows.

$$
\begin{bmatrix} \gamma_1 \\ \gamma_2 \\ \gamma_3 \\ \gamma_4 \\ \gamma_5 \\ \gamma_{12} \\ \gamma_{13} \\ \gamma_{23} \\ \gamma_{123} \end{bmatrix} =
\begin{bmatrix}
0 & 0 & 0 & 0 & 0 & 0 & 0 & 0 & 0 \\
0 & 0 & 0 & 0 & 0 & 0 & 0 & 0 & 0 \\
0 & 0 & 0 & 0 & 0 & 0 & 0 & 0 & 0 \\
0 & 0 & 0 & 0 & 0 & 0 & 0 & 0 & 0 \\
b_1 & b_2 & b_3 & 0 & 0 & b_4 & b_5 & b_6 & b_7 \\
0 & 0 & 0 & b_8 & 0 & 0 & 0 & 0 & 0 \\
0 & 0 & 0 & b_9 & 0 & 0 & 0 & 0 & 0 \\
0 & 0 & 0 & b_{10} & 0 & 0 & 0 & 0 & 0 \\
0 & 0 & 0 & b_{11} & 0 & 0 & 0 & 0 & 0
\end{bmatrix}
\begin{bmatrix} \gamma_1 \\ \gamma_2 \\ \gamma_3 \\ \gamma_4 \\ \gamma_5 \\ \gamma_{12} \\ \gamma_{13} \\ \gamma_{23} \\ \gamma_{123} \end{bmatrix} +
\begin{bmatrix} \gamma_1 \\ \gamma_2 \\ \gamma_3 \\ \gamma_4 \\ \zeta_5 \\ \zeta_{12} \\ \zeta_{13} \\ \zeta_{23} \\ \zeta_{123} \end{bmatrix}
$$

$$
\boldsymbol{\gamma} = \mathbf{B}' \boldsymbol{\gamma} + \boldsymbol{\zeta}. \tag{6.10}
$$

The weighted relation model is given in matrix notation, as follows.

$$
\begin{bmatrix} \gamma_1 \\ \gamma_2 \\ \gamma_3 \\ \gamma_4 \\ \gamma_5 \\ \gamma_{12} \\ \gamma_{13} \\ \gamma_{23} \\ \gamma_{123} \end{bmatrix} = \begin{bmatrix} w_1 & w_2 & 0 & 0 & 0 & 0 & 0 & 0 & 0 & 0 & 0 & 0 & 0 & 0 \\ 0 & 0 & w_3 & w_4 & 0 & 0 & 0 & 0 & 0 & 0 & 0 & 0 & 0 & 0 \\ 0 & 0 & 0 & 0 & w_5 & w_6 & 0 & 0 & 0 & 0 & 0 & 0 & 0 & 0 \\ 0 & 0 & 0 & 0 & 0 & 0 & w_7 & w_8 & 0 & 0 & 0 & 0 & 0 & 0 \\ 0 & 0 & 0 & 0 & 0 & 0 & 0 & 0 & w_9 & w_{10} & 0 & 0 & 0 & 0 \\ 0 & 0 & 0 & 0 & 0 & 0 & 0 & 0 & 0 & 0 & 1 & 0 & 0 & 0 \\ 0 & 0 & 0 & 0 & 0 & 0 & 0 & 0 & 0 & 0 & 0 & 1 & 0 & 0 \\ 0 & 0 & 0 & 0 & 0 & 0 & 0 & 0 & 0 & 0 & 0 & 0 & 1 & 0 \\ 0 & 0 & 0 & 0 & 0 & 0 & 0 & 0 & 0 & 0 & 0 & 0 & 0 & 1 \end{bmatrix} \begin{bmatrix} z_1 \\ z_2 \\ z_3 \\ z_4 \\ z_5 \\ z_6 \\ z_7 \\ z_8 \\ z_9 \\ z_{10} \\ \gamma_{12} \\ \gamma_{13} \\ \gamma_{23} \\ \gamma_{123} \end{bmatrix}
$$

$$\gamma = \mathbf{W}'\dot{\mathbf{z}}. \tag{6.11}$$

Again, these submodels for the second prototype model can be combined into Equation 6.4, where $J = 10$, $P_1 = 5$, and $P_2 = 4$. Thus, as stated earlier, Equation 6.4 can be used to examine various types of latent interaction effects.

6.1.2 Parameter Estimation

To estimate the unknown parameters in $\dot{\mathbf{W}}$, $\dot{\mathbf{C}}$, and $\mathbf{B}$, we seek to minimize the following least-squares optimization criterion

$$\phi = \sum_{i=1}^{N} \mathbf{e}_i' \mathbf{e}_i = \sum_{i=1}^{N} (\dot{\mathbf{V}}'\dot{\mathbf{z}}_i - \dot{\mathbf{A}}'\dot{\mathbf{W}}'\dot{\mathbf{z}}_i)'(\dot{\mathbf{V}}'\dot{\mathbf{z}}_i - \dot{\mathbf{A}}'\dot{\mathbf{W}}'\dot{\mathbf{z}}_i), \tag{6.12}$$

subject to the standardization constraint on each latent variable

$$\left(\sum_{i=1}^{N} \gamma_{ip}^{2} = N \right).$$

However, the standardization constraint should not be imposed on latent interaction terms because the product of standardized latent variables is not equal to the standardized product of the latent variables (e.g., Aiken and West 1991; Henseler and Chin 2010). This important issue was neglected in the study of Hwang et al. (2010), which suggested normalizing latent interaction terms.

An alternating least-squares algorithm was developed to minimize the criterion. This algorithm repeats two steps until convergence. In one step, the estimates of loadings and/or path coefficients in $\dot{\mathbf{A}}$ are updated in the least-square sense, given the estimates of weights. In the other step, the estimates of weights in $\mathbf{W}$ are updated in the least-squares sense, given the estimates of loadings and/or path coefficients. The two steps are virtually the same as those of the algorithm for generalized structured component analysis discussed in Appendix 2.1. A distinction is that when estimating weights in Equation 6.4, we should take into account that latent variables can share their weights with latent interaction terms, because latent interaction terms are equivalent to the products of latent variables that are weighted composites of indicators. For example, in the prototype model depicted in Figure 6.1, a latent interaction term (γ_{12}) is defined as the product of two latent variables (γ_1 and γ_2), each of which is the weighted composite of two indicators. This indicates that $\gamma_{12} = \gamma_1 \gamma_2 = (z_1 w_1 + z_2 w_2)(z_3 w_3 + z_4 w_4)$. Accordingly, we should consider that the latent interaction term is associated with the same weights for its two interacting latent variables, when estimating these weights. We provide a description of the alternating least-squares algorithm in Appendix 6.1.

The value of FIT is calculated as

$$\text{FIT} = 1 - \left(\frac{\phi}{\alpha} \right),$$

where

$$\alpha = \sum_{i=1}^{N} (\dot{\mathbf{V}}' \dot{\mathbf{z}}_i)'(\dot{\mathbf{V}}' \dot{\mathbf{z}}_i).$$

The value of AFIT is calculated based on the value of FIT, as described in Chapter 2. The bootstrap method is used to estimate the standard errors or confidence intervals of parameter estimates.

6.2 Probing of Latent Interaction Effects

We briefly describe how to probe statistically significant latent interaction effects. As in standard generalized structured component analysis, Hwang et al.'s (2010) approach enables the calculation of the individual scores of latent variables as the weighted composites of indicators. Given the latent variable scores, we can probe latent interaction effects in the same manner as in multiple linear regression (Aiken and West 1991; Cohen et al. 2003).

As an example, we consider the structural model in Figure 6.1, where $\gamma_{12} = \gamma_1\gamma_2$. The structural model can be re-expressed as

$$\gamma_3 = \gamma_1 b_1 + \gamma_2 b_2 + \gamma_1\gamma_2 b_3 + \zeta_3 \tag{6.13}$$

$$= \gamma_2 b_2 + (b_1 + \gamma_2 b_3)\gamma_1 + \zeta_3, \tag{6.14}$$

where $(b_1 + \gamma_2 b_3)$ is called the simple slope of γ_3 regressed on γ_1, which is a function of the moderator γ_2. In Equation 6.13, we chose γ_2 as a moderator and γ_1 as an exogenous variable. However, the latent interaction term is the product of two latent variables, that is, $\gamma_{12} = \gamma_1\gamma_2$, which is commutative. Thus, it is mathematically arbitrary which latent variable is chosen as the moderator over the other, when converting Equations 6.13 to 6.14. This should be decided from a substantive point of view. If b_3 is statistically significant, the latent interaction effect can be probed to examine how the relationship between γ_3 and γ_1 is varied conditionally upon the values of γ_2.

A well-known probing method is to categorize the original values of the moderator into a limited number of categories or groups and to examine the significance of the simple effect across the different categories of the moderator (e.g., Aiken and West 1991). This probing method requires discretization of a continuous variable into an arbitrary number of categories, thus tending to decrease statistical power (Edwards and Lambert 2007).

As suggested by Preacher, Rucker, and Hayes (2007), we instead utilize the Johnson–Neyman technique (Johnson and Neyman 1936) for probing a latent interaction effect, in combination with bootstrapping. As shown in Equation 6.14, the simple slope of γ_1 can be estimated at all individual values of the moderator γ_2, which are obtained as the weighted composites of its indicators and then considered fixed. The confidence intervals of each simple slope estimate can be obtained by means of bootstrapping. Specifically, let $\hat{b}_1$ and $\hat{b}_3$ denote the estimates of b_1 and b_3 obtained from a single bootstrap sample. Let γ_{2i} denote the ith value of γ_2 ($i = 1, ..., N$). Then, the simple slope estimate of γ_1 calculated at γ_{2i}, denoted by $s(i)$, is given as:

$$s(i) = \hat{b}_1 + \hat{b}_3\gamma_{2i}. \tag{6.15}$$

The sampling distribution of $s(i)$ is derived based on B bootstrap samples (e.g., $B = 100$). The confidence intervals of $s(i)$ continuously plotted over all N values of the moderator are named confidence bands (Bauer and Curran 2005; Preacher, Curran, and Bauer 2006; Rogosa 1981). By examining whether confidence bands do not include zero in a particular value of the moderator, a so-called region of significance can be detected, for which the simple slope turns out to be statistically significant. The region of significance shows where the relationship between γ_1 and γ_3 is moderated by γ_2.

6.3 Testing Latent Interaction Effects in Partial Least Squares Path Modeling

In partial least squares path modeling, there are three major approaches to testing latent interaction effects, depending on how to construct a latent interaction term. The first approach is to create new indicators that a latent interaction term underlies. This approach includes the product-indicator method (Chin et al. 1996, 2003) and the orthogonalizing method (Henseler and Chin 2010; Little, Bovaird, and Widaman 2006). The product-indicator method forms all possible pairwise products of indicators for interacting latent variables and uses the products as indicators for the latent interaction term of the interacting latent variables. For example, for the latent interaction term (γ_{12}) in the prototype model depicted in Figure 6.1, we create four pairwise products of two indicators for each of two exogenous latent variables (γ_1 and γ_2), as follows.

$$z_{13} = z_1 z_3$$

$$z_{14} = z_1 z_4$$

$$z_{23} = z_2 z_3$$

$$z_{24} = z_2 z_4. \tag{6.16}$$

These product indicators are subsequently used as the indicators for the latent interaction term, as illustrated in Figure 6.3.

The orthogonalizing method begins to form the same product indicators and then regresses them on all related indicators. For example, each of the four pairwise products in Equation 6.16 is regressed on all the indicators for the two interacting latent variables as follows.

$$z_{13} = b_{01} + b_{11}z_1 + b_{12}z_2 + b_{13}z_3 + b_{14}z_4 + \varepsilon_1$$

$$z_{14} = b_{02} + b_{21}z_1 + b_{22}z_2 + b_{23}z_3 + b_{24}z_4 + \varepsilon_2$$

$$z_{23} = b_{03} + b_{31}z_1 + b_{32}z_2 + b_{33}z_3 + b_{34}z_4 + \varepsilon_3$$

$$z_{24} = b_{04} + b_{41}z_1 + b_{42}z_2 + b_{43}z_3 + b_{44}z_4 + \varepsilon_4. \tag{6.17}$$

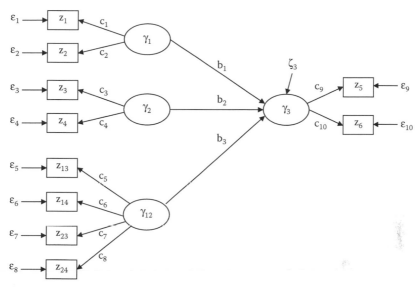

FIGURE 6.3
An example of the product-indicator method in partial least squares path modeling.

The residuals of the product indicators (ε's) are then used as the indicators for the latent interaction term. This indicates that the four product indicators in Figure 6.3 are replaced by the corresponding residual indicators. The residual indicators are orthogonal to the indicators for the interacting latent variables.

The construction of product or residual indicators is carried out independently of partial least squares path modeling. Thus, no technical refinement in partial least squares path modeling is needed to implement both methods. However, the two methods are appropriate for use only when all interacting latent variables are associated with reflective indicators (Chin et al. 2003; Henseler and Chin 2010). This is because a block of formative indicators for a latent variable are not assumed to stem from the same construct, and thus the pairwise multiplicative terms of two blocks of formative indicators do not necessarily serve as the indicators for an interaction between two constructs (Chin et al. 2003).

The second approach to testing a latent interaction effect in partial least squares path modeling is to use interacting latent variables themselves to construct their interaction term. That is, a latent interaction term is obtained as the product of interacting latent variables. This approach includes the two-stage method (Chin et al. 2003; also see Henseler and Fassott 2010). As its name suggests, this method follows two stages sequentially. In the first stage, latent variable scores are estimated via partial least squares path modeling, based on a model with no latent interactions included. In the second, a latent interaction term is calculated as the elementwise product of the latent variable scores and is added to a path-analytic model, in which the relationships among latent variables and latent interaction terms are specified, while treating them as indicators. For example, Figure 6.4 displays each stage of the method, which is applied to deal with the latent interaction term (γ_{12}) in Figure 6.1. As shown in the figure, in the first stage, the scores of three latent variables (γ_1, γ_2, and γ_3) are estimated based on a model with no latent interaction term, and then the latent interaction term is obtained as the product of two latent variables (γ_1 and γ_2). In the second stage, a path-analytic model that includes the three latent variables and latent interaction term as indicators is fitted to examine the effect of the latent interaction term. The path-analytic model is equivalent to the structural model in Figure 6.1, except for that all latent variables and latent interaction term are now indicators.

The two-stage method is straightforward to implement and does not call for any technical alteration to partial least squares path modeling. In addition, it can be used when any of interacting latent variables is formative. It is comparable to Hwang et al.'s (2010) approach in that it defines a latent interaction term as the product of interacting latent variables and does not require the construction of additional indicators for the latent interaction term.

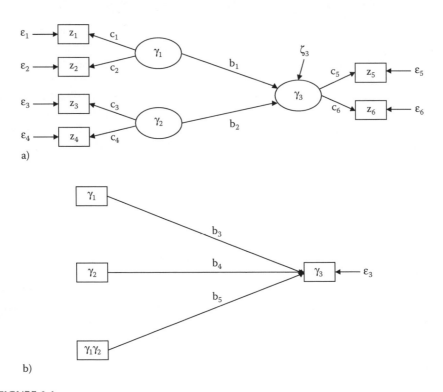

FIGURE 6.4
An example of the two-stage method in partial least squares path modeling. a) Stage 1. b) Stage 2.

Nonetheless, the two-stage method is a sequential procedure, where the first stage is carried out independently of the second one. The two stages adopt different optimization criteria for estimating the parameters of conceptually different models (e.g., the first stage considers a structural equation model with main-effect latent variables only, whereas the second is based on a path-analytic model with indicators only). Owing to the lack of coherence in the two stages, the method cannot assure that the solution from the first stage is optimal for the second stage. Moreover, how to obtain a solution in the first stage is irrelevant to obtaining a solution in the second stage. This suggests that in the first stage, we do not need to rely only on partial least squares path modeling for estimating the scores of latent variables. For example, a principal component analysis can be used to estimate the scores of latent variable for each block of indicators. However, in general, partial least squares path modeling can be more beneficial than a principal component analysis because it takes into account the relationships among latent variables, while estimating latent variable scores. In practice, Henseler and Fassott (2010) suggested using the two-stage method only when formative latent variables are involved in the construction of a latent interaction term.

The third approach is the hybrid method (Wold 1982). This method was originally introduced to deal with quadratic terms of latent variables in partial least squares path modeling. However, Henseler and Chin (2010) showed that it could also be adopted to accommodate latent interaction terms. Unlike the two approaches discussed earlier, the hybrid method aims to estimate a latent interaction term as well as latent variables simultaneously. This requires a modification of the first-stage algorithm of partial least squares path modeling, which we described in Appendix 2.3. Specifically, the following steps of the algorithm are refined to deal with a latent interaction term. In Step 1, each latent variable is estimated as a weighted composite of indicators. This step is then extended to obtain a latent interaction term as the product of interacting latent variables. In Step 2 (internal estimation), the so-called inner estimate for each latent variable is computed, which is equivalent to a weighted composite of its connected latent variables. The so-called inner weight can be calculated based on one of three schemes (i.e., centroid, factorial, and path weighting schemes). The second step is extended such that the interaction term is also used for the calculation of the inner estimate. In Step 3 (external estimation), the weights for each block of indicators are estimated based on either Mode A or Mode B. The third step remains unchanged because the latent interaction term does not involve indicators.

Compared to the first and second approaches, the generalized structured component analysis approach proposed by Hwang et al. (2010) represents a single-stage method that considers all latent variables and latent interaction terms simultaneously to estimate parameters in a given model. It can be used regardless of whether a latent variable is formative or reflective. In addition, it is efficient in dealing with higher-way interaction terms, which are likely to involve an unwieldy number of new indicators if the product- or residual-indicator method is adopted.

The hybrid method is similar to the two-stage method in that it regards a latent interaction term as the product of latent variables, and does not need to form additional indicators for the latent interaction term. Unlike the two-stage method, however, it integrates the estimation of a latent interaction term into the algorithm of partial least squares path modeling. This renders the hybrid method similar in spirit to Hwang et al.'s (2010) approach. A major concern with the hybrid method is that the modified algorithm seems to be less optimal to estimate a latent interaction term as well as latent variables. For example, when estimating weights for a latent variable, the fact that the same weights can be shared by a latent interaction term is not taken into consideration. Moreover, similar to the original algorithm of partial least squares path modeling, it is unclear which criterion is newly considered and optimized to estimate a latent interaction term. On the other hand, Hwang et al.'s (2010) approach employs a well-defined optimization criterion that is minimized to estimate all model parameters simultaneously.

6.4 Example

The present example is part of a longitudinal study conducted for investigating the relationships among materialism, stress, depression, and anxiety (Abela et al. 2008). In the study, 110 participants were repeatedly measured on their levels of materialism, stress, depression, and anxiety over several occasions (weeks). The Aspirations Index (ASPQ) (Kasser and Ryan 1996) was used to assess the levels of materialism. The ASPQ consists of 35 items, which are related to particular life goals. Participants were asked to rate how important each goal was for them on a 7-point scale (1 = "not at all important" to 7 = "very important"). The ASPQ items are grouped into six different types of goals: financial success, social recognition, attractive appearance, self-acceptance, affiliation, and community feeling. The first three types (financial success, social recognition, and attractive appearance) are considered extrinsic values, whereas the other three (self-acceptance, affiliation, and community feeling) are intrinsic values. To evaluate the relative importance of materialistic values to each participant, the average of a set of items for each extrinsic value was computed and then subtracted from the average of all items for the three intrinsic values. The resultant three difference scores served as indicators for the latent variable of *materialism*.

The Center for Epidemiological Studies Depression Scale (CES-D) (Radloff 1977) was used to measure each participant's depressive symptoms. The CES-D is a 20-item self-report measure designed for the general population, for which participants were to answer how often they had experienced a particular symptom over the past week, ranging from 1 (rarely or none of the time) to 4 (most or all of the time). The 62-item version of the Mood and Anxiety Symptom Questionnaire (MASQ) (Watson and Clark 1991) was used to assess both specific and nonspecific anxious symptoms, for each of which participants were to rate on a scale of 1 to 5 the extent to which they had felt this way during the past 24 hours. A 30-item abbreviated version of the General, Academic, Social Hassles Scale for Students (GASHSS) (Blankstein and Flett 1993) was used to measure participants' levels of stress. The GASHSS includes items assessing general hassles (8 items), academic hassles (10 items), and social hassles (12 items). Participants were asked to evaluate how persistent a given hassle was (i.e., its frequency and duration) over the past week (0 = "no hassle; not at all persistent" to 6 = "extremely persistent hassle; high frequency and/or duration").

In this example, each set of original items for three latent variables (*depression, anxiety,* and *stress*) was grouped to create three item parcels, which were subsequently used as indicators for each of the three latent variables, that is, three indicators per latent variable.

Figure 6.5 displays the structural model specified for the data. As shown in the model, materialism evaluated at the first time point (M) was assumed to influence depression at time points 3 and 5 (D3 and D5) and anxiety at time points 3 and 5 (A3 and A5). Materialism, often defined as "a set of centrally held beliefs about the importance of possessions in one's life" (Richins and Dawson 1992), could be considered a form of modernistic survival value (Egri and Ralston 2004). However, it was also found to have a negative impact on individual well-being over time (Kasser and Ryan 1993, 1996; Sheldon and Elliot 1999; Sheldon and Kasser 1998). Moreover, stress at time point 2 (S2) was hypothesized to influence depression and anxiety at time point 3 (D3 and A3). Similarly, stress at time point 4 (S4) was assumed

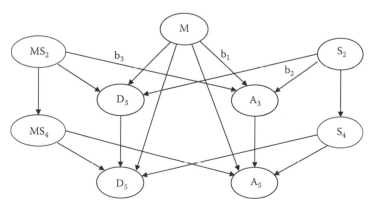

FIGURE 6.5
The structural model specified for the materialism data.

to affect depression and anxiety at time point 5 (D5 and A5). Furthermore, materialism was assumed to moderate the relationship between stress and depression and between stress and anxiety at later time points. Specifically, a latent interaction between materialism and stress at time point 2 (MS2) was hypothesized to have an effect on depression and anxiety at time point 3 (D3 and A3), whereas a latent interaction between materialism and stress at time point 4 (MS4) was to influence depression and anxiety at time point 5 (D5 and A5). These latent interaction effects were specified because materialistic individuals' self-esteem tends to be particularly vulnerable in the face of threat/stress; such individuals are likely to lack social support networks in times of stress; and individuals with extrinsic motivations (i.e., materialism) tend to perceive negative events as relatively more stressful. Depression and anxiety at time point 3 (D3 and A3) were specified to influence depression and anxiety at time point 5 (D5 and A5). Moreover, stress at time point 2 (S2) was assumed to have an impact on that at time point 4 (S4). Lastly, the latent interaction term of materialism and stress at time point 2 (MS2) was to influence that of materialism and stress at time point 4 (MS4).

We applied Hwang et al.'s (2010) approach to fit the model. The model provided FIT = 0.68 (SE = 0.02; 95% CI = 0.65–0.73) and AFIT = 0.67 (SE = 0.02; 95% CI = 0.64–0.72). This indicates that the model accounted for about 68% of the variance of all variables. Table 6.1 provides the estimates of weights and loadings for the indicators, along with their standard errors and 95% confidence intervals. We used 100 bootstrap samples to estimate the standard errors and 95% confidence intervals. All weight estimates were statistically significant. Moreover, all loading estimates were statistically significant and large.

Table 6.2 provides the estimates of path coefficients along with their standard errors and 95% confidence intervals. Materialism (M) had a statistically significant and positive impact on anxiety at time point 5 (A_5). Stress at time points 2 and 4 (S_2 and S_4) showed statistically significant and positive influences on depression at time points 3 and 5 (D_3 and D_5) and anxiety at time points 3 and 5 (A_3 and A_5). In addition, stress at time point 2 (S_2) had a statistically significant and positive effect on stress at time point 4 (S_4). Similarly, depression at time point 3 and anxiety at time point 3 (D_3 and A_3) exhibited statistically significant and positive effects on depression at time point 5 and anxiety at time point 5 (D_5 and A_5), respectively. Furthermore, the latent interaction between materialism and stress at time point 2 (MS_2) showed a positive and significant effect on anxiety at time point 3 (A_3). It also had a statistically significant and positive impact on the latent interaction between materialism and stress at time point 4 (MS_4). The latent interaction between materialism and stress at time point 4 (MS_4) had statistically significant and positive effects on depression and anxiety at time point 5 (D_5 and A_5).

TABLE 6.1

Estimates of Weights and Loadings along with Their Standard Errors and 95% Confidence Intervals

Latent Variable	Indicators	Weights			Loadings		
		Estimate	SE	95% CI	Estimate	SE	95% CI
Materialism (M)	z_1	0.54	0.09	0.33–0.67	0.94	0.02	0.86–0.96
	z_2	0.22	0.10	0.06–0.43	0.86	0.04	0.80–0.93
	z_3	0.35	0.06	0.22–0.46	0.85	0.04	0.78–0.92
Stress at time point 2 (S2)	z_4	0.32	0.07	0.21–0.48	0.88	0.03	0.82–0.93
	z_5	0.37	0.08	0.18–0.51	0.92	0.02	0.87–0.96
	z_6	0.42	0.09	0.21–0.59	0.91	0.04	0.83–0.96
Stress at time point 4 (S4)	z_7	0.35	0.07	0.23–0.51	0.89	0.03	0.84–0.94
	z_8	0.50	0.08	0.33–0.66	0.95	0.02	0.91–0.98
	z_9	0.25	0.10	0.02–0.45	0.86	0.05	0.76–0.93
Depression at time point 3 (D3)	z_{10}	0.35	0.04	0.26–0.42	0.95	0.01	0.93–0.97
	z_{11}	0.39	0.04	0.30–0.44	0.95	0.01	0.92–0.97
	z_{12}	0.32	0.04	0.24–0.43	0.95	0.01	0.92–0.97
Depression at time point 5 (D5)	z_{13}	0.33	0.04	0.25–0.41	0.93	0.02	0.89–0.96
	z_{11}	0.37	0.04	0.28–0.44	0.95	0.01	0.93–0.97
	z_{15}	0.36	0.04	0.28–0.43	0.95	0.01	0.93–0.97
Anxiety at time point 3 (A3)	z_{16}	0.37	0.03	0.31–0.45	0.88	0.02	0.82–0.92
	z_{17}	0.40	0.05	0.30–0.49	0.92	0.02	0.88–0.96
	z_{18}	0.34	0.05	0.24–0.42	0.92	0.02	0.87–0.95
Anxiety at time point 5 (A5)	z_{19}	0.38	0.03	0.33–0.45	0.91	0.02	0.87–0.95
	z_{20}	0.35	0.04	0.27–0.44	0.93	0.02	0.89–0.96
	z_{21}	0.36	0.04	0.26–0.44	0.91	0.03	0.86–0.95

SE, standard errors; CI, confidence intervals.

Figure 6.6 displays the simple slope estimates of anxiety at time point 3 (A_3) on stress at time point 2 (S_2) as a function of materialism, along with its confidence bands. The simple slopes were given by $b_2 + b_3M$, based on the following equation.

$$A_3 = b_1M + b_2S_2 + b_3MS_2 + \zeta_3$$

$$= b_1M + (b_2 + b_3M)S_2 + \zeta_3. \tag{6.18}$$

In the figure, the horizontal dotted line denotes the zero line, whereas the vertical straight line is the boundary of a region of significance. We can see that the simple slope estimates are statistically significantly different from zero for any value of materialism greater than −0.7. This indicates that

TABLE 6.2

Estimates of Path Coefficients along with Their Standard Errors
and 95% Confidence Intervals

	Estimate	SE	95% CI
$M \rightarrow D_3$	0.03	0.10	−0.13–0.25
$M \rightarrow D_5$	0.13	0.07	−0.01–0.26
$M \rightarrow A_3$	0.09	0.08	−0.11–0.24
$M \rightarrow A_5$	0.22	0.05	0.10–0.32
$S_2 \rightarrow S_4$	0.80	0.04	0.70–0.88
$S_2 \rightarrow D_3$	0.43	0.09	0.23–0.59
$S_2 \rightarrow A_3$	0.39	0.09	0.22–0.57
$S_4 \rightarrow D_5$	0.39	0.11	0.17–0.60
$S_4 \rightarrow A_5$	0.31	0.07	0.19–0.48
$D_3 \rightarrow D_5$	0.41	0.12	0.20–0.60
$A_3 \rightarrow A_5$	0.45	0.09	0.29–0.61
$MS_2 \rightarrow D_3$	0.07	0.13	−0.15–0.35
$MS_2 \rightarrow A_3$	0.19	0.08	0.05–0.38
$MS_2 \rightarrow MS_4$	0.78	0.08	0.65–0.92
$MS_4 \rightarrow D_5$	0.18	0.07	0.04–0.31
$MS_4 \rightarrow A_5$	0.27	0.06	0.16–0.39

SE, standard errors; CI, confidence intervals.

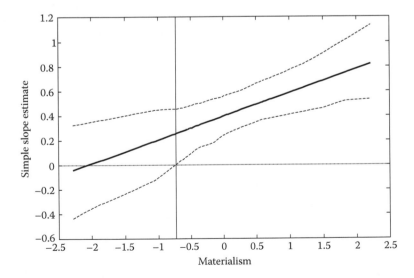

FIGURE 6.6
A plot of the simple slope estimates of anxiety at time point 3 on stress at time point 2 as a
function of materialism, along with 95% confidence bands.

materialism moderates the relationship between stress at time point 2 and anxiety at time point 3 solely over a certain range of its values.

6.5 Summary

We discussed Hwang et al.'s (2010) extension of generalized structured component analysis to examine various types of latent interaction effects. In their approach, a latent interaction term is obtained as the product of interacting latent variables, whose scores are uniquely determined as the weighted composites of indicators. Consequently, the approach does not need to create additional indicators for latent interaction terms. Moreover, it can easily deal with exogenous and endogenous, two- or higher-way latent interaction terms. Importantly, all model specification and estimation are carried out in a unified manner. That is, a single model is specified to take into account all latent interaction effects and a single optimization criterion is minimized to estimate parameters by means of an alternating least-squares algorithm. For illustrative purposes, we applied the approach to investigate the effects of materialism on the relationships between stress and depressive/anxious symptoms over time.

Partial least squares path modeling has adopted several approaches for testing a latent interaction effect. Although they are different in implementation, the indicator-based and two-stage approaches involve separate steps that are carried out independently of each other. The hybrid method is perhaps most comparable to Hwang et al.'s (2010) approach. However, it appears necessary to provide more theoretical or algorithmic justifications as to why and how this method works.

Appendix 6.1 The Alternating Least-Squares Algorithm for Generalized Structured Component Analysis with Latent Interactions

Let $\dot{Z} = \left[Z, \Gamma^{**} \right]$, where Z is an N by J matrix of indicators and Γ^{**} is an N by R matrix of latent interaction terms. Let $\dot{V} = [I^*, \dot{W}]$. We can re-write Equation 6.12 as

$$\phi = SS\left(\dot{Z}\dot{V} - \dot{Z}\dot{W}\dot{A} \right)$$

$$= SS\left(\Psi - \Gamma\dot{A} \right), \tag{A6.1}$$

where $\Psi = \dot{Z}\dot{V}$ and $\Gamma = \dot{Z}\dot{W}$.

An alternating least-squares algorithm was developed to minimize this criterion. This algorithm consists of the following two main steps.

Step 1: We update $\dot{A}$ for fixed $\dot{W}$. Minimizing Equation A6.1 with respect to $\dot{A}$ is equivalent to minimizing

$$\phi = SS(\Psi - \Gamma\dot{A})$$

$$= SS(vec(\Psi) - (I \otimes \Gamma)vec(\dot{A})). \qquad (A6.2)$$

Let **a** denote the vector formed by eliminating any zero elements from $vec(\dot{A})$. Let Φ denote the matrix of the columns of $I \otimes \Gamma$ corresponding to the zero elements in $vec(\dot{A})$. Then, the least-squares estimate of **a** is obtained by

$$\hat{a} = (\Phi'\Phi)^{-1}\Phi'vec(\Psi). \qquad (A6.3)$$

The updated $\dot{A}$ is reconstructed from $\hat{a}$.

Step 2: We update $\dot{W}$ for fixed $\dot{A}$. In Equation A6.1, Γ^{**} shares some columns of $\dot{W}$ because latent interaction terms are defined as the products of latent variables. Let w_p denote the pth column in $\dot{W}$, which is shared by the tth column in $\dot{V}$ and also by the rth column in Γ^{**}, where $t = J + p$ ($p = 1,...,P$; $r = 1,..., R$). Let **G** denote Γ whose columns involving w_p are the vectors of zeros. Let **H** denote Γ whose columns containing w_p are the vectors of zeros. Let m_1 and m_2 denote 1 by $J + P + R$ vectors whose elements are all zeros except the tth and kth elements being unity, respectively, where $k = J + P + r$. Let a_1 denote the pth row of $\dot{A}$. Let a_2 denote the sth row of $\dot{A}$, where $s = P + r$. Let **D** denote an N by N diagonal matrix consisting of the individual scores of a latent variable or a product of latent variables, which forms a latent interaction term but does not involve w_p. For instance, if $\gamma_{12} = \gamma_1 \circ \gamma_2 = Zw_1 \circ Zw_2$ and $p = 1$, then $D = diag(\gamma_2)$, where o indicates the elementwise multiplication of two vectors or the Hadamard product.

To update w_p, Equation A6.1 can generally be re-expressed as

$$\phi = SS(\Psi - \Gamma\dot{A})$$

$$= SS(vec(Zw_p m_1 + DZw_p m_2 + G) - vec(Zw_p a_1 + DZw_p a_2 + H))$$

$$= SS((m_1' \otimes Z)w_p + (m_2' \otimes DZ)w_p - (a_1' \otimes Z)w_p - (a_2' \otimes DZ)w_p - vec(H - G))$$

$$= SS(\Omega w_p - vec(\Pi)), \qquad (A6.4)$$

where $\Omega = (\mathbf{m}_1' \otimes \mathbf{Z}) + (\mathbf{m}_2' \otimes \mathbf{DZ}) - (\mathbf{a}_1' \otimes \mathbf{Z}) - (\mathbf{a}_2' \otimes \mathbf{DZ})$, and $\Pi = \mathbf{H} - \mathbf{G}$. Let $\boldsymbol{\theta}_p$ denote the vector formed by eliminating any zero elements from $\mathbf{w}_p$. Let Ξ denote the matrix formed by eliminating the columns of Ω corresponding to the zero elements in $\mathbf{w}_p$. Then, the least-squares estimate of $\boldsymbol{\theta}_p$ is obtained by

$$\hat{\boldsymbol{\theta}}_p = (\Xi'\Xi)^{-1}\Xi'\text{vec}(\Pi) \tag{A6.5}$$

The updated $\mathbf{w}_p$ is recovered from $\hat{\boldsymbol{\theta}}_p$, and subsequently multiplied by

$$\frac{\sqrt{N}}{\mathbf{w}_p'\mathbf{Z}'\mathbf{Zw}_p}$$

in order to satisfy the standardization constraint imposed on the pth latent variable.

7

Multilevel Generalized Structured Component Analysis

In practice, data are often hierarchically structured such that their individual-level cases are grouped within higher-level units. For example, adolescents' substance use can be measured across different urban areas nested within different provinces. The standardized test scores of students can be collected from different schools. In functional neuroimaging data, neuronal activities of participants, who come from different experimental groups, are recorded repeatedly over time in voxels. For such hierarchical data, generalized structured component analysis estimates parameters by aggregating the data across all individuals, under the assumption that they are independent. This implies that each parameter does not depend on higher-level units, ignoring any nested structure inherent to the data. However, the individual-level measures nested within the same higher-level unit are likely to be more similar than those in different units, thus leading to dependency among individual-level observations within the same unit. If parameters are estimated under the independence assumption, ignoring such potential dependency, this is likely to lead to inference errors (cf. Bryk and Raudenbush 1992; Snijders and Bosker 1999).

In addition to such a nested nature of the scores of observed variables or indicators, the scores of latent variables may also be hierarchically structured in the context of structural equation modeling. For example, in the National Longitudinal Survey of Youth data, antisocial behavior of children can be viewed as a latent variable, which is often obtained from mothers' responses to six items from the Behavior Problems Index (e.g., Curran 1998). The antisocial behavior of each child can be measured repeatedly across multiple time points. In the American customer satisfaction index (ACSI) model (Fornell et al. 1996), customer satisfaction is regarded as a focal latent variable, whose scores are measured for about 200 companies nested within different sectors of the US economy. Accordingly, it is desirable to take into account the nested structure of the individual scores of both latent variables and indicators in structural equation modeling.

Hierarchical linear models or multilevel models have been widely used to deal with hierarchical data (Bock 1989; Hox 1995; Bryk and Raudenbush 1992; Goldstein 1987). These models are typically designed to capture the nested structure of indicators in the context of regression analysis. Thus, they are not flexible enough to accommodate nested structures of both indicators and latent variables. Raudenbush and Bryk (2002) discussed some extensions of hierarchical linear models that involve latent variables. Nonetheless, they mainly addressed the issues of missing data and measurement error, where regression coefficients were regarded as latent variables. This approach is essentially a simple translation of hierarchical linear models into the framework of factor-based structural equation modeling (e.g., Bauer 2003; Curran 2003; Rovine and Molenaar 2000).

In this chapter, we discuss an extension of generalized structured component analysis, called *multilevel generalized structured component analysis*, to deal with nested structures of both indicators and latent variables. In multilevel generalized structured component analysis, both loadings and path coefficients are permitted to vary across higher-level units in order to address the nested structure of indicators and latent variables. As such, it enables us to take into account a nested structure in the measurement model between indicators and latent variables as well as in the structural model between latent variables. We provide a technical account and an empirical illustration of multilevel generalized structured component analysis.

7.1 Model Specification

In this section, for simplicity, we focus only on two-level generalized structured component analysis, where level-1 units (e.g., students) are nested within level-2 units (e.g., schools). We refer to level-1 units as individuals and level-2 units as groups, following Snijders and Bosker (1999, p. 38). An extension to three-level generalized structured component analysis is presented in Appendix 7.2.

We begin by presenting the standard generalized structured component analysis model, in which parameters are assumed not to vary across groups. Let N_g and G denote the number of individuals in each group, and the number of groups, respectively. Let $\mathbf{z}_{ig}$ and γ_{ig} denote J by 1 and P by 1 vectors of indicators and latent variables, respectively, for the ith individual belonging to the gth group ($i = 1,\ldots,N_g$; $g = 1,\ldots,G$). As discussed in Chapter 2, the measurement model can be given in matrix notation, as follows.

$$\mathbf{z}_{ig} = \mathbf{C}'\gamma_{ig} + \varepsilon_{ig}, \tag{7.1}$$

where ε_{ig} denotes a J by 1 vector of residuals for $\mathbf{z}_{ig}$. The structural model is given as

$$\gamma_{ig} = \mathbf{B}'\gamma_{ig} + \zeta_{ig}, \tag{7.2}$$

where ζ_{ig} denotes a J by 1 vector of residuals for γ_{ig}. The weighted relation model is given as

$$\gamma_{ig} = \mathbf{W}'\mathbf{z}_{ig}. \tag{7.3}$$

In Equations 7.1 and 7.2, loadings and path coefficients do not depend on the group, denoted by g. This is difficult to justify when data involve meaningful nested structures, unless each group consists of only one observation (Snijders and Bosker 1999, p. 40).

We now change Equations 7.1 and 7.2 to have loadings and path coefficients vary from group to group. The measurement model can be changed to

$$\mathbf{z}_{ig} = \mathbf{C}_g'\gamma_{ig} + \varepsilon_{ig}, \tag{7.4}$$

where $\mathbf{C}_g$ is a P by J matrix comprising group-dependent loadings (i.e., c_{jg}). This group-dependent loading matrix can be expressed as

$$\mathbf{C}_g' = \mathbf{L}' + \mathbf{L}_g', \tag{7.5}$$

where $\mathbf{L}$ is a P by J matrix consisting of average loadings, and $\mathbf{L}_g$ is a P by J matrix consisting of group-dependent deviations.

Likewise, the structural model can be changed to

$$\gamma_{ig} = \mathbf{B}_g'\gamma_{ig} + \zeta_{ig}, \tag{7.6}$$

where $\mathbf{B}_g$ is a P by P matrix consisting of group-dependent path coefficients. The $\mathbf{B}_g$ matrix can be split into two elements as follows.

$$\mathbf{B}_g' = \mathbf{Q}' + \mathbf{Q}_g', \tag{7.7}$$

where $\mathbf{Q}$ is a P by P matrix consisting of average path coefficients, and $\mathbf{Q}_g$ is a P by P matrix consisting of group-dependent deviations. The weighted relation model remains unchanged, because the scores of latent variables already depend on the group, as shown in Equation 7.3.

We use a prototype model to illustrate the above model specification. Figure 7.1 displays the prototype model, in which each of two latent variables underlies two indicators and one latent variable affects the other latent variable. Moreover, all loadings and path coefficients are specified to vary from group to group, which is denoted by the index g.

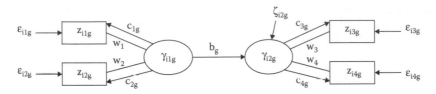

FIGURE 7.1
A prototype model for two-level generalized structured component analysis.

The measurement model for the prototype model is given as

$$z_{i1g} = \gamma_{i1g}c_{1g} + \varepsilon_{i1g}$$

$$z_{i2g} = \gamma_{i1g}c_{2g} + \varepsilon_{i2g}$$

$$z_{i3g} = \gamma_{i2g}c_{3g} + \varepsilon_{i3g}$$

$$z_{i4g} = \gamma_{i2g}c_{4g} + \varepsilon_{i4g}. \tag{7.8}$$

These equations can be expressed in matrix notation, as follows.

$$\begin{bmatrix} z_{i1g} \\ z_{i2g} \\ z_{i3g} \\ z_{i4g} \end{bmatrix} = \begin{bmatrix} c_{1g} & 0 \\ c_{2g} & 0 \\ 0 & c_{3g} \\ 0 & c_{4g} \end{bmatrix} \begin{bmatrix} \gamma_{i1g} \\ \gamma_{i2g} \end{bmatrix} + \begin{bmatrix} \varepsilon_{i1g} \\ \varepsilon_{i2g} \\ \varepsilon_{i3g} \\ \varepsilon_{i4g} \end{bmatrix}$$

$$\mathbf{z}_{ig} = \mathbf{C}_g{}'\boldsymbol{\gamma}_{ig} + \boldsymbol{\varepsilon}_{ig}, \tag{7.9}$$

Each group-dependent loading (c_{jg}) in Equation 7.8 or Equation 7.9 is the sum of an average loading, denoted by l_j, and a group-dependent deviation, denoted by l_{jg} ($j = 1, 2, 3, 4$). Thus, Equation 7.8 can be rewritten as

$$z_{i1g} = \gamma_{i1g}\left(l_1 + l_{1g}\right) + \varepsilon_{i1g}$$

$$z_{i2g} = \gamma_{i1g}\left(l_2 + l_{2g}\right) + \varepsilon_{i2g}$$

$$z_{i3g} = \gamma_{i2g}\left(l_3 + l_{3g}\right) + \varepsilon_{i3g}$$

$$z_{i4g} = \gamma_{i2g}\left(l_4 + l_{4g}\right) + \varepsilon_{i4g}. \tag{7.10}$$

Equivalently, in matrix notation,

$$\begin{bmatrix} z_{i1g} \\ z_{i2g} \\ z_{i3g} \\ z_{i4g} \end{bmatrix} = \left(\begin{bmatrix} l_1 & 0 \\ l_2 & 0 \\ 0 & l_3 \\ 0 & l_4 \end{bmatrix} + \begin{bmatrix} l_{1g} & 0 \\ l_{2g} & 0 \\ 0 & l_{3g} \\ 0 & l_{4g} \end{bmatrix} \right) \begin{bmatrix} \gamma_{i1g} \\ \gamma_{i2g} \end{bmatrix} + \begin{bmatrix} \varepsilon_{i1g} \\ \varepsilon_{i2g} \\ \varepsilon_{i3g} \\ \varepsilon_{i4g} \end{bmatrix}$$

$$\mathbf{z}_{ig} = \left[\mathbf{L}' + \mathbf{L}_g' \right] \gamma_{ig} + \varepsilon_{ig}, \tag{7.11}$$

where

$$\mathbf{L}' = \begin{bmatrix} l_1 & 0 \\ l_2 & 0 \\ 0 & l_3 \\ 0 & l_4 \end{bmatrix} \text{ and } \mathbf{L}_g' = \begin{bmatrix} l_{1g} & 0 \\ l_{2g} & 0 \\ 0 & l_{3g} \\ 0 & l_{4g} \end{bmatrix}.$$

The structural model for the prototype model is given as

$$\gamma_{i2g} = \gamma_{i1g} b_g + \zeta_{i2}$$

$$= \gamma_{i1g}(q + q_g) + \zeta_{i2}, \tag{7.12}$$

where b_g is the group-dependent path coefficient, which is equivalent to the sum of an average path coefficient (q) and a group-dependent deviation (q_g). We can express Equation 7.12 in matrix notation, as follows.

$$\begin{bmatrix} \gamma_{i1g} \\ \gamma_{i2g} \end{bmatrix} = \begin{bmatrix} 0 & 0 \\ b & 0 \end{bmatrix} \begin{bmatrix} \gamma_{i1g} \\ \gamma_{i2g} \end{bmatrix} + \begin{bmatrix} \gamma_{i1g} \\ \zeta_{i2g} \end{bmatrix}$$

$$= \left(\begin{bmatrix} 0 & 0 \\ q & 0 \end{bmatrix} + \begin{bmatrix} 0 & 0 \\ q_g & 0 \end{bmatrix} \right) \begin{bmatrix} \gamma_{i1g} \\ \gamma_{i2g} \end{bmatrix} + \begin{bmatrix} \gamma_{i1g} \\ \zeta_{i2g} \end{bmatrix}$$

$$\gamma_{ig} = \mathbf{B}_g' \gamma_{ig} + \zeta_{ig}$$

$$= \left[\mathbf{Q}' + \mathbf{Q}_g' \right] \gamma_{ig} + \zeta_{ig}, \tag{7.13}$$

where

$$\mathbf{Q}' = \begin{bmatrix} 0 & 0 \\ q & 0 \end{bmatrix} \text{ and } \mathbf{Q}_g' = \begin{bmatrix} 0 & 0 \\ q_g & 0 \end{bmatrix}.$$

The weighted relation model for the prototype model is given as:

$$\gamma_{i1g} = z_{i1g} w_1 + z_{i2g} w_2$$

$$\gamma_{i2g} = z_{i3g} w_3 + z_{i4g} w_4. \tag{7.14}$$

Equivalently, in matrix notation,

$$\begin{bmatrix} \gamma_{i1g} \\ \gamma_{i2g} \end{bmatrix} = \begin{bmatrix} w_1 & w_2 & 0 & 0 \\ 0 & 0 & w_3 & w_4 \end{bmatrix} \begin{bmatrix} z_{i1g} \\ z_{i2g} \\ z_{i3g} \\ z_{i4g} \end{bmatrix}$$

$$\gamma_{ig} = \mathbf{W}'\mathbf{z}_{ig}. \tag{7.15}$$

In multilevel generalized structured component analysis, the three sub-models are combined into a single model, as follows.

$$\begin{bmatrix} \mathbf{z}_{ig} \\ \gamma_{ig} \end{bmatrix} = \begin{bmatrix} \mathbf{C}_g' \\ \mathbf{B}_g' \end{bmatrix} \gamma_{ig} + \begin{bmatrix} \boldsymbol{\varepsilon}_{ig} \\ \boldsymbol{\zeta}_{ig} \end{bmatrix}$$

$$\begin{bmatrix} \mathbf{z}_{ig} \\ \mathbf{W}'\mathbf{z}_{ig} \end{bmatrix} = \begin{bmatrix} \mathbf{L}' + \mathbf{L}_g' \\ \mathbf{Q}' + \mathbf{Q}_g' \end{bmatrix} \mathbf{W}'\mathbf{z}_{ig} + \begin{bmatrix} \boldsymbol{\varepsilon}_{ig} \\ \boldsymbol{\zeta}_{ig} \end{bmatrix}$$

$$\begin{bmatrix} \mathbf{I} \\ \mathbf{W}' \end{bmatrix} \mathbf{z}_{ig} = \begin{bmatrix} \mathbf{L}' + \mathbf{L}_g' \\ \mathbf{Q}' + \mathbf{Q}_g' \end{bmatrix} \mathbf{W}'\mathbf{z}_{ig} + \begin{bmatrix} \boldsymbol{\varepsilon}_{ig} \\ \boldsymbol{\zeta}_{ig} \end{bmatrix}$$

$$\mathbf{V}'\mathbf{z}_{ig} = \mathbf{A}_g'\mathbf{W}'\mathbf{z}_{ig} + \mathbf{e}_{ig}, \tag{7.16}$$

where

$$\mathbf{A}_g' = \begin{bmatrix} \mathbf{C}_g' \\ \mathbf{B}_g' \end{bmatrix} = \begin{bmatrix} \mathbf{L}' + \mathbf{L}_g' \\ \mathbf{Q}' + \mathbf{Q}_g' \end{bmatrix}, \text{ and } \mathbf{e}_{ig} = \begin{bmatrix} \boldsymbol{\varepsilon}_{ig} \\ \boldsymbol{\zeta}_{ig} \end{bmatrix}.$$

This is the general form of the two-level generalized structured component analysis model.

In multilevel generalized structured component analysis, the group-dependent deviations of loadings and path coefficients in $\mathbf{L}_g$ and $\mathbf{Q}_g$ are treated as fixed parameters, rather than random variables that are typically assumed to be independently and identically distributed. This way of modeling the group-dependent deviations is comparable to the fixed-effects analysis of variance (ANOVA) model. In theory, this modeling seems restrictive because it regards the group units as unique entities with their own distinctive interpretations, such as gender, religious, or ethnic groups (Snijders and Bosker 1999, p. 42). Nonetheless, it is well suited to the distributional-free optimization procedure of generalized structured component analysis because it does not require any distributional assumptions on the group-dependent deviations. In practice, when the number of groups is small (e.g., $G < 10$), this approach can be a sensible choice over the random-effects modeling approach. Moreover, the number of individuals per group is large (e.g., $N_g \geq 100$), it can be used because it estimates the group-dependent deviations as precisely as the random-effects counterpart (Snijders and Bosker 1999, p. 44).

7.2 Parameter Estimation

Let $\mathbf{Z}_g$ denote an N_g by J matrix of indicators in the gth group. To estimate parameters, we seek to minimize the following least-squares criterion.

$$\phi = \sum_{g=1}^{G} \sum_{i=1}^{N_g} \text{SS}(\mathbf{V}'\mathbf{z}_{ig} - \mathbf{A}_g'\mathbf{W}'\mathbf{z}_{ig})$$

$$= \sum_{g=1}^{G} \text{SS}(\mathbf{Z}_g\mathbf{V} - \mathbf{Z}_g\mathbf{W}\mathbf{A}_g)$$

$$= \sum_{g=1}^{G} \text{SS}(\mathbf{Z}_g\mathbf{V} - \mathbf{Z}_g\mathbf{W}[\mathbf{C}_g, \mathbf{B}_g])$$

$$= \sum_{g=1}^{G} \text{SS}(\mathbf{Z}_g\mathbf{V} - \mathbf{Z}_g\mathbf{W}[\mathbf{L} + \mathbf{L}_g, \mathbf{Q} + \mathbf{Q}_g]) \tag{7.17}$$

with respect to $\mathbf{W}$, $\mathbf{L}$, $\mathbf{Q}$, $\mathbf{L}_g$, and $\mathbf{Q}_g$, subject to the constraints that the sum of the group-dependent deviations for each loading over groups is equal to zero, that is,

$$\sum_{g} l_{jg} = 0,$$

and the sum of the group-dependent deviations for each path coefficient over groups is equal to zero, that is,

$$\sum_{g} q_{kg} = 0$$

($k = 1, ..., K$, where K is the number of path coefficients). The constraints are imposed in the fixed-effect ANOVA model. In addition, the usual standardization constraint is imposed on each latent variable.

An alternating least-squares algorithm can be developed to minimize Equation 7.17. This algorithm can be viewed as a simple extension of the algorithm for generalized structured component analysis. We provide a description of the algorithm in Appendix 7.1.

As stated earlier, it is desirable to apply multilevel generalized structured component analysis when the number of individuals per group (N_g) is large. When N_g is much greater than the number of indicators, the alternating least-squares algorithm can be made more efficient computationally by adopting a procedure similar to that discussed in Appendix 2.1. That is, let

$\mathbf{Z}_g'\mathbf{Z}_g = \mathbf{R}_g\mathbf{R}_g'$ denote any square root decomposition of $\mathbf{Z}_g'\mathbf{Z}_g$. For example, $\mathbf{R}_g$ can be obtained by the Cholesky factorization or the QR decomposition. Then, minimizing Equation 7.17 is equivalent to minimizing

$$\phi = \sum_{g=1}^{G} SS(\mathbf{R}_g'(\mathbf{V} - \mathbf{WA}_g)). \tag{7.18}$$

Minimizing this criterion is computationally advantageous because the maximum size of $\mathbf{R}_g$ is equal to the number of indicators, which is much smaller than N_g. In addition, it is straightforward to extend the algorithm to estimate the parameters of a higher-level model than a two-level one. For example, as shown in Appendix 7.2, the three-level generalized structured component analysis model involves one more matrix of parameters, compared to the two-level model. The additional matrix of parameters can be easily updated by adding another step similar to step 2 or 3 in Appendix 7.1.

Multilevel generalized structured component analysis provides the value of FIT for a fitted model. The FIT is given by

$$FIT = 1 - \frac{\sum_{g=1}^{G} SS(\mathbf{Z}_g\mathbf{V} - \mathbf{Z}_g\mathbf{WA}_g)}{\sum_{g=1}^{G} SS(\mathbf{Z}_g\mathbf{V})}. \tag{7.19}$$

The value of AFIT can then be calculated based on the value of FIT. In addition, we can calculate the values of FIT_M and FIT_S, as follows.

$$FIT_M = 1 - \frac{\sum_{g=1}^{G} SS(\mathbf{Z}_g - \mathbf{Z}_g\mathbf{WC}_g)}{\sum_{g=1}^{G} SS(\mathbf{Z}_g)} \tag{7.20}$$

$$FIT_S = 1 - \frac{\sum_{g=1}^{G} SS(\mathbf{Z}_g\mathbf{W} - \mathbf{Z}_g\mathbf{WB}_g)}{\sum_{g=1}^{G} SS(\mathbf{Z}_g\mathbf{W})}. \tag{7.21}$$

As in generalized structured component analysis, multilevel generalized structured component analysis utilizes the bootstrap method in order to estimate the standard errors or confidence intervals of parameter estimates.

7.3 Example: The ACSI Data

The present example consists of customer-level measures of the 14 ACSI indicators for 13 American financial services companies, including banks and insurance companies. Thus, in this example, individual customers can be considered to be nested within different companies. The total number of customers, who have used services offered by one of the 13 companies, reached 3096. The number of customers within each company ranged from 100 to 400. The average number of customers was equal to 238.

A two-level model was specified for the data, in which loadings and the path coefficients were assumed to vary across companies. The specified model is displayed in Figure 7.2. For simplicity, the index i for each individual customer was omitted, and all the weights for the 14 indicators were not displayed in the figure. In the model, six latent variables were assumed to underlie 14 indicators in each group. The weight and loading for z_{12g} were fixed to one, because

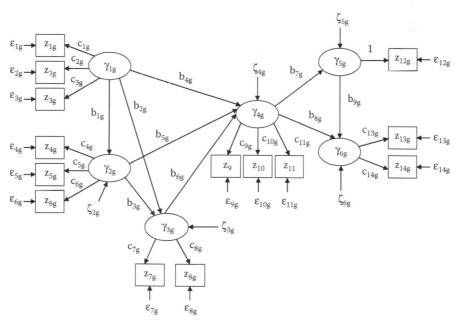

FIGURE 7.2
A two-level American customer satisfaction index model.

z_{12g} was the only indicator for the corresponding latent variable *customer complaints*, and its loading would remain the same as one across groups.

We applied multilevel generalized structured component analysis to fit the model to the data. The model provided that FIT = 0.63 (SE = 0.00, 95% CI = 0.62–0.64) and AFIT = 0.62 (SE = 0.00, 95% CI = 0.62–0.63). It also showed that FIT_M = 0.73 (SE = 0.00, 95% CI = 0.72–0.74) and FIT_S = 0.39 (SE = 0.01, 95% CI = 0.37–0.40). The number of bootstrap samples was 100.

Table 7.1 presents the estimates of group-invariant weights. All weight estimates turned out to be statistically significant. This suggests that all indicators contributed equally well to determining their latent variables.

Table 7.2 provides the estimates of average loadings. All these average loading estimates appeared large and were statistically significant. Table 7.2 also shows the between-group standard deviation of the value of each group-dependent loading in C_g, which is equivalent to the sum of the average loading and group-dependent deviation estimates. This between-group standard deviation is considered to reflect inter-group variability in each loading. All the between-group standard deviations were statistically significant. This suggests that there existed a substantial company-wise variation in each loading value.

Table 7.3 provides the estimates of average path coefficients. In general, the interpretations of these estimates appeared consistent with the relationships

TABLE 7.1

Estimates of Weights and Their 95% Confidence Intervals Obtained from the ACSI Data

Latent	Indicator	Estimate	95% CI
CE	z_1	0.42	0.40–0.44
	z_2	0.49	0.47–0.51
	z_3	0.35	0.33–0.37
PQ	z_4	0.44	0.43–0.46
	z_5	0.42	0.40–0.44
	z_6	0.28	0.26–0.29
PV	z_7	0.46	0.44–0.48
	z_8	0.59	0.57–0.60
CS	z_9	0.47	0.45–0.50
	z_{10}	0.31	0.28–0.33
	z_{11}	0.31	0.30–0.33
CL	z_{13}	0.64	0.62–0.65
	z_{14}	0.42	0.40–0.43

CC, customer complaints; CE, customer expectations; CI, confidence intervals; CL, customer loyalty; CS, customer satisfaction; PQ, perceived quality; PV, perceived value.

TABLE 7.2

Estimates of Average Loadings, the between-Group Standard Deviations of Group-Dependent Loadings, and Their 95% Confidence Intervals Obtained from the ASCI Data

Latent	Indicator	Average Loading		Between-Group Standard Deviation	
		Estimate	95% CI	Estimate	95% CI
CE	z_1	0.82	0.79–0.83	0.04	0.02–0.06
	z_2	0.87	0.85–0.88	0.05	0.02–0.05
	z_3	0.66	0.63–0.69	0.05	0.03–0.07
PQ	z_4	0.92	0.91–0.93	0.03	0.02–0.04
	z_5	0.91	0.90–0.92	0.03	0.02–0.04
	z_6	0.76	0.74–0.78	0.03	0.02–0.06
PV	z_7	0.95	0.94–0.96	0.02	0.01–0.03
	z_8	0.97	0.96–0.97	0.01	0.01–0.02
CS	z_9	0.95	0.94–0.95	0.02	0.01–0.03
	z_{10}	0.90	0.89–0.91	0.03	0.02–0.04
	z_{11}	0.88	0.86–0.89	0.03	0.02–0.05
CL	z_{13}	0.97	0.96–0.97	0.02	0.01–0.02
	z_{14}	0.92	0.91–0.93	0.04	0.02–0.04

CC, customer complaints, CE, customer expectations; CI, confidence intervals; CL, customer loyalty; CS, customer satisfaction; PQ, perceived quality; PV, perceived value.

TABLE 7.3

Estimates of Average Path Coefficients, the between-Group Standard Deviations of Group-Dependent Path Coefficients, and Their 95% Confidence Intervals Obtained from the ASCI Data

	Average Path Coefficient		Between-Group Standard Deviation	
	Estimate	95% CI	Estimate	95% CI
CE → PQ (b_1)	0.49	0.44–0.53	0.08	0.04–0.10
CE → PV (b_2)	0.15	0.12–0.18	0.06	0.04–0.09
PQ → PV (b_3)	0.66	0.63–0.69	0.06	0.03–0.10
CE → CS (b_4)	0.05	0.03–0.07	0.05	0.02–0.06
PQ → CS (b_5)	0.55	0.51–0.58	0.07	0.04–0.09
PV → CS (b_6)	0.39	0.36–0.43	0.07	0.04–0.08
CS → CC (b_7)	−0.39	−0.42 to −0.36	0.06	0.05–0.10
CS → CL (b_8)	0.71	0.69–0.73	0.07	0.03–0.08
CC → CL (b_9)	−0.07	−0.10 to −0.04	0.10	0.04–0.08

CC, customer complaints; CE, customer expectations; CI, confidence intervals; CL, customer loyalty; CS, customer satisfaction; PQ, perceived quality; PV, perceived value.

among the latent variables hypothesized in the ACSI model. That is, *customer expectations* had positive and statistically significant influences on *perceived quality, perceived value,* and *customer satisfaction.* In turn, *perceived quality* showed positive and statistically significant effects on *perceived value* and *customer satisfaction.* Moreover, *perceived value* exhibited a positive and statistically significant effect on *customer satisfaction. Customer satisfaction* had a positive and statistically significant impact on *customer loyalty,* while it had a negative and statistically significant impact on *customer complaints.* Finally, *customer complaints* had a negative and statistically significant effect on *customer loyalty.*

Table 7.3 also provides the between-group standard deviation of the value of each group-dependent path coefficient in $\mathbf{B}_g$, which is equivalent to the sum of the average path coefficient and group-dependant deviation estimates. All these standard deviations were statistically significant, indicating substantial differences in each of the path coefficient estimates across companies.

As stated above, multilevel generalized structured component analysis provides the values of group-dependent loadings and path coefficients, which are the sums of their estimated average and group-dependent deviations. As an example, Table 7.4 presents the values of all group-dependent path coefficients for each company. By looking into these values, we may examine specific group differences in each path coefficient.

TABLE 7.4

Values of Group-Dependent Path Coefficients for 13 Companies and Their 95% Confidence Intervals Obtained from the ASCI Data

Company	Paths	Estimate	95% CI
1	CE → PQ (b_1)	0.33	0.32–0.65
	CE → PV (b_2)	0.02	0.01–0.26
	PQ → PV (b_3)	0.67	0.52–0.79
	CE → CS (b_4)	0.04	−0.03–0.12
	PQ → CS (b_5)	0.63	0.45–0.68
	PV → CS (b_6)	0.32	0.29–0.48
	CS → CC (b_7)	−0.48	−0.50 to −0.23
	CS → CL (b_8)	0.80	0.59–0.82
	CC → CL (b_9)	−0.07	−0.22–0.04
2	CE → PQ (b_1)	0.61	0.37–0.64
	CE → PV (b_2)	0.12	−0.01–0.26
	PQ → PV (b_3)	0.60	0.55–0.78
	CE → CS (b_4)	−0.02	−0.03–0.12
	PQ → CS (b_5)	0.64	0.44–0.66
	PV → CS (b_6)	0.34	0.27–0.49

(Continued)

TABLE 7.4 (*Continued*)

Values of Group-Dependent Path Coefficients for 13 Companies and Their
95% Confidence Intervals Obtained from the ASCI Data

Company	Paths	Estimate	95% CI
	CS → CC (b_7)	−0.30	−0.54 to −0.27
	CS → CL (b_8)	0.71	0.59–0.81
	CC → CL (b_9)	−0.11	−0.18–0.05
3	CE → PQ (b_1)	0.53	0.33–0.61
	CE → PV (b_2)	0.22	0.02–0.29
	PQ → PV (b_3)	0.69	0.53–0.79
	CE → CS (b_4)	0.07	−0.05–0.13
	PQ → CS (b_5)	0.61	0.44–0.67
	PV → CS (b_6)	0.28	0.28–0.48
	CS → CC (b_7)	−0.40	−0.51 to −0.24
	CS → CL (b_8)	0.69	0.60–0.83
	CC → CL (b_9)	−0.10	−0.15–0.04
4	CE → PQ (b_1)	0.33	0.31–0.61
	CE → PV (b_2)	0.21	0.02–0.28
	PQ → PV (b_3)	0.63	0.48–0.80
	CE → CS (b_4)	−0.04	−0.04–0.13
	PQ → CS (b_5)	0.62	0.43–0.68
	PV → CS (b_6)	0.40	0.26–0.52
	CS → CC (b_7)	−0.46	−0.51 to −0.20
	CS → CL (b_8)	0.79	0.59–0.79
	CC → CL (b_9)	−0.07	−0.16–0.01
5	CE → PQ (b_1)	0.56	0.35–0.62
	CE → PV (b_2)	0.22	0.01–0.29
	PQ → PV (b_3)	0.52	0.52–0.81
	CE → CS (b_4)	0.08	−0.03–0.15
	PQ → CS (b_5)	0.51	0.38–0.67
	PV → CS (b_6)	0.42	0.24–0.49
	CS → CC (b_7)	−0.29	−0.54 to −0.27
	CS → CL (b_8)	0.82	0.55–0.83
	CC → CL (b_9)	0.23	−0.21–0.07
6	CE → PQ (b_1)	0.46	0.32–0.62
	CE → PV (b_2)	0.10	0.02–0.30
	PQ → PV (b_3)	0.68	0.49–0.79
	CE → CS (b_4)	0.05	−0.08–0.12
	PQ → CS (b_5)	0.51	0.41–0.66
	PV → CS (b_6)	0.45	0.26–0.54
	CS → CC (b_7)	−0.35	−0.49 to −0.23
	CS → CL (b_8)	0.69	0.58–0.84
	CC → CL (b_9)	−0.13	−0.21–0.04

(*Continued*)

TABLE 7.4 (*Continued*)

Values of Group-Dependent Path Coefficients for 13 Companies and Their 95% Confidence Intervals Obtained from the ASCI Data

Company	Paths	Estimate	95% CI
7	CE → PQ (b_1)	0.53	0.27–0.70
	CE → PV (b_2)	0.09	−0.01–0.28
	PQ → PV (b_3)	0.68	0.45–0.85
	CE → CS (b_4)	0.07	−0.06–0.16
	PQ → CS (b_5)	0.51	0.38–0.73
	PV → CS (b_6)	0.35	0.21–0.54
	CS → CC (b_7)	−0.36	−0.55 to −0.25
	CS → CL (b_8)	0.57	0.56–0.83
	CC → CL (b_9)	−0.09	−0.22–0.09
8	CE → PQ (b_1)	0.49	0.35–0.60
	CE → PV (b_2)	0.17	0.04–0.28
	PQ → PV (b_3)	0.64	0.50–0.78
	CE → CS (b_4)	0.06	−0.03–0.13
	PQ → CS (b_5)	0.38	0.40–0.70
	PV → CS (b_6)	0.57	0.27–0.54
	CS → CC (b_7)	−0.39	−0.52–0.24
	CS → CL (b_8)	0.75	0.59–0.80
	CC → CL (b_9)	−0.14	−0.17–0.05
9	CE → PQ (b_1)	0.54	0.34–0.60
	CE → PV (b_2)	0.16	0.06–0.25
	PQ → PV (b_3)	0.69	0.55–0.80
	CE → CS (b_4)	0.11	−0.03–0.13
	PQ → CS (b_5)	0.54	0.44–0.68
	PV → CS (b_6)	0.35	0.26–0.50
	CS → CC (b_7)	−0.35	−0.55–0.23
	CS → CL (b_8)	0.67	0.60–0.82
	CC → CL (b_9)	−0.06	−0.21–0.08
10	CE → PQ (b_1)	0.49	0.36–0.62
	CE → PV (b_2)	0.18	0.03–0.33
	PQ → PV (b_3)	0.71	0.50–0.80
	CE → CS (b_4)	−0.03	−0.03–0.13
	PQ → CS (b_5)	0.62	0.41–0.69
	PV → CS (b_6)	0.42	0.24–0.51
	CS → CC (b_7)	−0.47	−0.51 to −0.26
	CS → CL (b_8)	0.63	0.59–0.80
	CC → CL (b_9)	−0.19	−0.20–0.04
11	CE → PQ (b_1)	0.53	0.38–0.59
	CE → PV (b_2)	0.18	0.07–0.24

(*Continued*)

TABLE 7.4 (*Continued*)

Values of Group-Dependent Path Coefficients for 13 Companies and Their 95% Confidence Intervals Obtained from the ASCI Data

Company	Paths	Estimate	95% CI
	PQ → PV (b_3)	0.64	0.56–0.77
	CE → CS (b_4)	0.08	−0.02–0.12
	PQ → CS (b_5)	0.53	0.45–0.64
	PV → CS (b_6)	0.39	0.30–0.49
	CS → CC (b_7)	−0.37	−0.51–0.27
	CS → CL (b_8)	0.74	0.63–0.78
	CC → CL (b_9)	−0.04	−0.15–0.00
12	CE → PQ (b_1)	0.45	0.37–0.59
	CE → PV (b_2)	0.17	0.06–0.24
	PQ → PV (b_3)	0.66	0.58–0.76
	CE → CS (b_4)	0.15	−0.00–0.12
	PQ → CS (b_5)	0.50	0.44–0.64
	PV → CS (b_6)	0.39	0.27–0.48
	CS → CC (b_7)	−0.40	−0.47 to −0.27
	CS → CL (b_8)	0.66	0.60–0.78
	CC → CL (b_9)	−0.07	−0.18–0.00
13	CE → PQ (b_1)	0.50	0.38–0.63
	CE → PV (b_2)	0.06	0.05–0.24
	PQ → PV (b_3)	0.76	0.57–0.76
	CE → CS (b_4)	0.03	−0.03–0.13
	PQ → CS (b_5)	0.55	0.45–0.65
	PV → CS (b_6)	0.38	0.30–0.48
	CS → CC (b_7)	−0.46	−0.51 to −0.27
	CS → CL (b_8)	0.67	0.63–0.81
	CC → CL (b_9)	−0.10	−0.18–0.01

CC, customer complaints; CE, customer expectations; CI, confidence intervals; CL, customer loyalty; CS, customer satisfaction; PQ, perceived quality; PV, perceived value.

7.4 Summary

We discussed multilevel generalized structured component analysis that takes into account a nested structure of both indicators and latent variables by permitting loadings and path coefficients to vary across higher-level units. An alternating least-squares procedure can be used for parameter estimation. This technique was applied to two-level ACSI data, where customer-level measurements were clustered within different companies.

It provided average loading estimates between indicators and latent variables, as well as average path coefficient estimates between latent variables. These average estimates were found to validate the relationships hypothesized in the ACSI model. Moreover, the technique provided the values of group-dependent loadings and path coefficients. For the ACSI example, it seemed reasonable to apply multilevel generalized structured component analysis, because the number of individuals in each group was large (≥ 100).

Appendix 7.1 The Alternating Least-Squares Algorithm for Two-Level Multilevel Generalized Structured Component Analysis

To estimate parameters, we seek to minimize the following least-squares criterion.

$$\phi = \sum_{g=1}^{G} SS(\mathbf{Z}_g \mathbf{V} - \mathbf{Z}_g \mathbf{W} \mathbf{A}_g)$$

$$= \sum_{g=1}^{G} SS(\mathbf{Z}_g \mathbf{V} - \mathbf{Z}_g \mathbf{W}[\mathbf{L} + \mathbf{L}_g, \mathbf{Q} + \mathbf{Q}_g])$$

$$= \sum_{g=1}^{G} SS\left(\mathbf{Z}_g \mathbf{V} - \mathbf{Z}_g \mathbf{W}[\mathbf{L}, \mathbf{I}, \mathbf{Q}, \mathbf{I}] \begin{bmatrix} \mathbf{I} & \mathbf{0} \\ \mathbf{L}_g & \mathbf{0} \\ \mathbf{0} & \mathbf{I} \\ \mathbf{0} & \mathbf{Q}_g \end{bmatrix}\right)$$

$$= \sum_{g=1}^{G} SS(\mathbf{Z}_g \mathbf{V} - \mathbf{Z}_g \mathbf{W} \mathbf{M} \mathbf{H}_g), \qquad (A7.1)$$

with respect to $\mathbf{W}$, $\mathbf{L}$, $\mathbf{L}_g$, $\mathbf{Q}$, and $\mathbf{Q}_g$, where

$$\mathbf{M} = [\mathbf{L}, \mathbf{I}, \mathbf{Q}, \mathbf{I}], \ \mathbf{H}_g = \begin{bmatrix} \mathbf{I} & \mathbf{0} \\ \mathbf{L}_g & \mathbf{0} \\ \mathbf{0} & \mathbf{I} \\ \mathbf{0} & \mathbf{Q}_g \end{bmatrix},$$

and $\mathbf{I}$ is an identity matrix.

An alternating least-squares algorithm was developed to minimize this criterion. The algorithm repeats the following three main steps until convergence.

Step 1: Update $\mathbf{W}$ for fixed $\mathbf{L}$, $\mathbf{L}_g$, $\mathbf{Q}$, and $\mathbf{Q}_g$. This step is equivalent to minimizing

$$\phi = \sum_{g=1}^{G} \mathrm{SS}(\mathbf{Z}_g \mathbf{V} - \mathbf{Z}_g \mathbf{W} \mathbf{A}_g), \qquad (A7.2)$$

with respect to $\mathbf{W}$. This criterion is essentially equivalent to that for generalized structured component analysis. Thus, the same algorithm presented in Appendix 2.1 can be used to update $\mathbf{W}$.

Step 2: Update $\mathbf{L}$ and $\mathbf{Q}$ for fixed $\mathbf{W}$, $\mathbf{L}_g$, and $\mathbf{Q}_g$. Equation A7.1 can be rewritten as

$$\phi = \sum_{g=1}^{G} \mathrm{SS}(\mathrm{vec}(\mathbf{Z}_g \mathbf{V}) - (\mathbf{H}_g{}' \otimes \mathbf{Z}_g \mathbf{W}) \mathrm{vec}(\mathbf{M}))$$

$$= \sum_{g=1}^{G} \mathrm{SS}(\mathrm{vec}(\mathbf{Z}_g \mathbf{V}) - \tilde{\mathbf{\Omega}}_g \tilde{\mathbf{m}} - \mathbf{\Omega}_g \mathbf{m}), \qquad (A7.3)$$

where $\mathbf{\Omega}_g$ denotes the matrix formed by eliminating the columns of $\mathbf{H}_g{}' \otimes \mathbf{Z}_g \mathbf{W}$ corresponding to any zero elements in $\mathrm{vec}(\mathbf{M})$, $\mathbf{m}$ denotes a vector of free parameters in $\mathrm{vec}(\mathbf{M})$, $\tilde{\mathbf{\Omega}}_g$ denotes the matrix formed by eliminating the columns of $\mathbf{H}_g{}' \otimes \mathbf{Z}_g \mathbf{W}$ corresponding to ones in $\mathrm{vec}(\mathbf{M})$, and $\tilde{\mathbf{m}}$ denotes a vector of ones in $\mathrm{vec}(\mathbf{M})$. Then, the least-squares estimate of $\mathbf{m}$ is obtained by

$$\hat{\mathbf{m}} = \left\{ \sum_{g=1}^{G} \mathbf{\Omega}_g{}' \mathbf{\Omega}_g \right\}^{-1} \left\{ \sum_{g=1}^{G} \mathbf{\Omega}_g{}' \left(\mathrm{vec}(\mathbf{Z}_g \mathbf{V}) - \tilde{\mathbf{\Omega}}_g \tilde{\mathbf{m}} \right) \right\}. \qquad (A7.4)$$

The updated $\mathbf{M}$ is reconstructed from $\hat{\mathbf{m}}$.

Step 3: Update $\mathbf{L}_g$, and $\mathbf{Q}_g$ for fixed $\mathbf{W}$, $\mathbf{L}$, and $\mathbf{Q}$. Equation A7.1 can be rewritten as

$$\phi = \sum_{g=1}^{G} \mathrm{SS}(\mathrm{vec}(\mathbf{Z}_g \mathbf{V}) - (\mathbf{I} \otimes \mathbf{Z}_g \mathbf{W} \mathbf{M}) \mathrm{vec}(\mathbf{H}_g))$$

$$= \sum_{g=1}^{G} \mathrm{SS}(\mathrm{vec}(\mathbf{Z}_g \mathbf{V}) - \tilde{\mathbf{\Xi}}_g \tilde{\mathbf{h}}_g - \mathbf{\Xi}_g \mathbf{h}_g), \qquad (A7.5)$$

where $\boldsymbol{\Xi}_g$ denotes the matrix formed by eliminating the columns of $\mathbf{I} \otimes \mathbf{Z}_g \mathbf{WM}$ corresponding to any zero elements in $\mathrm{vec}(\mathbf{H}_g)$, $\mathbf{h}_g$ denotes a vector of free parameters in $\mathrm{vec}(\mathbf{H}_g)$, $\tilde{\boldsymbol{\Xi}}_g$ denotes the matrix formed by eliminating the columns of $\mathbf{I} \otimes \mathbf{Z}_g \mathbf{WM}$ corresponding to ones in $\mathrm{vec}(\mathbf{H}_g)$, and $\tilde{\mathbf{h}}_g$ denotes a vector of ones in $\mathrm{vec}(\mathbf{H}_g)$. Then, the least-squares estimate of $\mathbf{h}_g$ is obtained by

$$\hat{\mathbf{h}}_g = \left(\boldsymbol{\Xi}_g{}' \boldsymbol{\Xi}_g \right)^{-1} \boldsymbol{\Xi}_g{}' \left(\mathrm{vec}(\mathbf{Z}_g \mathbf{V}) - \tilde{\boldsymbol{\Xi}}_g \tilde{\mathbf{h}}_g \right). \tag{A7.6}$$

The updated $\mathbf{H}_g$ is reconstructed from $\tilde{\mathbf{h}}_g$

Appendix 7.2 The Three-Level Generalized Structured Component Analysis Model

Let $\mathbf{z}_{igs}$ denote a J by 1 vector of indicators for the ith level-1 unit in the gth level-2 unit nested within the sth level-3 unit ($i = 1,..., N_{gs}$; $g = 1,..., G_s$; $s = 1,..., S$). Let $\boldsymbol{\gamma}_{igs}$ denote a P by 1 vector of latent variables for the ith level-1 unit in the gth level-2 unit nested within the sth level-3 unit. Let $\mathbf{C}_{gs}{}'$ and $\mathbf{B}_{gs}{}'$ denote matrices of loadings and path coefficients, respectively, which depend on both level-2 and level-3 units. Let $\mathbf{L}_s{}'$ and $\mathbf{L}_{gs}{}'$ denote matrices consisting of level-2 average loadings and level-2 deviations, respectively. Let $\boldsymbol{\Lambda}'$ and $\boldsymbol{\Lambda}_s{}'$ denote matrices of level-3 average loadings and level-3 deviations, respectively. Let $\mathbf{Q}_s{}'$ and $\mathbf{Q}_{gs}{}'$ denote matrices consisting of level-2 average path coefficients and level-2 deviations, respectively. Let $\boldsymbol{\Delta}'$ and $\boldsymbol{\Delta}_s{}'$ denote matrices of level-3 average path coefficients and level-3 deviations, respectively.

The measurement model for three-level generalized structured component analysis can be written as

$$\mathbf{z}_{igs} = \mathbf{C}_{gs}{}' \boldsymbol{\gamma}_{igs} + \boldsymbol{\varepsilon}_{igs}. \tag{A7.7}$$

The $\mathbf{C}_{gs}{}'$ matrix can be split into

$$\mathbf{C}_{gs}{}' = \mathbf{L}_s{}' + \mathbf{L}_{gs}{}'. \tag{A7.8}$$

Similarly, the $\mathbf{L}_s{}'$ matrix is decomposed into

$$\mathbf{L}_s{}' = \boldsymbol{\Lambda}' + \boldsymbol{\Lambda}_s{}'. \tag{A7.9}$$

The structural model for three-level generalized structured component analysis can be written as

$$\gamma_{igs} = \mathbf{B}_{gs}{}'\gamma_{igs} + \zeta_{igs}. \tag{A7.10}$$

The $\mathbf{B}_{gs}{}'$ matrix can be expressed as

$$\mathbf{B}_{gs}{}' = \mathbf{Q}_s{}' + \mathbf{Q}_{gs}{}'. \tag{A7.11}$$

The $\mathbf{Q}_s{}'$ matrix is decomposed into

$$\mathbf{Q}_s{}' = \mathbf{\Delta}' + \mathbf{\Delta}_s{}'. \tag{A7.12}$$

As in the two-level model, the weighted relation model remains unchanged, as follows.

$$\gamma_{igs} = \mathbf{W}'\mathbf{z}_{igs}. \tag{A7.13}$$

We can combine the three submodels into a single model, as follows.

$$\begin{bmatrix} \mathbf{z}_{igs} \\ \gamma_{igs} \end{bmatrix} = \begin{bmatrix} \mathbf{C}_{gs}{}' \\ \mathbf{B}_{gs}{}' \end{bmatrix} \gamma_{igs} + \begin{bmatrix} \varepsilon_{igs} \\ \zeta_{igs} \end{bmatrix}$$

$$\begin{bmatrix} \mathbf{z}_{igs} \\ \mathbf{W}'\mathbf{z}_{igs} \end{bmatrix} = \begin{bmatrix} \mathbf{L}_s{}' + \mathbf{L}_{gs}{}' \\ \mathbf{Q}_s{}' + \mathbf{Q}_{gs}{}' \end{bmatrix} \mathbf{W}'\mathbf{z}_{igs} + \begin{bmatrix} \varepsilon_{igs} \\ \zeta_{igs} \end{bmatrix}$$

$$\begin{bmatrix} \mathbf{I} \\ \mathbf{W}' \end{bmatrix} \mathbf{z}_{igs} = \begin{bmatrix} \mathbf{\Delta}' + \mathbf{\Delta}_s{}' + \mathbf{L}_{gs}{}' \\ \mathbf{\Delta}' + \mathbf{\Delta}_s{}' + \mathbf{Q}_{gs}{}' \end{bmatrix} \mathbf{W}'\mathbf{z}_{igs} + \begin{bmatrix} \varepsilon_{igs} \\ \zeta_{igs} \end{bmatrix}$$

$$\mathbf{V}'\mathbf{z}_{igs} = \mathbf{A}'_{gs}\mathbf{W}'\mathbf{z}_{igs} + \mathbf{e}_{igs}, \tag{A7.14}$$

where

$$\mathbf{A}_{gs}{}' = \begin{bmatrix} \mathbf{C}_{gs}{}' \\ \mathbf{B}_{gs}{}' \end{bmatrix} = \begin{bmatrix} \mathbf{L}_s{}' + \mathbf{L}_{gs}{}' \\ \mathbf{Q}_s{}' + \mathbf{Q}_{gs}{}' \end{bmatrix} = \begin{bmatrix} \mathbf{\Delta}' + \mathbf{\Delta}_s{}' + \mathbf{L}_{gs}{}' \\ \mathbf{\Delta}' + \mathbf{\Delta}_s{}' + \mathbf{Q}_{gs}{}' \end{bmatrix}, \text{ and } \mathbf{e}_{igs} = \begin{bmatrix} \varepsilon_{igs} \\ \zeta_{igs} \end{bmatrix}.$$

This is the general form of the three-level multilevel generalized structured component analysis model.

Let $\mathbf{Z}_{gs}$ denote an N_{gs} by J matrix of indicators in the gth level-2 unit nested within the sth level-3 unit. To estimate parameters in Equation A7.14, we seek to minimize the following least-squares criterion.

$$\phi = \sum_{s=1}^{S}\sum_{g=1}^{G_s}\sum_{i=1}^{N_{gs}} SS(\mathbf{V}'\mathbf{z}_{igs} - \mathbf{A}_{gs}'\mathbf{W}'\mathbf{z}_{igs})$$

$$= \sum_{s=1}^{S}\sum_{g=1}^{G_s} SS(\mathbf{Z}_{gs}\mathbf{V} - \mathbf{Z}_{gs}\mathbf{W}\mathbf{A}_{gs})$$

$$= \sum_{s=1}^{S}\sum_{g=1}^{G_s} SS(\mathbf{Z}_g\mathbf{V} - \mathbf{Z}_g\mathbf{W}[\mathbf{\Lambda} + \mathbf{\Lambda}_s + \mathbf{L}_{gs}, \mathbf{\Delta} + \mathbf{\Delta}_s + \mathbf{Q}_{gs}])$$

$$= \sum_{s=1}^{S}\sum_{g=1}^{G_s} SS \left(\mathbf{Z}_{gs}\mathbf{V} - \mathbf{Z}_{gs}\mathbf{W}[\mathbf{\Lambda},\mathbf{I},\mathbf{I},\mathbf{\Delta},\mathbf{I},\mathbf{I}] \begin{bmatrix} \mathbf{I} & 0 & 0 & 0 & 0 & 0 \\ 0 & \mathbf{\Lambda}_s & 0 & 0 & 0 & 0 \\ 0 & 0 & \mathbf{I} & 0 & 0 & 0 \\ 0 & 0 & 0 & \mathbf{I} & 0 & 0 \\ 0 & 0 & 0 & 0 & \mathbf{\Delta}_s & 0 \\ 0 & 0 & 0 & 0 & 0 & \mathbf{I} \end{bmatrix} \times \begin{bmatrix} \mathbf{I} & 0 \\ \mathbf{I} & 0 \\ \mathbf{L}_{gs} & 0 \\ 0 & \mathbf{I} \\ 0 & \mathbf{I} \\ 0 & \mathbf{Q}_{gs} \end{bmatrix} \right)$$

$$= \sum_{s=1}^{S}\sum_{g=1}^{G_s} SS(\mathbf{Z}_{gs}\mathbf{V} - \mathbf{Z}_{gs}\mathbf{W}\mathbf{M}\mathbf{P}_s\mathbf{H}_{gs}), \tag{A7.15}$$

where

$$\mathbf{M} = [\mathbf{\Lambda}, \mathbf{I}, \mathbf{I}, \mathbf{\Delta}, \mathbf{I}, \mathbf{I}], \mathbf{P}_s = \begin{bmatrix} \mathbf{I} & 0 & 0 & 0 & 0 & 0 \\ 0 & \mathbf{\Lambda}_s & 0 & 0 & 0 & 0 \\ 0 & 0 & \mathbf{I} & 0 & 0 & 0 \\ 0 & 0 & 0 & \mathbf{I} & 0 & 0 \\ 0 & 0 & 0 & 0 & \mathbf{\Delta}_s & 0 \\ 0 & 0 & 0 & 0 & 0 & \mathbf{I} \end{bmatrix}, \text{ and } \mathbf{H}_{gs} = \begin{bmatrix} \mathbf{I} & 0 \\ \mathbf{I} & 0 \\ \mathbf{L}_{gs} & 0 \\ 0 & \mathbf{I} \\ 0 & \mathbf{I} \\ 0 & \mathbf{Q}_{gs} \end{bmatrix}.$$

8

Regularized Generalized Structured Component Analysis

Multicollinearity, also termed collinearity or ill-conditioning, generally refers to a data problem that two or more *predictor* variables are highly correlated or linearly dependent; one of them can be nearly a linear combination of other predictor variables (cf. Belsley, Kuh, and Welsch 1980, Chapter 3). In multiple regression, this problem represents high correlations or dependency between predictor variables. From a practical standpoint, multicollinearity makes it difficult to interpret the unique influence of a given predictor variable on the dependent variable because there exists redundant information between the predictor variable and other predictor variables (Stevens 2009, p. 74). However, a direct consequence of multicollinearity is rather computational. That is, multicollinearity renders the cross-products matrix of predictor variables (nearly) singular, so that the cross products matrix of predictor variables cannot be inverted (i.e., the matrix inversion problem) or the calculation of the inverted matrix becomes less accurate. This in turn affects least-squares estimation, which requires the calculation of the inversed cross-products matrix. Another consequence is that the variance of a parameter estimate tends to be large. An inflated variance is likely to lead to inferential error, for example, by underestimating the t statistic of its parameter estimate. It can also cause the mean square error of the estimate to be high, thereby leading the estimate to be deviated farther from the true parameter on average.

Various heuristics or procedures have been used to detect potential multicollinearity in multiple regression. For example, if the signs of regression coefficient estimates are different from those expected, one may be suspicious of the presence of multicollinearity. Moreover, the correlation matrix of predictor variables can be used to inspect whether any pairwise correlation is high in absolute value (say, 0.9 or above). Another procedure is to examine tolerance, which is 1 minus the squared multiple correlation of a predictor variable regressed on all the other predictor variables. An index related closely to tolerance is the variance inflation factor (VIF) (Chatterjee and Price 1977), which is calculated as 1/tolerance. Although there exists no rule of thumb for their values, a VIF value greater than 10, or equivalently, a tolerance value less than 0.1 often serves as evidence to raise some concern (Myers 1990, p. 369). However, Belsley, Kuh, and Welsch (1980, Chapter 3) criticized the use of these traditional heuristics in that they lacked theoretical justification (i.e., unexpected signs of regression coefficient estimates) or were unable to diagnose whether more than two predictor variables are collinear. Instead, they suggested examining two diagnostic indexes at the same time. One is the condition index that is equal to the ratio of the largest singular value to each successive singular value, whereas the other is the variance-decomposition proportion associated with each predictor variable for each dimension. If for a given dimension, the condition index is greater than 30 and the variance-decomposition proportions of at least two variables are greater than 0.5, this may point to multicollinearity, although these criteria should be considered situation-dependent.

The presence of multicollinearity does not influence prediction. In other words, it does not affect the calculation of the predicted scores of the dependent variable as a weighted composite of predictor variables (Belsley, Kuh, and Welsch 1980, p. 116). Thus, as long as the objective of the study centers solely on prediction, we can disregard multicollinearity. On the other hand, we need to address multicollinearity, when we are interested in interpreting individual regression coefficient estimates. There are several ways for dealing with multicollinearity. One is to eliminate one of the redundant variables or delete the variable with the highest variance-decomposition proportion (Tabachnick and Fidell 2012, p. 91). However, we can lose potentially valuable information by dropping variables. Moreover, omitted variables may lead to biased estimates of the remaining variables (e.g., Clarke 2005). Another option is to combine redundant variables into a composite variable such as their mean or sum. Although the composite variable can preserve information in its original predictor variables, it does not tell how each of the original variables is related to the dependent variable. Perhaps, the most direct and general solution is to collect more observations, so that predictor variables can be more independent (Belsley, Kuh, and Welsch 1980, p. 193). This will be of help in yielding a more accurate parameter estimate and reducing its variance. However, collecting more observations is often impractical. Lastly, a shrinkage or data-reduction approach can be adopted to deal with multicollinearity. This approach aims to resolve the matrix inversion problem by adding further information to the cross-products matrix or reducing the dimension of the matrix. For example, ridge regression (Hoerl and Kennard 1970), principal component regression (e.g., Hastie et al. 2001, p. 66), or partial least squares regression (e.g., Frank and Friedman 1993; Höskuldsson 1988; Wold 1966) can be considered to deal with multicollinearity in multiple regression.

Multicollinearity has been discussed and studied mainly in the context of multiple regression. However, it is also a prevalent problem in structural equation modeling (Marsh et al. 2004). Unlike multiple regression that involves observed predictor variables only, structural equation modeling generally involves two sets of predictor variables: *observed predictor* variables and *latent predictor* variables. For example, when a formative relationship between indicators and their latent variables is specified in the measurement model, the indicators may be considered observed predictor variables. In addition, any latent variable affecting other latent variables in the structural model is a latent predictor variable. Thus, two types of multicollinearity should be considered in structural equation modeling. One is multicollinearity among latent predictor variables (Grapentine 2000; Grewal, Cote, and Baumgartner 2004; Jagpal 1982; Kaplan 1994) and the other is multicollinearity among formative indicators for a single latent variable (Temme, Kreis, and Hildebrandt 2006). The former type is illustrated in Figure 8.1, where three latent variables have influences on an endogenous latent variable. Multicollinearity can arise from high correlations between two or all of the exogenous latent variables. The latter type is illustrated in Figure 8.2, in which four indicators form a latent variable. Multicollinearity can occur because of high correlations between the formative indicators for each latent variable. If a model involves reflective indicators only, the former type of multicollinearity is a main concern. On the other hand, if a model includes reflective and formative indicators at the same time, both types of multicollinearity can occur.

In this chapter, we discuss how to deal with the two types of multicollinearity in generalized structured component analysis. As mentioned earlier, there are several general options for dealing with multicollinearity. We can apply them to generalized structured component analysis. For example, we can remove some of the redundant formative indicators or latent predictor variables. We can create a composite variable of redundant formative indicators, or a second-order latent variable that underlies highly correlated

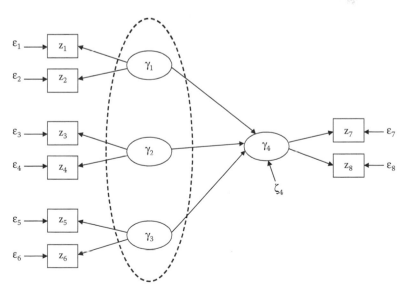

FIGURE 8.1
Latent variables in a dashed circle are highly correlated.

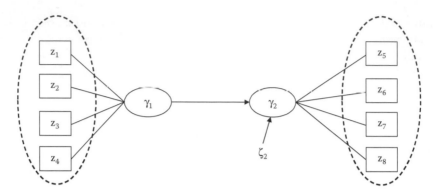

FIGURE 8.2
Formative indicators for each latent variable in a dashed circle are highly correlated.

first-order latent predictor variables (Kenny 2011). We can also consider collecting more observations. However, omitting variables or creating composite variables or second-order latent variables leads to the specification of a new model that is different substantively from the original model. Adding more observations can be costly.

We thus focus on a recent extension of generalized structured component analysis that aims to deal with both types of multicollinearity, while maintaining the same model specification as well as the same data. The extension is called *regularized generalized structured component analysis* (Hwang 2009), in which a ridge penalty term (Hoerl and Kennard 1970) is imposed on each set of parameters. In the sections that follow, we shall provide a technical account and an empirical application of regularized generalized structured component analysis.

8.1 Ridge Regression

Technically, regularized generalized structured component analysis can be regarded as an adaptation of ridge regression (Hoerl and Kennard 1970; also see Vinod 1978; Vinod and Ullah 1981) to generalized structured component analysis. We thus begin by reviewing ridge regression in brevity.

Let us consider a linear regression model

$$\mathbf{y} = \mathbf{X}\boldsymbol{\beta} + \boldsymbol{\varepsilon}, \tag{8.1}$$

where $\mathbf{y}$ is an N by 1 vector of the dependent variable, $\mathbf{X}$ is an N by J matrix of predictor variables, $\boldsymbol{\beta}$ is a J by 1 vector consisting of regression coefficients, and ε is an *N* by 1 vector of residuals. The ordinary least-squares estimate of β is then obtained by

$$\hat{\beta}_{ols} = (\mathbf{X}'\mathbf{X})^{-1}\mathbf{X}'\mathbf{y}. \tag{8.2}$$

In the presence of multicollinearity in **X**, the cross-products matrix **X'X** in Equation 8.2 can be (nearly) singular, so that the calculation of the inverse of **X'X** is not possible or inaccurate.

To resolve the so-called matrix inversion problem, in ridge regression, the estimate of β is obtained by

$$\hat{\beta}_{ridge} = (\mathbf{X}'\mathbf{X} + \lambda\mathbf{I})^{-1}\mathbf{X}'\mathbf{y}, \tag{8.3}$$

where the constant λ is called the ridge parameter, which is typically a small positive value, and **I** is an identity matrix of order *J*. By adding the ridge parameter to each diagonal element of **X'X**, the resultant cross-products matrix may be inverted. This is the main motivation of ridge regression.

Obtaining the ridge estimate of β in Equation 8.3 indicates that we seek to minimize the following penalized least-squares criterion

$$\varphi_1 = SS(\mathbf{y} - \mathbf{X}\beta) + \lambda SS(\beta). \tag{8.4}$$

This can be viewed as an L_2-norm penalized least-squares criterion because the sum of squared regression coefficients (SS(β)) is called an L_2-norm or quadratic penalty (also see Le Cessie and Van Houwelingen 1992; Lee and Silvapulle 1988). This penalty term takes into account the magnitudes of the regression coefficients. If the values of the regression coefficients in absolute value are large, the penalty term becomes large. The ridge parameter (λ) plays a role in balancing out the relative importance of the penalty term in estimation of the regression coefficients. If the ridge parameter value becomes large, then a greater penalty is imposed on the sizes of the regression coefficient estimates, thus shrinking them toward zero. In this regard, ridge regression can be viewed as a shrinkage approach to dealing with multicollinearity.

To give further insight into how ridge regression works, we can consider an alternative way of obtaining the ridge estimate of β in Equation 8.3. That is, the same problem can be solved by minimizing the following constrained least-squares criterion

$$\varphi_2 = SS(\mathbf{y} - \mathbf{X}\beta), \text{ subject to } SS(\beta) \le \tau \tag{8.5}$$

(Hastie et al. 2009, p. 63). In Equation 8.5, the size constraint (SS(β) ≤ τ) is explicitly imposed on the regression coefficients. It is known that there is a one-to-one correspondence between the ridge parameter (λ) in Equation 8.4 and τ in Equation 8.5 (Hastie et al. 2009, p. 63). As stated earlier, when two predictor variables are highly correlated, a very large positive regression coefficient of one predictor variable can be offset by a very large negative regression coefficient of the other predictor variable. By imposing the size constraint, ridge regression keeps the magnitudes of the regression coefficients within a certain range.

The ordinary least-squares estimator ($\hat{\beta}_{ols}$) in Equation 8.2 serves as the minimum-variance estimator among all linear unbiased estimators, that is, the best linear unbiased estimator (BLUE), under mild assumptions that the residuals have expectation zero, and are uncorrelated and homoscedastic. However, it can become poorly determined and involve large variance, when the cross-products matrix $\mathbf{X'X}$ is (nearly) singular under the presence of multicollinearity. On the other hand, the ridge estimator ($\hat{\beta}_{ridge}$) in Equation 8.3 is biased yet is more robust to multicollinearity, thereby leading to smaller variance.

The accuracy of a parameter estimate is evaluated by the squared Euclidean distance between the estimate and parameter. If we take the expected value of the squared distance over sample replications, we obtain mean squared error (MSE), given by

$$\text{MSE} = E\left[(\hat{\beta} - \beta)^2\right] = E\left(\hat{\beta} - E(\hat{\beta})\right)^2 + \left(E(\hat{\beta}) - \beta\right)^2, \tag{8.6}$$

where $\hat{\beta}$ and β denote an estimate and its parameter, respectively. As shown in Equation 8.6, the mean square error is the sum of two properties. The first term in Equation 8.6 is equivalent to variance (i.e., the expected squared distance between the estimate and its mean), whereas the second is squared bias (the squared distance between the parameter and the mean of the estimate). The smaller the mean square error of an estimate, the closer it is to the parameter on average.

The ordinary least-squares estimator is unbiased (i.e., zero bias), but it may exhibit large variance particularly in the presence of multicollinearity. On the other hand, the ridge estimator involves a small amount of bias but tends to be associated with a smaller variance. If the variance is sufficiently small, the ridge estimator is likely to show a smaller mean square error than the ordinary least-squares estimator. This suggests that despite the fact that it is a biased estimator; the ridge estimator can be more accurate on average than the ordinary least-squares estimator. In fact, it is known that within a certain range of the ridge parameters, the ridge estimator always exhibits a smaller mean square error than the ordinary least-squares estimator, regardless of whether multicollinearity is present (Hoerl and Kennard 1970). Importantly, this tendency becomes salient in the presence of multicollinearity (Takane and Hwang 2007).

8.2 Regularized Generalized Structured Component Analysis

Generalized structured component analysis involves three sets of parameters: weights (**W**), path coefficients (**B**), and loadings (**C**). In the presence of multicollinearity, ordinary least-squares estimation is likely to result in less accurate estimates of these parameters with large variances. To address this issue, regularized generalized structured component analysis was developed in which a ridge penalty is imposed on each set of parameters.

In regularized generalized structured component analysis, we seek to minimize the following penalized least-squares criterion

$$\phi = SS(\mathbf{ZV} - \mathbf{ZWA}) + \lambda_1 SS(\mathbf{W}) + \lambda_2 SS(\mathbf{C}) + \lambda_3 SS(\mathbf{B}), \tag{8.7}$$

with respect to **W**, **C**, and **B**, subject to the standardization constraint on each latent variable. In Equation 8.7, λ_1, λ_2, and λ_3 denote the prescribed ridge parameters for each ridge penalty term that is added to penalize the magnitude of each set of parameter estimates. It is easy to see that Equation 8.7 reduces to the ordinary least-squares criterion for generalized structured component analysis, when all the ridge parameters are set to zero. By minimizing Equation 8.7 under nonzero ridge parameters, we may mitigate the adverse influences of multicollinearity on all sets of the estimates simultaneously. This in turn helps deal with the two types of multicollinearity at once. For example, when there exists multicollinearity between formative indicators, the ridge penalty SS(**W**) tends to be large, thereby leading to a nonzero value of λ_1. On the other hand, when multicollinearity is present in latent predictor variables, the ridge penalty SS(**B**) tends to be large, thereby leading to a nonzero value of λ_3.

An iterative algorithm, named the *alternating regularized least-squares algorithm* (Hwang 2009), was developed to minimize Equation 8.7. This algorithm is a regularized version of the alternating least-squares algorithm for generalized structured component analysis. It repeats three main steps, given the values of λ_1, λ_2, and λ_3. As described in detail in the Appendix, each step of the algorithm is essentially equivalent to solving a ridge regression problem. This algorithm is versatile, so that it can be adopted for minimizing various penalized least-squares criteria with an L_2-norm or quadratic penalty (e.g., Hwang et al. 2012; Hwang et al. in press).

Prior to applying the alternating regularized least-squares algorithm, we should determine the values of λ_1, λ_2, and λ_3. We can apply K-fold cross validation (e.g., Hastie et al. 2001, p. 214) to choose their values. In general, the entire dataset is divided into K subsets in K-fold cross validation. One of the K subsets is taken as a validation or test sample, whereas the remaining $K - 1$ subsets are used as a calibration or training sample. We estimate parameters based on the calibration sample, and then apply the parameter estimates to the validation sample in order to calculate prediction error. We repeat this procedure K times, changing the validation and calibration samples systematically. Finally, we compute the cross-validation estimate of prediction error over all K validation samples.

Specifically, let $\mathbf{Z}^{(k)}$ denote the kth validation sample of $\mathbf{Z}$ ($k = 1, \cdots, K$). Let $\mathbf{Z}^{(-k)}$ denote the calibration sample that remained after removing $\mathbf{Z}^{(k)}$ from $\mathbf{Z}$. Under given values of λ_1, λ_2, and λ_3, we fit a specified model to $\mathbf{Z}^{(-k)}$, and obtain the estimates of $\mathbf{W}$, $\mathbf{C}$, and $\mathbf{B}$, denoted by $\hat{\mathbf{W}}^{(-k)}$, $\hat{\mathbf{C}}^{(-k)}$, and $\hat{\mathbf{B}}^{(-k)}$, respectively. Then, the cross-validation estimate of prediction error, denoted by CV, under the values of λ_1, λ_2, and λ_3 is calculated by

$$CV(\lambda_1,\lambda_2,\lambda_3)=\frac{1}{K}\sum_{k=1}^{K} SS(\mathbf{Z}^{(k)}\hat{\mathbf{V}}^{(-k)} - \mathbf{Z}^{(k)}\hat{\mathbf{W}}^{(-k)}\hat{\mathbf{A}}^{(-k)}), \tag{8.8}$$

where $\hat{\mathbf{A}}^{(-k)} = [\mathbf{C}^{(-k)}, \hat{\mathbf{B}}^{(-k)}]$. The values of Equation 8.8 are repeatedly calculated with different values of λ_1, λ_2, and λ_3. The values of λ_1, λ_2, and λ_3 associated with the smallest value of Equation 8.8 can be chosen as the optimal ones. In setting the number of subsets (K), it is important to consider sample size. If sample size is large, large K still gives rise to a calibration sample of sufficiently large size, so that it may result in a less biased estimate of prediction error. In practice, $K = 5$ or 10 is typically employed (Hastie et al. 2001, p. 216).

As in its nonregularized counterpart, regularized generalized structured component analysis provides the same measures of overall model fit. It employs the bootstrap method to estimate the standard errors (SE) or confidence intervals (CI) of parameter estimates.

Partial least squares path modeling has provided remedies for the two types of multicollinearity. As described in Appendix 2.3, in the first stage of the algorithm for partial least squares path modeling, the weights $\mathbf{w}_p$ for a single latent variable γ_p are updated based on Mode A or Mode B (see the third step, called the external estimation step). In Mode A, $\mathbf{w}_p$ is updated by

$$\hat{\mathbf{w}}_p = \mathbf{Z}_p'\mathbf{q}_p, \tag{8.9}$$

where $\mathbf{Z}_p$ and $\mathbf{q}_p$ indicate a matrix of indicators and the inner estimate for γ_p, respectively. On the other hand, in Mode B, $\mathbf{w}_p$ is updated by

$$\hat{\mathbf{w}}_p = (\mathbf{Z}_p'\mathbf{Z}_p)^{-1}\mathbf{Z}_p'\mathbf{q}_p. \tag{8.10}$$

Thus, multicollinearity among formative indicators affects the estimation of weights under Mode B only, which involves the inverse of a cross-products matrix. In the presence of such multicollinearity, it was suggested applying partial least squares regression to estimate the weights in Equation 8.10. This procedure is called *Mode PLS* (Esposito Vinzi 2008, 2009; Esposito Vinzi and Russolillo 2010).

Multicollinearity in exogenous latent variables is likely to influence the least-squares estimation of path coefficients in the second stage of the algorithm. To address this type of multicollinearity, it was again suggested using partial least squares regression for the estimation of path coefficients (Esposito Vinzi, Trinchera, and Amato 2010).

Partial least squares regression is a well-known data-reduction technique where a series of mutually orthogonal components of predictor variables are extracted to account for the variance of dependent variables as much as possible. Consequently, partial least squares regression serves as a useful tool for dealing with multicollinearity. It thus seems a natural choice to use partial least squares regression to address the two types of multicollinearity in partial least squares path modeling. Despite the practical merit, it is unclear which criterion is being optimized by partial least squares regression and whether some sort of convergence can be ensured in the first stage of the algorithm. Furthermore, technically, there is no particular reason to use partial least squares regression only. We can certainly consider ridge regression or other data-reduction technique, such as principal component regression, to deal with multicollinearity in partial least squares path modeling.

8.3 Example

The present example is company-level data from the American Customer Satisfaction Index (ACSI) (Fornell et al. 1996) database collected in 2002. The sample size was 152 companies in total. Table 8.1 presents the measurement model for the ACSI, in which all six latent variables underlie 14 indicators. Figure 8.3 displays the ACSI model. In the model, the symbol "+" or "−" in parentheses on a path indicated a positive or negative relationship between two latent variables, respectively. The hypothesized relationships among the six latent variables were well derived from previous theories (see Fornell et al. 1996).

We first applied nonregularized generalized structured component analysis to fit the model. We used 100 bootstrap samples to estimate SE and 95% CI. The model showed that FIT = 0.83 (SE = 0.01, 95% CI = 0.82–0.86), AFIT = 0.83 (SE = 0.01, 95% CI = 0.82–0.85), GFI = 1.00 (SE = 0.01, 95% CI = 1.00–1.00), and SRMR = 0.08 (SE = 0.01, 95% CI = 0.07–0.10). This indicates that the model accounted for about 83% of the total variance of all variables. GFI was virtually equal to 1, whereas SRMR was small. Moreover, it was found that $FIT_M = 0.95$ (SE = 0.01, 95% CI = 0.94–0.96) and $FIT_S = 0.57$ (SE = 0.02, 95% CI = 0.53–0.62). This indicates that the measurement model of the ACSI accounted for about

TABLE 8.1

Estimates of Weights and Loadings and Their Standard Errors and 95% Confidence Intervals Obtained from (Nonregularized) Generalized Structured Component Analysis

Latent[a]	Indicator[b]	Weights			Loadings		
		Estimate	SE	95% CI	Estimate	SE	95% CI
CE	z_1	0.29	0.03	0.24–0.34	0.95	0.01	0.93–0.96
	z_2	0.34	0.03	0.29–0.40	0.97	0.01	0.96–0.98
	z_3	0.41	0.02	0.37–0.44	0.95	0.01	0.94–0.96
PQ	z_4	0.41	0.03	0.35–0.46	0.98	0.00	0.97–0.99
	z_5	0.39	0.03	0.34–0.48	0.98	0.01	0.96–0.99
	z_6	0.22	0.03	0.17–0.27	0.96	0.01	0.95–0.97
PV	z_7	0.06	0.05	−0.03–0.15	0.95	0.01	0.93–0.97
	z_8	0.94	0.04	0.86–1.03	1.00	0.00	1.00–1.00
CS	z_9	0.56	0.05	0.48–0.67	0.99	0.00	0.98–0.99
	z_{10}	0.13	0.05	0.01–0.22	0.97	0.00	0.97–0.98
	z_{11}	0.33	0.02	0.31–0.37	0.95	0.01	0.93–0.96
CC	z_{12}	1	0	1–1	1	0	1–1
CL	z_{13}	0.39	0.05	0.29–0.49	0.98	0.00	0.96–0.99
	z_{14}	0.62	0.05	0.53–0.73	0.99	0.00	0.98–1.00

SE, standard errors; CI, confidence intervals.

[a] CC, customer complaints; CE, customer expectations; CL, customer loyalty; CS, customer satisfaction; PQ, perceived quality; PV, perceived value.

[b] z_1, Customer expectations about overall quality; z_2, Customer expectations about reliability; z_3, Customer expectations about customization; z_4, Overall quality; z_5, Reliability; z_6, Customization; z_7, Price given quality; z_8, Quality given price; z_9, Overall customer satisfaction; z_{10}, Confirmation of expectations; z_{11}, Distance to ideal product or service; z_{12}, Formal or informal complaint behavior; z_{13}, Repurchase intention; z_{14}, Price tolerance.

95% of the total variance of all indicators, whereas the structural model explained about 57% of the total variance of all latent variables.

Table 8.1 provides the estimates of weights and loadings obtained from nonregularized generalized structured component analysis. All weight estimates turned out to be statistically significant, except for the weight estimate for z_7 ("price given quality"). Table 8.2 shows the correlation matrix of the 14 indicators. The correlation between z_7 and z_8 ("quality given price") for *perceived value* appeared fairly high ($r = 0.94$). This might lead to the small, statistically nonsignificant weight estimate for z_7. Besides this correlation, several bivariate correlations also appeared quite large. All loading estimates were large and statistically significant.

Table 8.3 gives the path coefficient estimates obtained from nonregularized generalized structured component analysis. The interpretations of the path coefficient estimates seem to be generally consistent with the relationships

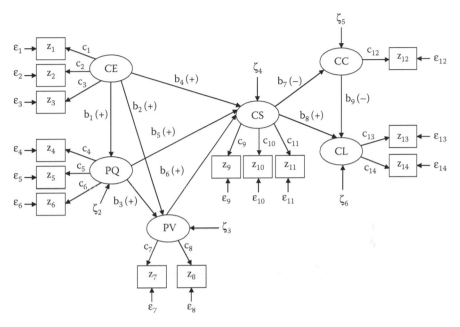

FIGURE 8.3
A path diagram of the American customer satisfaction index model

among the latent variables displayed in Figure 8.3. Nonetheless, *customer expectations* had negative and statistically nonsignificant impacts on *perceived value* and *customer satisfaction*. Clearly, the signs and nonsignificance of the two path coefficient estimates were inconsistent with the relationships hypothesized in the ACSI. In addition, *customer expectations* was quite highly correlated with *perceived quality* ($r = 0.91$), as shown in Table 8.4.

We applied regularized generalized structured component analysis to the same data. We used 100 bootstrap samples to estimate SE and 95% CI. We adopted five fold cross validation, varying each value of λ_1, λ_2, and λ_3 from 0 to 5 in the common logarithmic scale. We chose $\lambda_1 = 1$, $\lambda_2 = 0$, and $\lambda_3 = 10$ because these values produced the smallest cross-validation estimate of prediction error.

Regularized generalized structured component analysis provided FIT = 0.83 (SE = 0.01, 95% CI = 0.82–0.85), AFIT = 0.83 (SE = 0.01, 95% CI = 0.81–0.85), GFI = 0.99 (SE = 0.00, 95% CI = 0.98–0.99), SRMR = 0.07 (SE = 0.01, 95% CI = 0.06–0.08), FIT_M = 0.95 (SE = 0.01, 95% CI = 0.94–0.96), and FIT_S = 0.57 (SE = 0.02, 95% CI = 0.53–0.62). The values of these fit measures were almost identical to those obtained under the nonregularized counterpart.

Table 8.5 provides the estimates of weights and loadings obtained from regularized generalized structured component analysis. These estimates were quite similar to those obtained from nonregularized generalized structured component analysis, leading to essentially the same interpretations. However, it is noticeable that the weight estimate for z_7 became larger than that

Generalized Structured Component Analysis

TABLE 8.2

Correlation Matrix of the 14 Indicators in the ACSI Data

		CE			PQ			PV		CS			CC	CL	
	z_1	z_2	z_3	z_4	z_5	z_6	z_7	z_8	z_9	z_{10}	z_{11}	z_{12}	z_{13}	z_{14}
CE z_1	1	0.92	0.83	0.84	0.83	0.76	0.62	0.74	0.80	0.80	0.79	-0.10	0.34	0.40
z_2		1	0.87	0.83	0.90	0.82	0.64	0.76	0.84	0.84	0.80	-0.13	0.33	0.41
z_3			1	0.87	0.88	0.91	0.63	0.78	0.85	0.83	0.80	-0.24	0.41	0.45
PQ z_4				1	0.93	0.91	0.74	0.89	0.96	0.94	0.90	-0.46	0.65	0.68
z_5					1	0.92	0.72	0.86	0.95	0.94	0.87	-0.35	0.56	0.60
z_6						1	0.67	0.82	0.91	0.89	0.84	-0.41	0.52	0.55
PV z_7							1	0.94	0.84	0.83	0.67	-0.35	0.48	0.57
z_8								1	0.95	0.93	0.82	-0.41	0.58	0.65
CS z_9									1	0.98	0.89	-0.47	0.62	0.68
z_{10}										1	0.88	-0.40	0.60	0.65
z_{11}											1	-1.46	0.71	0.71
CC z_{12}												1	-0.64	-0.68
CL z_{13}													1	0.93
z_{14}														1

CC, customer complaints; CE, customer expectations; CL, customer loyalty; CS, customer satisfaction; PQ, perceived quality; PV, perceived value.

TABLE 8.3

Estimates of Path Coefficients and Their Standard Errors and 95% Confidence Intervals Obtained from (Nonregularized) Generalized Structured Component Analysis

	Estimate	SE	95% CI
CE → PQ (b_1)	0.92	0.01	0.90–0.94
CE → PV (b_2)	−0.08	0.12	−0.37–0.10
CE → CS (b_4)	−0.04	0.04	−0.11–0.03
PQ → PV (b_3)	0.96	0.12	0.76–1.23
PQ → CS (b_5)	0.73	0.04	0.64–0.82
PV → CS (b_6)	0.31	0.03	0.25–0.37
CS → CC (b_7)	−0.47	0.07	−0.59 to −0.32
CS → CL (b_8)	0.49	0.05	0.42–0.61
CC → CL (b_9)	−0.44	0.05	−0.53 to −0.33

CC, customer complaints; CE, customer expectations; CI, confidence intervals; CL, customer loyalty; CS, customer satisfaction; PQ, perceived quality; PV, perceived value; SE, standard errors.

TABLE 8.4

Correlation Matrix of the Six Latent Variables in the ACSI Obtained from (Nonregularized) Generalized Structured Component Analysis

	CE	PQ	PV	CS	CC	CL
CE	1	0.91	0.74	0.88	−0.16	0.42
PQ		1	0.82	0.97	−0.42	0.62
PV			1	0.88	−0.39	0.59
CS				1	−0.46	0.69
CC					1	−0.67
CL						1

CC, customer complaints; CE, customer expectations; CL, customer loyalty; CS, customer satisfaction; PQ, perceived quality; PV, perceived value.

obtained from nonregularized generalized structured component analysis, and turned out to be statistically significant. Table 8.6 presents the estimates of path coefficients obtained from regularized generalized structured component analysis. It was shown that *customer expectations* had positive and statistically significant effects on both *perceived value* and *customer satisfaction*. These seemed to support the hypothesized relationships among the three latent variables in the ACSI. Although these two path coefficients were specified in the ACSI model based on previous marketing theories, keeping them in the model has been questioned because their estimates were found to be statistically nonsignificant in some empirical studies (e.g., Johnson et al. 2001). However, as suggested in our analysis, such nonsignificant path coefficient estimates might be due to the presence of multicollinearity in the data, not due to their theoretical misspecification.

TABLE 8.5

Estimates of Weights and Loadings and Their Standard Errors and 95% Confidence Intervals Obtained from Regularized Generalized Structured Component Analysis

Latent	Indicator	Weights			Loadings		
		Estimate	SE	95% CI	Estimate	SE	95% CI
CE	z_1	0.29	0.02	0.25–0.35	0.95	0.01	0.93–0.96
	z_2	0.34	0.03	0.30–0.40	0.97	0.00	0.96–0.98
	z_3	0.41	0.02	0.36–0.44	0.95	0.01	0.93–0.97
PQ	z_4	0.40	0.02	0.35–0.44	0.98	0.00	0.97–0.99
	z_5	0.38	0.03	0.34–0.44	0.98	0.01	0.97–0.99
	z_6	0.24	0.02	0.21–0.28	0.96	0.01	0.95–0.97
PV	z_7	0.08	0.04	0.00–0.16	0.95	0.01	0.93–0.97
	z_8	0.92	0.04	0.85–1.00	1.00	0.00	1.00–1.00
CS	z_9	0.54	0.04	0.46–0.60	0.99	0.00	0.98–0.99
	z_{10}	0.15	0.03	0.09–0.22	0.97	0.00	0.96–0.98
	z_{11}	0.34	0.02	0.30–0.37	0.95	0.01	0.93–0.97
CC	z_{12}	1	0	1–1	1	0	1–1
CL	z_{13}	0.40	0.04	0.30–0.47	0.98	0.00	0.97–0.98
	z_{14}	0.61	0.04	0.55–0.71	0.99	0.00	0.98–1.00

CC, customer complaints; CE, customer expectations; CI, confidence intervals; CL, customer loyalty; CS, customer satisfaction; PQ, perceived quality; PV, perceived value; SE, standard errors.

TABLE 8.6

Estimates of Path Coefficients and Their Standard Errors and 95% Confidence Intervals Obtained from Regularized Generalized Structured Component Analysis

	Estimate	SE	95% CI
CE → PQ (b_1)	0.86	0.01	0.84–0.88
CE → PV (b_2)	0.13	0.06	0.03–0.29
CE → CS (b_4)	0.12	0.02	0.07–0.18
PQ → PV (b_3)	0.71	0.07	0.55–0.81
PQ → CS (b_5)	0.52	0.03	0.47–0.58
PV → CS (b_6)	0.35	0.03	0.29–0.39
CS → CC (b_7)	−0.44	0.08	−0.57 to 0.26
CS → CL (b_8)	0.47	0.05	0.36–0.57
CC → CL (b_9)	−0.42	0.04	−0.50 to 0.32

CC, customer complaints; CE, customer expectations; CI, confidence intervals; CL, customer loyalty; CS, customer satisfaction; PQ, perceived quality; PV, perceived value; SE, standard errors.

8.4 Summary

In practice, generalized structured component analysis can suffer from two types of multicollinearity. We discussed regularized generalized structured component analysis, in which generalized structured component analysis was combined with ridge-type regularization in a unified framework to address both types of multicollinearity. A regularized extension of the alternating least-squares algorithm for generalized structured component analysis was developed to obtain the ridge least-squares estimates of parameters. This algorithm can be used for regularized versions of various statistical techniques that build on least-squares optimization (cf. ten Berge 1993). We can use cross validation to determine the values of the ridge parameters in a systematic fashion.

We applied regularized generalized structured component analysis to a company-level ACSI dataset. Generalized structured component analysis yielded several substantively counter-intuitive and statistically nonsignificant parameter estimates in the presence of multicollinearity. On the other hand, regularized generalized structured component analysis resulted in parameter estimates that were all statistically significant and congruent with the relationships hypothesized in the ACSI.

In practice, regularized generalized structured component analysis may be used in lieu of regular, nonregularized generalized structured component analysis. In the presence of multicollinearity, regularized generalized structured component analysis will likely result in more accurate estimates than generalized structured component analysis by taking nonzero ridge parameters. On the other hand, when no multicollinearity is present, regularized generalized structured component analysis will become equivalent to generalized structured component analysis by setting all ridge parameters to zeros. The only concern is that regularized generalized structured component analysis can be more costly computationally than generalized structured component analysis, because it involves cross validation for selection of the ridge parameters.

Appendix 8.1 The Alternating Regularized Least-Squares Algorithm for Regularized Generalized Structured Component Analysis

Hwang (2009) proposed an alternating regularized least-squares algorithm to minimize Equation 8.7. The algorithm repeats the following three main steps, given the values of λ_1, λ_2, and λ_3.

Step 1: $\mathbf{C}$ is updated for fixed $\mathbf{W}$ and $\mathbf{B}$. Let $\mathbf{A} = [\mathbf{C}^* + \mathbf{B}^*]$, where $\mathbf{C}^* = [\mathbf{C}, 0]$ and $\mathbf{B}^* = [0, \mathbf{B}]$. Then, we can re-express Equation 8.7 as

$$\phi = SS(\mathbf{ZV} - \mathbf{ZW}[\mathbf{C}^* + \mathbf{B}^*]) + \lambda_1 SS(\mathbf{W}) + \lambda_2 SS(\mathbf{C}^*) + \lambda_3 SS(\mathbf{B}^*). \quad (A8.1)$$

Minimizing Equation A8.1 with respect to $\mathbf{C}$ is equivalent to minimizing

$$\begin{aligned}
\phi_1 &= SS(\mathbf{ZV} - \mathbf{ZWB}^* - \mathbf{ZWC}^*) + \lambda_2 SS(\mathbf{C}^*) \\
&= SS(\text{vec}(\mathbf{T}) - (\mathbf{I} \otimes \mathbf{ZW})\text{vec}(\mathbf{C}^*)) + \lambda_2 SS(\text{vec}(\mathbf{C}^*)) \\
&= SS(\text{vec}(\mathbf{T}) - \mathbf{\Omega c}) + \lambda_2 \mathbf{c}'\mathbf{c} \quad\quad\quad (A8.2)
\end{aligned}$$

where $\mathbf{T} = \mathbf{ZV} - \mathbf{ZWB}^*$, $\mathbf{c}$ denotes the vector consisting of any nonzero elements in $\text{vec}(\mathbf{C}^*)$, and $\mathbf{\Omega}$ denotes the matrix of the columns of $\mathbf{I} \otimes \mathbf{ZW}$ corresponding to the nonzero elements in $\text{vec}(\mathbf{C}^*)$. Then, the ridge least-squares estimate of $\mathbf{c}$ is obtained by

$$\hat{\mathbf{c}} = (\mathbf{\Omega}'\mathbf{\Omega} + \lambda_2 \mathbf{I})^{-1} \mathbf{\Omega}' \text{vec}(\mathbf{T}). \quad (A8.3)$$

The updated $\mathbf{C}$ is reconstructed from $\hat{\mathbf{c}}$.

Step 2: $\mathbf{B}$ is updated for fixed $\mathbf{W}$ and $\mathbf{C}$. This is equivalent to minimizing

$$\begin{aligned}
\phi_2 &= SS(\mathbf{ZV} - \mathbf{ZWC}^* - \mathbf{ZWB}^*) + \lambda_3 SS(\mathbf{B}^*) \\
&= SS(\text{vec}(\mathbf{Y}) - (\mathbf{I} \otimes \mathbf{ZW})\text{vec}(\mathbf{B}^*)) + \lambda_3 SS(\text{vec}(\mathbf{B}^*)) \\
&= SS(\text{vec}(\mathbf{Y}) - \mathbf{\Phi b}) + \lambda_3 SS \mathbf{b}'\mathbf{b} \quad\quad\quad (A8.4)
\end{aligned}$$

with respect to $\mathbf{B}$, where $\mathbf{Y} = \mathbf{ZV} - \mathbf{ZWC}^*$, $\mathbf{b}$ denotes the vector of any nonzero elements in $\text{vec}(\mathbf{B}^*)$, and $\mathbf{\Phi}$ denotes the matrix consisting of the columns of $\mathbf{I} \otimes \mathbf{ZW}$ corresponding to the nonzero elements in $\text{vec}(\mathbf{B}^*)$. Then, the ridge least-squares estimate of $\mathbf{b}$ is obtained by

$$\hat{\mathbf{b}} = (\mathbf{\Phi}'\mathbf{\Phi} + \lambda_3 \mathbf{I})^{-1} \mathbf{\Phi}' \text{vec}(\mathbf{Y}). \quad (A8.5)$$

The updated $\mathbf{B}$ is reconstructed from $\hat{\mathbf{b}}$.

Step 3: $\mathbf{W}$ is updated for fixed $\mathbf{B}$ and $\mathbf{C}$ (or equivalently $\mathbf{A}$). This is equivalent to minimizing

$$\phi_3 = SS(\mathbf{ZV} - \mathbf{ZWA}) + \lambda_1 SS(\mathbf{W}), \quad (A8.6)$$

with respect to **W**. Let $\mathbf{w}_p$ denote the pth column of unknown component weights in **W**, which is shared by the tth column in **V**, where $t = J + p$ ($p = 1,..., P$). Let $\mathbf{\Lambda} = \mathbf{WA}$. Let $\mathbf{V}_{(-t)}$ denote **V** whose tth column is a vector of zeros. Let $\mathbf{V}_{(t)}$ denote **V** whose columns are all zero vectors except the tth column. Let $\mathbf{\Lambda}_{(-p)}$ denote a product matrix of **W** whose pth column is a vector of zeros and **A** whose pth row is a zero vector. Let $\mathbf{\Lambda}_{(p)}$ denote a product matrix of **W** whose columns are all zero vectors except the pth column and **A** whose rows are all zero vectors except the pth row. Let $\mathbf{m}_{(t)}$ denote a 1 by $J + P$ vector whose elements are all zeros except the tth element being unity. Let $\mathbf{a}_{(p)}$ denote the pth row of **A**. To update $\mathbf{w}_p$, Equation A8.6 can be rewritten as

$$\phi_3 = \sum_{p=1}^{P} \mathrm{SS}((\boldsymbol{\alpha}' \otimes \mathbf{Z})\mathbf{w}_p - \mathrm{vec}(\mathbf{Z\Delta})) + \lambda_1 \sum_{p-1}^{P} \mathbf{w}_p' \mathbf{w}_p, \tag{A8.7}$$

where $\boldsymbol{\alpha} = \mathbf{m}_{(t)} - \mathbf{a}_{(p)}$, and $\boldsymbol{\Delta} = \mathbf{\Lambda}_{(-p)} - \mathbf{V}_{(t)}$

Let $\boldsymbol{\theta}_p$ denote the vector formed by eliminating any zero elements from $\mathbf{w}_p$. Let $\boldsymbol{\Xi}$ denote the matrix formed by eliminating the columns of $\boldsymbol{\alpha}' \otimes \mathbf{Z}$ corresponding to the zero elements in $\mathbf{w}_p$. Then, the ridge least-squares estimate of $\boldsymbol{\theta}_p$ is obtained by

$$\hat{\boldsymbol{\theta}}_p = (\boldsymbol{\Xi}'\boldsymbol{\Xi} + \lambda_1 \mathbf{I})^{-1} \boldsymbol{\Xi}' \mathrm{vec}(\mathbf{Z\Delta}). \tag{A8.8}$$

The updated $\mathbf{w}_p$ is recovered from $\hat{\boldsymbol{\theta}}_{(p)}$, and subsequently multiplied by

$$\frac{\sqrt{N}}{\mathbf{w}_p' \mathbf{Z}' \mathbf{Z} \mathbf{w}_p}$$

in order to satisfy the standardization constraint imposed on the pth latent variable.

9

Lasso Generalized Structured Component Analysis

Generalized structured component analysis is a component-based approach to structural equation modeling. That is, it defines latent variables as weighted composites or components of indicators, thereby permitting the provision of unique latent variable scores. At the same time, however, this characteristic may lead to a potential issue that the quality and interpretability of latent variables depend directly on which indicators are used. For example, it may be difficult to interpret a latent variable clearly when it is constructed from a large number of indicators. In addition, possessing irrelevant or uninformative indicators is likely to obscure or distort the meaning of a latent variable. Furthermore, the use of too many indicators for a single latent variable tends to increase the likelihood of having redundant indicators, thus resulting in less stable parameter estimates, as discussed in Chapter 8. Accordingly, it is important to select an appropriate set of indicators for each latent variable in the measurement model of generalized structured component analysis.

It is also important to select an optimal set of latent variables in the structural model. For example, eliminating uninformative latent variables is of use in easing interpretations of a given structural model, uncovering the most relevant features and structures of the phenomena under study (e.g., Baumann, Albert, and von Korff 2002). Thus, it is crucial to specify a simple yet highly predictive structural model with relevant latent variables only.

In linear regression, several methods are available to select a subset of predictor variables. A popular method is sequential procedures that include forward, backward, and stepwise selection (e.g., Stevens 2009, pp. 76–77). Forward selection begins with the null model with no predictor variables, and sequentially adds a predictor variable into the model, which most improves model fit at each stage. Adding a predictor variable continues until it improves model fit substantially. The improvement in fit is typically calculated based on the F statistic for comparing two nested models. Backward selection proceeds in the reverse direction. It starts with the full model with all predictor variables, and sequentially removes a predictor variable from the model, which least influences model fit. Eliminating a predictor variable continues until no substantial decline in fit is found. Statistical/stepwise selection is a hybrid procedure of forward and backward selection. It proceeds as in forward selection, but removes a predictor variable at any stage, if it does not improve fit substantially.

The sequential procedures are a discrete process of variable section, in which predictor variables are either retained or discarded (Hastie et al. 2009, p. 61). This discrete selection process can be of use in improving model interpretability by considering only a subset of predictor variables. However, it can often involve high variance. More importantly, although they are useful in linear regression, it seems difficult to apply them to generalized structured component analysis for reasons. First, unlike linear regression, it is difficult to specify a dependent variable in selecting a subset of indicators for each latent variable in the measurement model. This is also the case in factor analysis models, principal component analysis models, or some covariance structure models (Kano 2002). In the structural model, each dependent latent variable can be used for selection of its predictor latent variables. However, it can be cumbersome to apply the sequential procedures for each dependent latent variable in the structural model. Second, there is no easy-to-use procedure for examining improvement in fit between nested models. In linear regression, the F statistic is used for such testing. However, generalized structured component analysis is a distribution-free approach, so that it is not equipped with such a parametric test statistic. The R-squared value of each dependent latent variable may be used for comparing different nested structural models in combination with the bootstrap method. However, this procedure can be burdensome computationally because it is applied to each pair of nested models, for which a number of bootstrap samples are generated. No such index is yet available for selecting a subset of indicators for each latent variable, which can be used for both formative and reflective indicators. In the same vein, there exists no index similar to Mallows' (1973) C_p, which is used for variable selection in linear regression.

Another well-known method for variable selection in linear regression is to use a shrinkage method such as the lasso (least absolute shrinkage and selection operator) (Tibshirani 1996). As in ridge regression (Hoerl and Kennard 1970), the lasso imposes a size constraint on predictor variables. Unlike ridge regression, however, the lasso tends to make some of the regression coefficient estimates exact zero, thereby leading to selection of a subset of predictor variables in an automatic manner. The lasso is a more continuous process, and tends to show a lesser degree of variability (Hastie et al. 2001, p. 59).

The lasso is particularly attractive for variable selection in generalized structured component analysis. First, it can be used for selecting subsets of indicators and latent variables simultaneously. For example, we may examine which weight estimates become zeros to select a subset of formative indicators, whereas we may look into which loading estimates become zeros to select a subset of reflective indicators. Similarly, we may investigate which path coefficient estimates are shrunk to zeros to select a subset of latent variables. Second, it does not require fitting different (nested) models repeatedly. The whole model can be fitted to the data only once in order to select subsets of indicators and latent variables in an automatic manner. Third, it is technically straightforward to combine the lasso into generalized structured component analysis in a unified manner.

In this chapter, we discuss *lasso generalized structured component analysis* that combines generalized structured component analysis with the lasso so as to address the issue of variable selection for both indicators and latent variables. We begin by reviewing the lasso to facilitate an understanding of the development of lasso generalized structured component analysis. We then provide technical accounts and an empirical application of lasso generalized structured component analysis.

9.1 Lasso Regression

Let us consider a linear regression model

$$ \mathbf{y} = \mathbf{X}\boldsymbol{\beta} + \boldsymbol{\varepsilon}, \tag{9.1} $$

where $\mathbf{y}$ is an N by 1 vector of the dependent variable, $\mathbf{X}$ is an N by J matrix of predictor variables, $\boldsymbol{\beta} = [\beta_1, \dots, \beta_J]'$ is a J by 1 vector consisting of regression coefficients, and $\boldsymbol{\varepsilon}$ is an N by 1 vector of residuals.

As with ridge regression (Hoerl and Kennard 1970), the lasso is a shrinkage method that imposes a size constraint to shrink regression coefficients. In the lasso, the following constrained least-squares criterion is minimized.

$$ \varphi_1 = \frac{1}{2}SS(\mathbf{y} - \mathbf{X}\boldsymbol{\beta}), \text{ subject to } \sum_{j=1}^{J} |\beta_j| \leq \tau \tag{9.2} $$

or equivalently,

$$ \varphi_1 = \frac{1}{2}SS(\mathbf{y} - \mathbf{X}\boldsymbol{\beta}), \text{ subject to } \mathbf{1}' \, | \, \boldsymbol{\beta} \, | \leq \tau, \tag{9.3} $$

where $\mathbf{1}$ is a J by 1 vector of ones. The sum of the absolute values of regression coefficients

$$ \left(\sum_{j=1}^{J} |\beta_j| \right) $$

is called the L_1 lasso penalty. In Equation 9.2, a size constraint, that is, the lasso penalty is equal to or less than some constant τ, is explicitly imposed on regression coefficients. If τ is greater than the sum of the ordinary least-squares estimates of regression coefficients in absolute value, the constraint has no effect, and then lasso estimates will be equivalent to their least-squares counterparts. On the other hand, if τ is sufficiently small, the constraint will make some of the regression coefficient estimates exact zero. This is different from ridge regression that shrinks regression coefficient estimates toward zero but never sets them to exact zero. Another difference is that ridge regression tends to shrink the regression coefficient estimates of correlated predictor variables toward each other, whereas the lasso tends to choose one coefficient estimate and ignore the other coefficient estimate (Ng 2012). Thus, the lasso performs shrinkage like ridge regression, but at the same time, performs in a manner similar to the procedure of selecting a subset of predictor variables (Berk 2008, p. 64).

Minimizing Equation 9.2 is also equivalent to minimizing the following penalized least-squares criterion

$$\varphi_2 = SS(\mathbf{y} - \mathbf{X}\boldsymbol{\beta}) + \lambda \mathbf{1}' \mid \boldsymbol{\beta} \mid. \tag{9.4}$$

There is a one-to-one correspondence between τ and λ. As the value of λ increases, the regression coefficients tend to be shrunk toward zeros. Unlike the ridge penalty, the lasso penalty makes the problem in Equation 9.4 non-linear. A quadratic programming as well as other specific algorithms can be used to minimize the criterion (e.g., Efron et al. 2004; Friedman et al. 2007; Osborne, Presnell, and Thurlach 2000).

9.2 Lasso Generalized Structured Component Analysis

Lasso generalized structured component analysis aims to select subsets of both indicators and latent variables by minimizing the following penalized least-squares criterion

$$\phi = \frac{1}{2} SS(\mathbf{Z}\mathbf{V} - \mathbf{Z}\mathbf{W}\mathbf{A}) + \lambda_1 \mathbf{1}' \mid \mathbf{W} \mid \mathbf{1} + \lambda_2 \mathbf{1}' \mid \mathbf{C} \mid \mathbf{1} + \lambda_3 \mathbf{1}' \mid \mathbf{B} \mid \mathbf{1} \tag{9.5}$$

with respect to $\mathbf{W}$, $\mathbf{C}$, and $\mathbf{B}$, subject to the standardization constraint on each latent variable, where $\mathbf{1}$ is a vector of ones of appropriate order, and λ_1, λ_2, and λ_3 are non-negative penalizing parameters. In Equation 9.5, the L_1 lasso penalty of each set of parameters plays a role in shrinking parameter estimates toward zero, and possibly setting some of them to exact zero. Specifically, the first and second penalty terms (i.e., $\mathbf{1}' \mid \mathbf{W} \mid \mathbf{1}$ and $\mathbf{1}' \mid \mathbf{C} \mid \mathbf{1}$) are added to select a subset of indicators by penalizing weight and loading estimates, whereas the last penalty term (i.e., $\mathbf{1}'|\mathbf{B}|\mathbf{1}$) is used for selecting a subset of latent variables by penalizing path coefficient estimates.

An iterative algorithm is developed to minimize Equation 9.5 with respect to the unknown parameters ($\mathbf{W}$, $\mathbf{C}$, and $\mathbf{B}$). Given the predetermined values of λ_1, λ_2, and λ_3, this algorithm alternates three main steps until convergence. Each step is virtually equivalent to minimizing the penalized least-squares criterion for the lasso in Equation 9.4. The coordinate-descent algorithm developed by Friedman et al. (2007) can be adopted for each step, which is fast and easy-to-implement. We provide a description of the so-called alternating coordinate-descent algorithm in the Appendix.

As with regularized generalized structured component analysis discussed in Chapter 8, K-fold cross validation can be used to determine the values of λ_1, λ_2, and λ_3 prior to the above estimation procedure. Again, the values of λ_1, λ_2, and λ_3 yielding the smallest cross-validation estimate of prediction error can be chosen as the final ones.

Lasso generalized structured component analysis provides the same measures of overall model fit as those provided in generalized structured component analysis. However, if some of the weight estimates become zero, this indicates that the corresponding indicators are excluded from the model. Thus, in such cases, we eliminate these indicators from the calculation of the overall fit measures. Likewise, if all path coefficient estimates of a latent variable are equal to zeros, this implies that the latent variable is not relevant to the phenomenon under study. Thus, both the latent variable and its indicators are also removed from the model. Lasso generalized structured component analysis uses the bootstrap method to estimate the standard errors or confidence intervals of parameter estimates.

9.3 Example: The Company-Level ACSI Data

The present example is the company-level ACSI data that were analyzed in Chapter 8. The sample size was 152 companies in total. As depicted in Figure 9.1, the ACSI model involves 14 reflective indicators for six latent variables.

We applied (nonregularized) generalized structured component analysis and lasso generalized structured component analysis to fit the model to the data. We used 100 bootstrap samples to estimate standard errors (SE) and 95% confidence intervals (CI) in both analyses. For lasso generalized structured component analysis, we chose $\lambda_1 = 0$, $\lambda_2 = 0$, and $\lambda_3 = 0.05$ based on five fold cross validation, in which each value of the penalizing parameters varied from 0 to 0.2 by 0.05. As also reported in Chapter 8, generalized structured component analysis provided that FIT = 0.83 (SE = 0.01, 95%

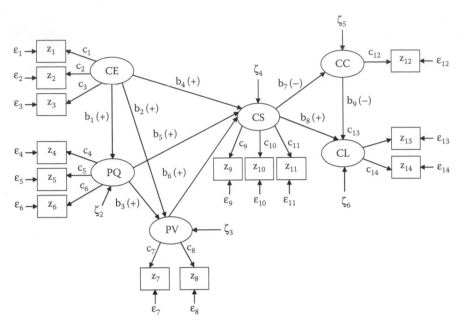

FIGURE 9.1
A path diagram of the American customer satisfaction index model.

CI = 0.82–0.86), AFIT = 0.83 (SE = 0.01, 95% CI = 0.82–0.85), GFI = 1.00 (SE = 0.01, 95% CI = 1.00–1.00), and SRMR = 0.08 (SE = 0.01, 95% CI = 0.07–0.10). On the other hand, lasso generalized structured component analysis provided that FIT = 0.83 (SE = 0.01, 95% CI = 0.81–0.85), AFIT = 0.83 (SE = 0.01, 95% CI = 0.81–0.85), GFI = 0.98 (SE = 0.00, 95% CI = 0.98–0.98), and SRMR = 0.06 (SE = 0.01, 95% CI = 0.06–0.08). This indicates that both analyses exhibited quite similar levels of model fit.

Tables 9.1 and 9.2 show the weight and loading estimates obtained from the two analyses. The two analyses resulted in quite similar weight and loading estimates, leading to the same interpretations. This was expected because no lasso penalties were imposed on the weight and loading estimates in lasso generalized structured component analysis. A difference is that the weight estimate for z_7 ("price given quality") turned out to be statistically significant in lasso generalized structured component analysis, whereas it was not statistically significant in generalized structured component analysis.

Tables 9.3 and 9.4 show the path coefficient estimates obtained from both analyses. As reported in Chapter 8, generalized structured component analysis showed that *customer expectations* had negative but statistically nonsignificant effects on *perceived value* and *customer satisfaction*. Lasso structured component analysis shrank the same path coefficient estimates to exact zero by imposing a nonzero lasso penalty. As discussed in Chapter 8, researchers have raised concerns with the addition of these two path coefficients to the ACSI model because their estimates were sometimes found to be statistically

TABLE 9.1

Estimates of Weights and Loadings and Their Standard Errors and 95% Confidence Intervals Obtained from (Nonregularized) Generalized Structured Component Analysis

Latent[a]	Indicator[b]	Weights			Loadings		
		Estimate	SE	95% CI	Estimate	SE	95% CI
CE	z_1	0.29	0.03	0.24–0.34	0.95	0.01	0.93–0.96
	z_2	0.34	0.03	0.29–0.40	0.97	0.01	0.96–0.98
	z_3	0.41	0.02	0.37–0.44	0.95	0.01	0.94–0.96
PQ	z_4	0.41	0.03	0.35–0.46	0.98	0.00	0.97–0.99
	z_5	0.39	0.03	0.34–0.48	0.98	0.01	0.96–0.99
	z_6	0.22	0.03	0.17–0.27	0.96	0.01	0.95–0.97
PV	z_7	0.06	0.05	−0.03–0.15	0.95	0.01	0.93–0.97
	z_8	0.94	0.04	0.86–1.03	1.00	0.00	1.00–1.00
CS	z_9	0.56	0.05	0.48–0.67	0.99	0.00	0.98–0.99
	z_{10}	0.13	0.05	0.01–0.22	0.97	0.00	0.97–0.98
	z_{11}	0.33	0.02	0.31–0.37	0.95	0.01	0.93–0.96
CC	z_{12}	1	0	1–1	1	0	1–1
CL	z_{13}	0.39	0.05	0.29–0.49	0.98	0.00	0.96–0.99
	z_{14}	0.62	0.05	0.53–0.73	0.99	0.00	0.98–1.00

[a] CC, customer complaints; CE, customer expectations; CI, confidence intervals; CL, customer loyalty; CS, customer satisfaction; PQ, perceived quality; PV, perceived value; SE, standard errors.

[b] z_1, customer expectations about overall quality; z_2, customer expectations about reliability; z_3, customer expectations about customization; z_4, overall quality; z_5, reliability; z_6, customization; z_7, price given quality; z_8, quality given price; z_9, overall customer satisfaction; z_{10}, confirmation of expectations; z_{11}, distance to ideal product or service; z_{12}, formal or informal complaint behavior; z_{13}, repurchase intention; z_{14}, price tolerance.

nonsignificant (e.g., Johnson et al. 2001). We pointed out in Chapter 8 that the statistical nonsignificance of their estimates might be mainly due to the presence of multicollinearity among latent variables. In general, keeping them in the model appears more consistent with marketing theories (Fornell et al. 1996). We used the present example merely to demonstrate the feasibility of lasso generalized structured component analysis for selecting a subset of latent variables in real data.

9.4 Summary

In generalized structured component analysis, the quality and interpretability of latent variables are likely affected by the usefulness and relevance of their indicators, as latent variables are obtained as components. Thus, it is an important issue to select an appropriate set of indicators in the measurement

TABLE 9.2

Estimates of Weights and Loadings and Their Standard Errors and 95% Confidence Intervals Obtained from Lasso Generalized Structured Component Analysis

Latent	Indicator	Weights			Loadings		
		Estimate	SE	95% CI	Estimate	SE	95% CI
CE	z_1	0.29	0.03	0.24–0.36	0.95	0.01	0.93–0.96
	z_2	0.34	0.03	0.29–0.40	0.97	0.01	0.96–0.98
	z_3	0.41	0.02	0.36–0.44	0.95	0.01	0.93–0.97
PQ	z_4	0.41	0.02	0.36–0.45	0.98	0.00	0.97–0.98
	z_5	0.39	0.03	0.35–0.46	0.98	0.01	0.97–0.99
	z_6	0.22	0.02	0.19–0.28	0.96	0.01	0.95–0.97
PV	z_7	0.08	0.04	0.01–0.15	0.95	0.01	0.93–0.97
	z_8	0.93	0.03	0.86–0.99	1.00	0.00	1.00–1.00
CS	z_9	0.55	0.05	0.47–0.67	0.99	0.00	0.98–0.99
	z_{10}	0.14	0.05	0.03–0.22	0.97	0.01	0.96–0.98
	z_{11}	0.34	0.02	0.30–0.37	0.95	0.01	0.92–0.97
CC	z_{12}	1	0	1–1	1	0	1–1
CL	z_{13}	0.40	0.04	0.30–0.47	0.98	0.00	0.97–0.98
	z_{14}	0.62	0.04	0.55–0.71	0.99	0.00	0.98–1.00

CC, customer complaints; CE, customer expectations; CI, confidence intervals; CL, customer loyalty; CS, customer satisfaction; PQ, perceived quality; PV, perceived value; SE, standard errors.

TABLE 9.3

Estimates of Path Coefficients and Their Standard Errors and 95% Confidence Intervals Obtained from (Nonregularized) Generalized Structured Component Analysis

	Estimate	SE	95% CI
CE → PQ (b_1)	0.92	0.01	0.90–0.94
CE → PV (b_2)	−0.08	0.12	−0.37–0.10
CE → CS (b_4)	−0.04	0.04	−0.11–0.03
PQ → PV (b_3)	0.96	0.12	0.76–1.23
PQ → CS (b_5)	0.73	0.04	0.64–0.82
PV → CS (b_6)	0.31	0.03	0.25–0.37
CS → CC (b_7)	−0.47	0.07	−0.59 to −0.32
CS → CL (b_8)	0.49	0.05	0.42–0.61
CC → CL (b_9)	−0.44	0.05	−0.53 to −0.33

CC, customer complaints; CE, customer expectations; CI, confidence intervals; CL, customer loyalty; CS, customer satisfaction; PQ, perceived quality; PV, perceived value; SE, standard errors.

TABLE 9.4

Estimates of Path Coefficients and Their Standard Errors and 95%
Confidence Intervals Obtained from Lasso Generalized
Structured Component Analysis

	Estimate	SE	95% CI
CE → PQ (b_1)	0.87	0.01	0.85–0.89
CE → PV (b_2)	0	—	—
CE → CS (b_4)	0	—	—
PQ → PV (b_3)	0.83	0.07	0.71–0.99
PQ → CS (b_5)	0.67	0.03	0.60–0.73
PV → CS (b_6)	0.29	0.04	0.22–0.36
CS → CC (b_7)	−0.42	0.08	−0.57 to −0.26
CS → CL (b_8)	0.46	0.06	0.34–0.56
CC → CL (b_9)	−0.41	0.05	−0.52 to −0.30

CC, customer complaints; CE, customer expectations; CI, confidence
intervals; CL, customer loyalty; CS, customer satisfaction; PQ, perceived
quality; PV, perceived value; SE, standard errors.

model, particularly when a large number of indicators are used for constructing a latent variable. In addition, it may be necessary to select a suitable set of latent variables to facilitate the simplicity, interpretability, and predictive power of the structural model. In this chapter, we discuss that generalized structured component analysis can be combined with the lasso in a unified framework in order to address the issue of variable selection in both indicators and latent variables. An alternating coordinate-descent algorithm was developed for parameter estimation of lasso generalized structured component analysis, in which each set of parameters was updated by minimizing an L_1-penalized linear regression via the coordinate-descent algorithm. An application was provided to illustrate the usefulness of lasso generalized structured component analysis.

Lasso generalized structured component analysis is still in an early stage of development. Thus, it can be further extended and refined to improve its data-analytic scope and capability. A potentially intriguing extension may be to utilize the elastic net for variable selection, instead of the lasso. The elastic net (Zou and Hastie 2005) was proposed as an alternative to the lasso when there are more variables than observations. The elastic net takes into consideration both ridge-type of regularization and the lasso. Another possibility may be to consider the group lasso (Yuan and Lin 2006), which was developed for selecting grouped variables to accurately predict the dependent variable in linear regression. The lasso was designed to select predictor variables individually. However, it is sometimes more desirable to choose a group of predictor variables. For example, in a multifactor analysis of variance, a group of dummy variables represents a factor. In this case, selecting a factor, instead of individual variables, will be of more interest, and the group lasso may select important factors more efficiently than the lasso. In generalized structured component analysis, a latent variable is also constructed by a block of indicators. Thus, selection of latent variables may be achieved more efficiently by adopting the group lasso.

Appendix 9.1 The Alternating Coordinate-Descent Algorithm for Lasso Generalized Structured Component Analysis

Let $\mathbf{A} = [\mathbf{C}^* + \mathbf{B}^*]$, where $\mathbf{C}^* = [\mathbf{C}, 0]$ and $\mathbf{B}^* = [0, \mathbf{B}]$. Then, we can rewrite Equation 9.5 as

$$\phi = \frac{1}{2}\mathrm{SS}(\mathbf{ZV} - \mathbf{ZW}[\mathbf{C}^* + \mathbf{B}^*]) + \lambda_1 \mathbf{1}'\big|\mathbf{W}\big|\mathbf{1} + \lambda_2 \mathbf{1}'\big|\mathbf{C}^*\big|\mathbf{1} + \lambda_3 \mathbf{1}'\big|\mathbf{B}^*\big|\mathbf{1}. \quad (A9.1)$$

Given the prescribed values of λ_1, λ_2, and λ_3, the alternating coordinate-descent algorithm repeats the following three steps until convergence.

Step 1: $\mathbf{C}$ is updated for fixed $\mathbf{W}$ and $\mathbf{B}$. Minimizing Equation A9.1 with respect to $\mathbf{C}$ is equivalent to minimizing

$$\phi_1 = \frac{1}{2}\mathrm{SS}([\mathbf{ZV} - \mathbf{ZWB}^*] - \mathbf{ZWC}^*) + \lambda_2 \mathbf{1}'\big|\mathbf{C}^*\big|\mathbf{1}$$

$$= \frac{1}{2}\mathrm{SS}(\mathbf{T} - \mathbf{ZWC}^*) + \lambda_2 \mathbf{1}'\big|\mathbf{C}^*\big|\mathbf{1}$$

$$= \frac{1}{2}\mathrm{SS}(\mathrm{vec}(\mathbf{T}) - (\mathbf{I} \otimes \mathbf{ZW})\mathrm{vec}(\mathbf{C}^*)) + \lambda_2 \mathbf{1}'\big|\mathbf{C}^*\big|\mathbf{1} \quad (A9.2)$$

with respect to $\mathbf{C}^*$, where $\mathbf{T} = \mathbf{ZV} - \mathbf{ZWB}^*$. Let $\mathbf{c}$ denote a vector formed by eliminating any zero elements in $\mathrm{vec}(\mathbf{C}^*)$. Let $\boldsymbol{\Omega}$ indicate a matrix formed by eliminating the columns of $\mathbf{I} \otimes \mathbf{ZW}$ corresponding to the zero elements in $\mathrm{vec}(\mathbf{C}^*)$. Let $\mathbf{t} = \mathrm{vec}(\mathbf{T})$. Equation A9.2 can then be rewritten as

$$\phi_1 = \frac{1}{2}\mathrm{SS}(\mathbf{t} - \boldsymbol{\Omega}\mathbf{c}) + \lambda_2 \mathbf{1}'\big|\mathbf{c}\big|. \quad (A9.3)$$

This is of the same form as the penalized version of the lasso problem given in Equation 9.4. We use the coordinate-descent algorithm (Friedman et al. 2007) to minimize Equation A9.3. In the algorithm, each element of $\mathbf{c}$, denoted by c_k, is updated, holding the other elements constant. Let $\mathbf{c}_{(-k)}$ denote $\mathbf{c}$ whose kth element is zero. Then, we can have $\boldsymbol{\Omega}\mathbf{c} = \boldsymbol{\Omega}\mathbf{c}_{(-k)} + \boldsymbol{\omega}_k c_k$, where $\boldsymbol{\omega}_k$ is the kth column of $\boldsymbol{\Omega}$. Equation A9.3 can be rewritten as

$$\phi_1 = \frac{1}{2}SS([\mathbf{t} - \mathbf{\Omega}\mathbf{c}_{(-k)}] - \boldsymbol{\omega}_k c_k) + \lambda_2 \mathbf{1}' |\mathbf{c}_{(-k)}| + \lambda_2 |c_k|$$

$$= \frac{1}{2}\sum_{i=1}^{n}(f_i - \omega_{ik}c_k)^2 + \lambda_2 \mathbf{1}' |\mathbf{c}_{(-k)}| + \lambda_2 |c_k|, \tag{A9.4}$$

where f_i is the ith element of $\mathbf{t} - \mathbf{\Omega}\mathbf{c}_{(-k)}$, ω_{ik} is the ith element of $\boldsymbol{\omega}_k$, and $n = N(J + P)$. Minimizing Equation A9.4 with respect to c_k is equivalent to minimizing

$$\phi_1^* = \frac{1}{2}\sum_{i=1}^{n}(f_i - \omega_{ik}c_k)^2 + \lambda_2 |c_k|. \tag{A9.5}$$

The lasso estimate of c_k is obtained by the coordinate-descent algorithm, as follows. If $c_k > 0$, by solving $\partial\phi_1^* / \partial c_k = 0$, the estimate of c_k is obtained by

$$c_k = \frac{\sum_{i=n}^{n}\omega_{ik}f_i}{\sum_{i=1}^{n}\omega_{ik}^2} - \frac{\lambda_2}{\sum_{i=1}^{n}\omega_{ik}^2} \tag{A9.6}$$

The first term of the right-hand side of Equation A9.6 is the ordinary least-squares estimate of c_k, denoted by $\hat{c}_k$. It is easy to see that the updated c_k is obtained by shrinking $\hat{c}_k$ by

$\lambda_2 / \sum_{i=1}^{n}\omega_{ik}^2$. Because we assume $c_k > 0$, $\hat{c}_k - \lambda_2 / \sum_{i=1}^{n}\omega_{ik}^2$

should also be greater than 0. If

$$\hat{c}_k < -\lambda_2 / \sum_{i=1}^{n}\omega_{ik}^2,$$

Equation A9.5 has its minimum value at $c_k = 0$. Consequently, c_k is updated by Equation A9.6, if $\hat{c}_k < 0$ and

$$|\hat{c}_k| > \lambda_2 / \sum_{i=1}^{n}\omega_{ik}^2;$$

otherwise $c_k = 0$. Similarly, if $c_k < 0$, solving $\partial \varphi_1^* / \partial c_k = 0$ gives

$$c_k = \hat{c}_k + \frac{\lambda_2}{\sum\limits_{i=1}^{n} \omega_{ik}^2}. \tag{A9.7}$$

Again, the updated c_k is obtained by shrinking $\hat{c}_k$ toward zero by

$\lambda_2 / \sum\limits_{i=1}^{n} \omega_{ik}^2$. Because we assume $c_k < 0$, $c_k = \hat{c}_k + \lambda_2 / \sum\limits_{i=1}^{n} \omega_{ik}^2$

should also be smaller than 0. If

$$\hat{c}_k < -\lambda_2 / \sum\limits_{i=1}^{n} \omega_{ik}^2,$$

Equation A9.5 is minimized at $c_k = 0$. Thus, c_k is updated by Equation A9.7, if $\hat{c}_k < 0$ and

$$|\hat{c}_k| > \lambda_2 / \sum\limits_{i=1}^{n} \omega_{ik}^2;$$

otherwise $c_k = 0$. In sum, c_k is updated as follows.

$$c_k = \begin{cases} \hat{c}_k - \lambda_2 / \sum\limits_{i=1}^{n} \omega_{ik}^2, & \text{if } \hat{c}_k > 0 \text{ and } \lambda_2 / \sum\limits_{i=1}^{n} \omega_{ik}^2 < |\hat{c}_k| \\[2ex] \hat{c}_k + \lambda_2 / \sum\limits_{i=1}^{n} \omega_{ik}^2, & \text{if } \hat{c}_k < 0 \text{ and } \lambda_2 / \sum\limits_{i=1}^{n} \omega_{ik}^2 < |\hat{c}_k| \\[2ex] 0, & \text{if } \lambda_2 / \sum\limits_{i=1}^{n} \omega_{ik}^2 \geq |\hat{c}_k| \end{cases} \tag{A9.8}$$

In other words, if $|\hat{c}_k|$ is greater than the threshold,

$\lambda_2 / \sum_{i=1}^{n} \omega_{ik}^2$, it is shrunken toward zero by $\lambda_2 / \sum_{i=1}^{n} \omega_{ik}^2$,

and if $|\hat{c}_k|$ does not exceed the threshold, it is set to zero. Each element of **c** continues to be updated until convergence. The updated **C** is then constructed from **c**.

Step 2: **B** is updated for fixed **W** and **C**. Minimizing Equation A9.1 with respect to **B** is equivalent to minimizing

$$\phi_2 = \frac{1}{2} SS([ZV - ZWC^*] - ZWB^*) + \lambda_3 1' |B^*| 1$$

$$= \frac{1}{2} SS(L - ZWB^*) + \lambda_3 1' |B^*| 1$$

$$= \frac{1}{2} SS(\text{vec}(L) - (I \otimes ZW)\text{vec}(B^*)) + \lambda_3 1' |B^*| 1 \qquad (A9.9)$$

with respect to **B***, where **L** = **ZV** − **ZWC***. Let **b** denote a vector formed by removing any zero elements in vec(**B***). Let **H** indicate a matrix formed by eliminating the columns of **I** ⊗ **ZW** corresponding to the zero elements in vec(**B***). Let **g** = vec(**L**). Equation A9.9 can then be rewritten as

$$\phi_2 = \frac{1}{2} SS(g - Hb) + \lambda_3 1' |b|. \qquad (A9.10)$$

As in Step 1, each element of **b**, denoted by b_k, can be updated based on the coordinate-descent algorithm, as follows.

$$b_k = \begin{cases} \hat{b}_k - \lambda_3 / \sum_{i=1}^{n} \eta_{ik}^2, & \text{if } \hat{b}_k > 0 \text{ and } \lambda_3 / \sum_{i=1}^{n} \eta_{ik}^2 < |\hat{b}_k| \\ \\ \hat{b}_k + \lambda_3 / \sum_{i=1}^{n} \eta_{ik}^2, & \text{if } \hat{b}_k < 0 \text{ and } \lambda_3 / \sum_{i=1}^{n} \eta_{ik}^2 < |\hat{b}_k| \\ \\ 0, & \text{if } \lambda_3 / \sum_{i=1}^{n} \eta_{ik}^2 \geq |\hat{b}_k| \end{cases} \qquad (A9.11)$$

where η_{ik} is the (i,k) element of matrix **H** and $\hat{b}_k$ is the ordinary least-squares estimate of b_k. The updated **B** is subsequently constructed from **b**.

Step 3: $\mathbf{W}$ is updated for fixed $\mathbf{B}$ and $\mathbf{C}$ (or equivalently for fixed $\mathbf{A}$). Minimizing Equation A9.1 with respect to $\mathbf{W}$ is equivalent to minimizing

$$\phi_3 = \frac{1}{2}\text{SS}(\mathbf{ZV} - \mathbf{ZWA}) + \lambda_1 \mathbf{1}'|\mathbf{W}|\mathbf{1}. \tag{A9.12}$$

Let $\mathbf{w}_p$ denote the pth column of $\mathbf{W}$, which is shared by the tth column in $\mathbf{V}$, where $t = J + p$ ($p = 1,..., P$). Let $\mathbf{\Lambda} = \mathbf{WA}$. Let $\mathbf{V}_{(-t)}$ denote $\mathbf{V}$ whose tth column is a vector of zeros. Let $\mathbf{V}_{(t)}$ denote $\mathbf{V}$ whose columns are all zero vectors except the tth column. Let $\mathbf{\Lambda}_{(-p)}$ denote a product matrix of $\mathbf{W}$ whose pth column is a vector of zeros and $\mathbf{A}$ whose pth row is a zero vector. Let $\mathbf{\Lambda}_{(p)}$ denote a product matrix of $\mathbf{W}$ whose columns are all zero vectors except the pth column and $\mathbf{A}$ whose rows are all zero vectors except the pth row. Let $\mathbf{m}_{(t)}$ denote a 1 by $J + P$ vector whose elements are all zeros except the tth element being unity. Let $\mathbf{a}_{(p)}$ denote the pth row of $\mathbf{A}$. To update $\mathbf{w}_p$, we can minimize

$$\phi_3^* = \frac{1}{2}\text{SS}((\mathbf{\alpha}' \otimes \mathbf{Z})\mathbf{w}_p - \text{vec}(\mathbf{Z\Delta})) + \lambda_1 \mathbf{1}'|\mathbf{w}_p|, \tag{A9.13}$$

where $\mathbf{\alpha} = \mathbf{m}_{(t)} - \mathbf{a}_{(p)}$ and $\mathbf{\Delta} = \mathbf{\Lambda}_{(-p)} - \mathbf{V}_{(-t)}$

Let $\mathbf{\theta}_p$ denote the vector formed by eliminating any zero elements from $\mathbf{w}_p$. Let $\mathbf{\Xi}$ denote the matrix formed by eliminating the columns of $\mathbf{I} \otimes \mathbf{ZW}$ corresponding to the zero elements in $\mathbf{w}_p$. Let $\mathbf{u} = \text{vec}(\mathbf{Z\Delta})$. Then, Equation A9.13 can be re-expressed as

$$\phi_3^* = \frac{1}{2}\text{SS}(\mathbf{\Xi}\mathbf{\theta}_p - \mathbf{u}) + \lambda_1 \mathbf{1}'|\mathbf{\theta}_p|, \tag{A9.14}$$

Again, each element of $\mathbf{\theta}_p$, denoted by θ_{kp}, can be updated as follows.

$$\theta_{kp} = \begin{cases} \hat{\theta}_{kp} - \lambda_1 / \sum_{i=1}^{n} \xi_{ik}^2, & \text{if } \hat{\theta}_{kp} > 0 \text{ and } \lambda_1 / \sum_{i=1}^{n} \xi_{ik}^2 < |\hat{\theta}_{kp}| \\[2ex] \hat{\theta}_{kp} + \lambda_1 / \sum_{i=1}^{n} \xi_{ik}^2, & \text{if } \hat{\theta}_{kp} < 0 \text{ and } \lambda_1 / \sum_{i=1}^{n} \xi_{ik}^2 < |\hat{\theta}_{kp}| \\[2ex] 0, & \text{if } \lambda_1 / \sum_{i=1}^{n} \xi_{ik}^2 \geq |\hat{\theta}_{kp}| \end{cases} \tag{A9.15}$$

where ξ_{ik} is the (i,k) element of matrix Ξ and $\hat{\theta}_{kp}$ is the ordinary least-squares estimate of θ_{kp}. The updated $\mathbf{w}_p$ is constructed from θ_p, and subsequently multiplied by

$$\sqrt{\frac{N}{\mathbf{w}_p'\mathbf{Z}'\mathbf{Z}\mathbf{w}_p}}$$

in order to satisfy the standardization constraint imposed on the pth latent variable.

10

Dynamic Generalized Structured Component Analysis

10.1 Introduction

Up to now, we have largely assumed that the data we analyze by generalized structured component analysis consist of cases (subjects) by variables matrices, in which the cases can usually be assumed to be statistically independent. However, this is not always the case. In this chapter, we discuss an extension of generalized structured component analysis to analyzing time series data, which often arise from repeated measurements of a quantity or quantities over time. Time series data are temporally (serially) correlated. To account for the correlations, we incorporate multivariate autoregressive models into generalized structured component analysis. The proposed method is called dynamic generalized structured component analysis. The word "dynamic" here refers to a process in which a state of a variable at a particular time may be influenced by a state of the same or other variables at previous times. The bootstrap procedure needs a special attention to deal with the temporal correlations. In this chapter, we treat time as a discrete variable. See Suk and Hwang (2014), and the next chapter (Chapter 11) for an extension of generalized structured component analysis, which treats the time variable as continuous. The method in Chapter 11 also treats time as an extension of variables (corresponding to the columns of a data matrix), whereas the approach in this chapter treats observed time points as an extension of cases (corresponding to the rows).

Dynamic generalized structured component analysis is applicable to any kinds of time series data. In this chapter, however, we focus mainly on the analysis of brain connectivity in functional neuroimaging studies (Friston 1994). In these studies, directional influences among various parts of the brain are investigated based on functional magnetic resonance imaging (fMRI) data (Huettel, Song, and McCarthy 2004). The fMRI data record changes in blood oxygenation over scans (time points), called blood oxygen level-dependent (BOLD) signals, while a subject is presented with stimuli or asked to perform a task. The basic element of spatial measurements in fMRI is referred to as a voxel. A region of interest (ROI) is defined by a collection of voxels that show a common pattern of activations in BOLD signals, and is often used as a unit of analysis. That is, a number of specific brain regions (ROIs) are preselected based on a hypothesis about their importance in performing a task, and then their relationships are modeled and tested.

To get an idea of what the fMRI data are like, let us look at Figure 10.1, in which BOLD signals are plotted over several hundred time points in three ROIs, primary visual cortex (V1), middle temporal area (V5), and superior parietal cortex (SPC). These ROIs are deemed important in performing a visual task. In this figure, BOLD signals at five voxels are superposed on top of each other within each ROI. It can be observed that the BOLD signals go up and down rather irregularly over time. Note also that there are quite a bit of variations among voxels within ROIs. Nonetheless, it seems that there are common variations across voxels within ROIs. We are interested in these common variations and the relationships among them. In dynamic generalized structured component analysis, BOLD signals in voxels in a given ROI correspond to a block of observed variables, and common activities in ROIs correspond to latent variables. We investigate how the latent variables are related to the observed BOLD signals, and how they are related to each other and to stimulus inputs. Later, we will analyze the data, part of which is shown in Figure 10.1, where a more detailed description of the data is given.

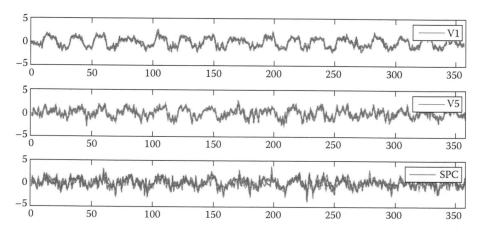

FIGURE 10.1
Time series plots of five voxels in each of three regions of interests, primary visual cortex (V1), middle temporal area (V5), and superior parietal cortex (SPC).

10.2 The Method

10.2.1 The Model

As in other structural equation models, dynamic generalized structured component analysis consists of the measurement and structural models. The measurement model specifies hypothesized relationships between observed and latent variables. This part of the proposed model remains essentially the same as in the original generalized structured component analysis model (Hwang and Takane 2004). The structural model, on the other hand, specifies hypothesized relationships among latent variables. This part of the proposed model has many new features to capture the dynamic nature of time series data, and to accommodate input variables in brain imaging studies. Specifically, dynamic generalized structured component analysis has a mechanism for examining relationships between latent variables involving different time points (called time-lagged effects) by incorporating a multivariate autoregressive model. In addition, dynamic generalized structured component analysis is able to investigate direct and modulating effects of input variables on specific latent variables and on connections between the latent variables, respectively. Note, however, that the proposed method is designed to analyze multiple-time and multiple-voxel data collected from a single subject at a time. Later we consider some extensions of the method to multiple-subject data.

Let us begin with a brief description of the measurement model. Let Z_i ($i = 1, ..., p$) denote a T by v_i matrix of observations, where i indexes a latent variable (ROI), T indicates the number of time points, v_i is the number of observed variables (voxels) for latent variable i, and p is the total number of latent variables (ROIs). (In conventional generalized structured component analysis, the number of cases (subjects) replaces T.) The matrix Z_i is assumed to be columnwise standardized. As in conventional generalized structured component analysis, the weighted relation model is specified to define a latent variable, γ_i, as an exact linear combination of the observed variables. It is assumed that each latent variable is scaled to have unit variance. Let w_i denote a v_i-element vector of component weights. Then, the weighted relation model is

$$\gamma_i = Z_i w_i \tag{10.1}$$

for $i = 1, ..., p$. As noted earlier, the measurement model specifies the relationship between the observed variables Z_i and the latent variable γ_i. Let c_i denote the vector of weights applied to γ_i to best approximate Z_i. Then, the measurement model for latent variable i is stated as

$$Z_i = \gamma_i c_i' + E_i^{(M)} = Z_i w_i c_i' + E_i^{(M)} \tag{10.2}$$

for $i = 1, \ldots, p$, where $\mathbf{E}_i^{(M)}$ is the matrix of disturbance terms. As noted earlier, $\mathbf{Z}_i$ and γ_i are standardized, and $\mathbf{w}_i$ and $\mathbf{c}_i$ are scaled accordingly.

It is convenient to write Equations 10.1 and 10.2 for $i = 1, \ldots, p$ as a single equation. Define a row block matrix $\mathbf{Z}$ by $\mathbf{Z} = [\mathbf{Z}_1, \mathbf{Z}_2, \ldots, \mathbf{Z}_p]$. The total number of columns in $\mathbf{Z}$ is denoted by

$$V = \sum_{i=1}^{p} v_i.$$

Also, define a block diagonal matrix $\mathbf{W}$ with $\mathbf{w}_i$ as the ith diagonal block, a similar block diagonal matrix $\mathbf{C}$ with $\mathbf{c}_i'$ as the ith diagonal block, and a row block matrix with γ_i as the ith column vector, $\Gamma = [\gamma_1, \ldots, \gamma_p]$. Then, Equations 10.1 and 10.2 can collectively be written as

$$\Gamma = \mathbf{ZW}, \tag{10.3}$$

and

$$\mathbf{Z} = \Gamma\mathbf{C} + \mathbf{E}^{(M)} = \mathbf{ZWC} + \mathbf{E}^{(M)}, \tag{10.4}$$

where $\mathbf{E}^{(M)} = [\mathbf{E}_1^{(M)}, \cdots, \mathbf{E}_p^{(M)}]$.

We now specify the structural model, for which several more matrices have to be introduced. Let $\mathbf{u}_j$ represent the T-element vector of the jth input variable ($j = 1, \ldots, J$). It is assumed that each $\mathbf{u}_j$ is *a priori* standardized. An example of input variable time series is depicted in Figure 10.2. Although input variables can be any kind of time series vectors of interest, here we have an input vector simulating a hemodynamic response function. Specifically, we first generated a delta function (0 = no stimulus, 1 = stimulus) with an experimental stimulus presented every 15th time point starting at time 5 (in the upper panel), and then the input was convolved with a gamma function to model the hemodynamic response (in the bottom panel). We used the software Statistical Parametric Mapping (SPM; Wellcome Institute of Cognitive Neurology, London, UK, http://www.fil.ion.ucl.ac.uk/spm) for the convolution.

It is assumed that in the general case, there are J input vectors, which we denote collectively by the row block matrix $\mathbf{U} = [\mathbf{u}_1, \ldots, \mathbf{u}_J]$. Let diag($\mathbf{u}_j$) be the diagonal matrix with the elements of $\mathbf{u}_j$ as its diagonal elements. This diagonal matrix plays an important role in defining interaction effects between input variables and latent variables.

Shift matrices are now introduced to capture time-lagged effects. However, we define a series of shift matrices, so that both contemporaneous and lagged effects can be treated in a unified manner. Note that the contemporaneous effects denote concurrent relations between latent variables at the same time

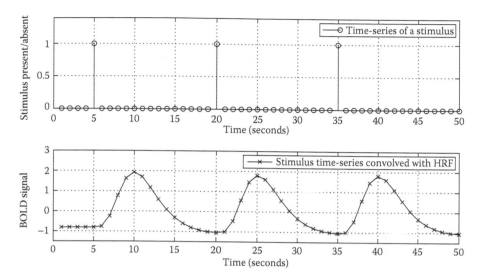

FIGURE 10.2
An example of a stimulus effect. A delta function (in the upper panel) and the convoluted input variable using the hemodynamic response function.

point. Specifically, the shift matrix with time lag 0 ($S_0 = I_T$, where I_T is the identity matrix of order T) represents contemporaneous effects, whereas S_l ($l = 1, ..., L$) represents time-lagged effects of order l. The matrix S_1, for example, is used to obtain the effect of time $t - 1$ on time t. This matrix looks like:

$$S_l = \begin{bmatrix} 0 & 0 & \cdots & 0 & 0 \\ 1 & 0 & \cdots & 0 & 0 \\ 0 & 1 & \cdots & 0 & 0 \\ \vdots & \vdots & \ddots & \vdots & \vdots \\ 0 & 0 & \cdots & 1 & 0 \end{bmatrix}. \tag{10.5}$$

The matrix above is obtained by down shifting the first $T - 1$ rows of S_0 to row 2 through row T of S_1 and by filling the first row by a zero vector. In general, $S_l = (S_1)^l$ for $l \geq 1$. Premultiplying Γ by S_l shifts down the rows of Γ by l rows and defines the matrix of the effect of latent variables at time $t - l$ on time t.

The generic structural model of dynamic generalized structured component analysis can be stated as

$$\Gamma = \sum_{l=0}^{L1} S_l \Gamma A_l + \sum_{l=0}^{L2} S_l UD_l + \sum_{l=0}^{L3} \sum_{j=1}^{J} S_l \text{diag}(u_j) \Gamma M + E^{(S)}, \tag{10.6}$$

where the $\mathbf{A}_l$ ($l = 0, \ldots, L_1$, where L_1 is the maximum time lags for between-ROI effects) are square matrices of order p of path coefficients for connections between latent variables of varying lags, the $\mathbf{D}_l$ ($l = 0, \ldots, L_2$, where L_2 is the maximum time lags for direct stimulus effects) are J by p matrices of path coefficients for the direct effects of input variables on latent variables, the $\mathbf{M}_{lj}$ ($l = 0, \ldots, L_3$, where L_3 is the maximum time lags for modulating effects of stimuli, and $j = 1, \ldots, J$) are square matrices of order p of path coefficients for modulating effects of input variables on connections between latent variables, and $\mathbf{E}^{(S)}$ is the matrix of prediction errors. These matrices typically have many prescribed zero elements representing the hypothesis that certain effects are 0. There are two kinds of zeros. One is called structural zeros, which always take the value of zero. An example of this is diagonal elements of $\mathbf{A}_0$, which are always set to zero (i.e., $\mathbf{A}_0$ is hollow). This implies that there are no contemporaneous effects of ROIs on themselves, which is logically impossible. The other is optional zeros. An example of this is off-diagonal elements of $\mathbf{A}_1$, which may be set to zeros, implying that the lag 1 between-ROI effects are purely autoregressive. There are no time-lagged effects between different ROIs (i.e., no cross-regressive effects). Free elements in the parameter matrices, on the other hand, represent the hypothesis that the corresponding effects are nonzero and are to be estimated from data. The first term in Equation 10.6 thus represents the contemporaneous and lagged effects among latent variables, the second term the direct effects of input variables on latent variables, and the third term the modulating effects of input variables on connections between latent variables. Note that Equation 10.6 reduces to the structural model in conventional generalized structured component analysis when there is only the first term and $L_1 = 0$.

For illustration, the structural model (the path diagram) for the data presented in Figure 10.1 is depicted in Figure 10.3. The three latent variables, γ_1, γ_2, and γ_3, correspond to V1, V5, and SPC, respectively. In the hypothesized model, the three latent variables are all bidirectionally connected. Contemporaneous effects between latent variables are indicated by solid arrows. Path coefficients indicating the strength of these effects are denoted by a_1 through a_6. We also assume that there are autoregressive effects of lag 1 of the latent variables on themselves. The time-lagged effects are indicated by dashed arrows. The path coefficients associated with the autoregressive effects are denoted by a_7 through a_9. Figure 10.3 also indicates that there are three input variables. To get an idea of how they look, the time series of the three input variables are plotted in Figure 10.4. A more detailed explanation of the stimuli will be given in the example section (Section 10.3.1). It is hypothesized that (1) the first input variable ($\mathbf{u}_1$) has a direct effect on γ_1 (with the path coefficient denoted as d_1), (2) the second input variable ($\mathbf{u}_2$) modulates the connection from γ_1 to γ_2 (with the path coefficient denoted as m_1), and (3) the third input variable ($\mathbf{u}_3$) modulates the connection from γ_3 to γ_2 (with the path coefficient denoted as m_2). The modulation effect here means that connectivity between latent variables is temporarily enhanced or

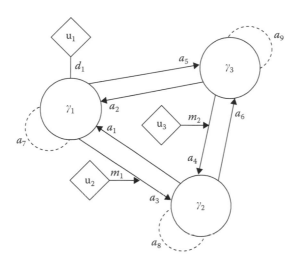

FIGURE 10.3
A structural model for the data set, part of which is displayed in Figure 10.1.

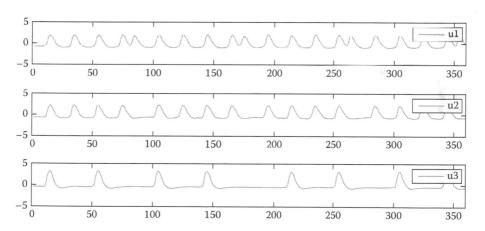

FIGURE 10.4
Time series plots of three input variables in the structural model depicted in Figure 10.3.

reduced as a result of an input variable effect. The models in this diagram can be written as

$$\gamma_1 = \gamma_2 a_1 + \gamma_3 a_2 + \mathbf{S}_1 \gamma_1 a_7 + \mathbf{u}_1 d_1 + \mathbf{e}_1^{(S)} \tag{10.7}$$

$$\gamma_2 = \gamma_1 a_3 + \gamma_3 a_4 + \mathbf{S}_1 \gamma_2 a_8 + \text{diag}(\mathbf{u}_2)\gamma_1 m_1 + \text{diag}(\mathbf{u}_3)\gamma_3 m_2 + \mathbf{e}_2^{(S)}, \tag{10.8}$$

and

$$\gamma_3 = \gamma_1 a_5 + \gamma_2 a_6 + S_1 \gamma_3 a_9 + e_3^{(S)}, \tag{10.9}$$

or collectively as

$$\Gamma = \Gamma A_0 + S_1 \Gamma A_0 + u_1 d_{10} + \mathrm{diag}(u_2) \Gamma M_{02} + \mathrm{diag}(u_3) \Gamma M_{03} + E^{(S)}, \tag{10.10}$$

where

$$A_0 = \begin{bmatrix} 0 & a_3 & a_5 \\ a_1 & 0 & a_6 \\ a_2 & a_4 & 0 \end{bmatrix},$$

$$A_1 = \begin{bmatrix} a_7 & 0 & 0 \\ 0 & a_8 & 0 \\ 0 & 0 & a_9 \end{bmatrix},$$

$$d_{01} = \begin{bmatrix} d_1 & 0 & 0 \end{bmatrix}$$

(where d_{01} is the first row of D_0),

$$M_{02} = \begin{bmatrix} 0 & m_1 & 0 \\ 0 & 0 & 0 \\ 0 & 0 & 0 \end{bmatrix},$$

and

$$M_{02} = \begin{bmatrix} 0 & 0 & 0 \\ 0 & 0 & 0 \\ 0 & m_2 & 0 \end{bmatrix}.$$

In general, the rows of A_l and M_{lj} represent latent variables exerting influence, whereas the columns represent latent variables being influenced. The rows of D_l, on the other hand, represent input variables exerting influence on latent variables corresponding to the columns. Later we analyze the data presented in Figure 10.1 by fitting the model displayed in Figure 10.3 with stimulus effects depicted in Figure 10.4.

The FIT is defined in essentially the same way as in conventional generalized structured component analysis. The numerator of the second term in Equation 2.20 should be replaced by the minimized value of φ defined in Equation 10.10, whereas the denominator remains exactly the same, which is equal to $TV + p$ in the current notation. The AFIT is similarly defined to Equation 2.21, except that there usually are many more unknown parameters to be estimated in dynamic generalized structured component analysis, which should be taken into account in calculating d_1. In the example given above, the number of path coefficients alone amounts to 12 = 6 (contemporaneous bidirectional effects) + 3 (autoregressive effects of lag 1) + 1 (a direct stimulus effect) + 2 (modulating stimulus effects), which constitutes part of G along with the number of component weights and loadings.

10.2.2 Parameter Estimation

As in generalized structured component analysis, a least-squares (LS) criterion is used for parameter estimation. Specifically,

$$\varphi = SS(\mathbf{E}) = tr(\mathbf{E}'\mathbf{E}) \tag{10.11}$$

is minimized with respect to model parameters, where $\mathbf{E} = [\mathbf{E}^{(M)}, \mathbf{E}^{(S)}]$. An alternating LS algorithm (e.g., de Leeuw, Young, and Takane 1976) is used to find LS estimates of parameters. The alternating LS algorithm is particularly attractive in the present context because dynamic generalized structured component analysis has two natural subsets of parameters; $\mathbf{w}_i$'s that define latent variables and all other parameters (component loadings; $\mathbf{c}_i$'s, and path coefficients; $\mathbf{A}_l$'s, $\mathbf{D}_l$'s, and $\mathbf{M}_{lj}$'s). The proposed algorithm thus consists of two major steps:

Step I. In the first step, component loadings, $\mathbf{c}_i$'s, and path coefficients, $\mathbf{A}_l$'s, $\mathbf{D}_l$'s, and $\mathbf{M}_{lj}$'s, are updated while $\mathbf{w}_i$'s are fixed.

Step II. Component weights $\mathbf{w}_i$'s are updated while $\mathbf{c}_i$'s, $\mathbf{A}_l$'s, $\mathbf{D}_l$'s, and $\mathbf{M}_{lj}$'s are fixed.

Details of these two steps will be given in Appendix 10.1. Initial estimates of $\mathbf{w}_i$'s may be randomly generated or obtained by applying principal component analysis to $\mathbf{Z}_i$'s ($i = 1, ..., p$).

10.2.3 A Special Bootstrap Method

In generalized structured component analysis, a bootstrap method (Efron 1982) is used for assessing the reliability of parameter estimates, since it is impossible to obtain analytic expressions for standard errors of parameter estimates. However, the standard bootstrap method used in conventional generalized structured component analysis is not appropriate for time series data because it does not take into account the time order of observations, and consequently it results in incorrect standard error estimates (Lahiri 2003). In dynamic generalized structured component analysis, a modified moving block bootstrap method (Jung et al. 2012) is employed to deal with the problem.

In the conventional bootstrap method, individual cases corresponding to the rows of the data matrix $\mathbf{Z}$ are randomly sampled with replacement. In the modified moving block bootstrap method, on the other hand, blocks of cases are sampled to maintain the time sequence of observations within blocks. The block size is determined as the sequence of time points over which serial correlations are deemed non-negligible. These correlations are supposed to arise from autoregressive and other time-lagged effects included in the structural model. To account for these effects at a particular time point, information about preceding time points is necessary, which is retained by the previous observations within blocks. This is the essential idea behind sampling blocks in the modified moving block bootstrap method.

The modified moving block bootstrap method essentially boils down to the following simplified procedure. If a certain row of the data matrix $\mathbf{Z}$ is sampled in a bootstrap sample, the same rows of $\mathbf{S}_l\mathbf{Z}$ ($l = 1, \ldots, L_1$), $\mathbf{S}_l\mathbf{U}$ ($l = 0, \ldots, L_2$), and $\mathbf{S}_l\mathrm{diag}(\mathbf{u}_j)\mathbf{Z}$ ($l = 0, \ldots, L_3; j = 1, \ldots, J$), as they appear in the structural model, should also be sampled. These auxiliary data provide crucial information to derive a proper prediction of $\boldsymbol{\Gamma}$ for the bootstrap sample. The rest of the bootstrap procedure proceeds in essentially the same manner as in the conventional bootstrap method. An important thing is that the specified structural model account for most, if not all, of the serial correlations present in the data. In the examples of analysis of fMRI data to be reported in this chapter, it is always assumed that first-order time-lag effects are sufficient to "explain away" all the serial correlations. In the fMRI data, the time resolution is typically not very high, and thus inclusions of higher order time-lag effects are not likely to be substantively important. There is also some evidence in the literature (e.g., Gates et al. 2011) that serial correlations beyond order 1 are negligible, once the first-order time-lag effects are taken into account.

10.3 Examples of Application to Real Functional Neuroimaging Data

In this section, we report analyses of two real data sets by dynamic generalized structured component analysis to demonstrate its use in empirical research.

10.3.1 The Attention to Visual Motion Data

The first example pertains to the "attention to visual motion" data available from the SPM web site (http://www.fil.ion.ucl.ac.uk/spm/data/attention/).

As noted earlier, there are three ROIs in the data set: V1 (the primary visual cortex), V5 (middle temporal area), and SPC (superior parietal cortex). BOLD signals for the three ROIs were extracted using the SPM software. Specifically, all activated voxels over a threshold level corresponding to an experimental condition were first identified using SPM, and then BOLD signals for a ROI were selected with an 8 mm sphere centered on the global maxima within a ROI (for more details, see Friston et al. 2007). A total of 92 voxels were selected as multiple indicators for the three ROIs and used in dynamic generalized structured component analysis: 54 voxels for V1, 24 voxels for V5, and 14 voxels for SPC. The BOLD signals were extracted from each of 360 scans ($T = 360$) with the time interval of 3.22 seconds.

Part of the data set is displayed in Figure 10.1. As noted earlier, the time series plots are displayed only for five voxels in each ROI. Although there is some variability among the voxels within a ROI, there was also some common variability. This common variability is our main focus of analysis. It is considered representative of neuronal activities in the ROI and is captured as a latent variable corresponding to the ROI. We are interested in its relationships with activities in other ROIs (contemporaneous effects among ROIs), with its previous states (autoregressive effects), and with stimulus inputs (effects of experimental stimuli).

In the experiment, subjects performed a visual motion processing task under four different experimental conditions while undergoing fMRI records. The four experimental conditions were fixation, static (nonmoving dots), no attention (moving dots but no attention required), and attention. Three experimental stimuli (i.e., input variables) were created by combining the four conditions: photic ($\mathbf{u}_1$, comprising all conditions with visual inputs), motion ($\mathbf{u}_2$, including all conditions with moving dots), and attention ($\mathbf{u}_3$, including the only conditions with attention to moving dots). The time series of these three experimental inputs are presented in Figure 10.4.

The structural model used in this analysis is depicted in Figure 10.3. (See also Equation 10.7 through 10.10.) Recall that the three ROIs in the hypothesized model were fully and bidirectionally connected with each other. It was also assumed that there were autoregressive effects of lag 1 of ROIs on themselves. Furthermore, it was hypothesized that (1) the experimental input "photic" ($\mathbf{u}_1$) had a direct effect on V1 (γ_1), (2) the experimental manipulation "motion" ($\mathbf{u}_2$) modulated the connection from V1 (γ_1) to V5 (γ_2), and (3) the experimental manipulation "attention" ($\mathbf{u}_3$) modulated the connection from SPC (γ_3) to V5 (γ_2).

Dynamic generalized structured component analysis was applied to fit the specified model to the data. The overall fit of the model was found to be FIT = 0.767 and AFIT = 0.766, indicating that the specified model accounted for about 77% of the total variance of all observed and latent variables. Table 10.1 shows estimates of path coefficients, standard errors, and the corresponding p-values. Bootstrap standard errors were calculated based on 500 bootstrap samples using the special bootstrap method described in

TABLE 10.1

Estimates of Path Coefficients, Standard Errors, and
p-Values in the Attention to Visual Motion Study

Coefficient	Path	Estimate	SE	p-Value
a_1	V1→V1	0.35	0.03	0
a_2	SPC→V1	0.15	0.04	0
a_3	V1→V5	0.60	0.06	0
a_4	SPC→V5	0.31	0.05	0
a_5	V1→SPC	0.34	0.07	0
a_6	V5→SPC	0.40	0.06	0
a_7	V1→V1	0.33	0.03	0
a_8	V5→V5	−0.04	0.06	0.18
a_9	SPC→SPC	0.05	0.05	0.14
d_1	V1	0.27	0.03	0
m_1	V1→V5	−0.03	0.04	0.18
m_2	SPC→V5	0.02	0.05	0.29

SE, standard errors; V1, primary visual cortex; V5, middle temporal
area; SPC, superior parietal cortex.

Section 10.2.3. All contemporaneous effects (a_1 through a_6) were statistically significant at $\alpha = 0.05$. Only one autoregressive effect (a_7) turned out to be statistically significant, whereas the other two autoregressive effects (a_8 on V1 and a_9 on SPC) were not. The direct effect (d_1) of $\mathbf{u}_1$ on V1 was statistically significant. However, neither of the two modulating effects of experimental stimuli was statistically significant.

To confirm some of these effects, we display, in Figure 10.5, the time series of $\mathbf{u}_1$ (the top panel), the three latent variables representing the three ROIs (the second, fourth, and sixth panels), and the interaction effects between $\mathbf{u}_2$ and V1 on V5, and between $\mathbf{u}_3$ and SPC on V5 (the third and fifth panels, respectively). The time series plot of $\mathbf{u}_1$ looks fairly highly correlated with V1 ($r = 0.73$). This confirms the statistically significant direct effect of $\mathbf{u}_1$ on V1. The modulating effects of $\mathbf{u}_2$ and $\mathbf{u}_3$ seem only negligibly correlated with V5 ($r = -0.10$ and $r = 0.06$, respectively). Thus, it makes intuitive sense that these modulating effects are not statistically significant.

Component loadings indicating the relationships between ROIs (latent variables) and their respective indicators (i.e., observed BOLD signals) were found to be fairly strong and homogeneous. Most of them were positive and greater than 0.80.

10.3.2 The Memory Data

In the second example, we again considered a fully and bidirectionally connected structural model. The number of ROIs was, however, increased from three to seven, and there was no stimulus input. This model is a popular

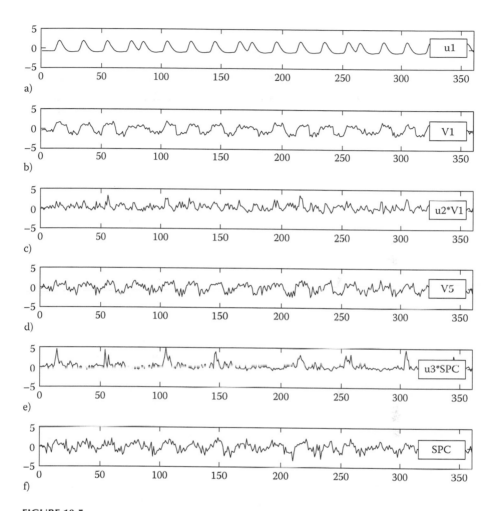

FIGURE 10.5

The time series of u_1 (the top panel), and the three latent variables (the second, fourth, and sixth panels), and the interactions between u_2 and γ_1, and between u_3 and γ_3 on γ_2 (the third and fifth panels).

brain network analysis model (Smith et al. 2010), which is also known as being difficult to fit by factor-based methods of structural equation models (e.g., Gates et al. 2011).

In this example, we used the data collected as part of a larger study examining changes in topological patterns of large-scale functional brain networks during memory tasks (Grady et al. 2006) available at the fMRI Data Center (http://www.fmirdc.org). In the study, participants performed four encoding and two recognition tasks. The encoding tasks involved the presentation of pictures or words in either perceptual or semantic conditions.

After the encoding tasks, participants performed two recognition tasks. In each recognition condition, participants responded as to whether each stimulus was familiar or novel.

Using the data, Wang et al. (2010) found several network hubs in association cortex regions using a correlation-based network analysis. Specifically, in their analysis, the fMRI data were first parceled into 90 cortical and subcortical regions (ROIs) using an automated anatomical labeling brain atlas. Then, a representative time series with 154 data points was extracted for each ROI by averaging the BOLD signals over all voxels within the ROI. They then conducted a correlation-based network analysis to reveal alterations in topology of functional brain networks during the performance of memory tasks.

For the present analysis, we used the data extracted from one adult belonging to an older adult group (the mean age of this group was 74.4 years) during a recognition task. We adopted the network hubs of older adults in a recognition task, revealed by Wang et al. (2010), and constructed a hypothesized structural model. Specifically, the network hubs consisted of seven ROIs in the left-hemisphere: insula (INS), median cingulate and paracingulate gyri (DCG), hippocampus (HIP), middle occipital gyrus (MOG), precuneus (PCUN), thalamus (THA), and middle temporal gyrus (MTG). The number of voxels was 542 in INS, 575 in DCG, 268 in HIP, 968 in MOG, 1064 in PCUN, 305 in THA, and 1487 in MTG. There were simply too many voxels (observed variables) to be processed for each ROI. So they were aggregated into five distinct signals for each ROI, capturing representative patterns of BOLD signals within the ROI. We calculated the centroids of clusters, which were used as indicators of ROIs in dynamic generalized structured component analysis. For the cluster analysis, the K-means algorithm in MATLAB was used (www.mathworks.com).

Dynamic generalized structured component analysis was applied to fit the specified model to the data. The specified model provided FIT = 0.714 and AFIT = 0.709, indicating that it accounted for about 71% of the total variance of all observed and latent variables. Table 10.2 shows the estimates of path coefficients, their standard errors, and p-values. Note that the diagonal portion of the table indicates the autoregressive effects of ROIs on themselves. As the table indicates, a majority of path coefficients are found to be statistically significant. In this way, we arrive at a new model (*albeit* tentative) that is different from the initially postulated. Figure 10.6 depicts the path diagram (the structural model) that we have arrived at as a result of the analysis.

It is also possible (although not demonstrated here) to test the statistical significance of any contrasts between parameters of interest. For example, $a_{0,ij}$ indicates the contemporaneous effect of ROI i on ROI j, whereas $a_{0,ji}$ indicates the effect of ROI j on ROI i. It is intriguing to test if there is a significant difference between the two. We may then bootstrap $a_{0,ij} - a_{0,ji}$ to see if the difference is significantly positive, negative, or neither.

TABLE 10.2

Estimates of Path Coefficients, (Top) Their Standard Errors (Middle), and *p*-Values (Bottom) for the Memory Study

	INS	DCG	HIP	MOG	PCUN	THA	MTG
INS	0.36	−0.29	−0.17	0.02	0.20	0.14	0.05
	(0.09)	(0.05)	(0.06)	(0.06)	(0.05)	(0.06)	(0.06)
	0	0	0	**0.30	0	0.00	**0.22
DCG	−0.74	0.14	−0.71	−0.10	0.64	0.35	0.24
	(0.12)	(0.06)	(0.09)	(0.08)	(0.08)	(0.10)	(0.11)
	0	0.01	0	**0.11	0	0.00	0.01
HIP	−0.39	−0.68	0.12	0.05	0.17	0.41	−0.08
	(0.14)	(0.08)	(0.07)	(0.08)	(0.10)	(0.09)	(0.11)
	0.00	0	**0.07	**0.23	0.02	0	**0.23
MOG	−0.09	−0.14	0.17	0.37	0.30	−0.26	0.65
	(0.16)	(0.10)	(0.11)	(0.05)	(0.10)	(0.08)	(0.09)
	**0.30	0.02	**0.05	**0	**0	**0.00	**0
PCUN	0.84	0.54	0.25	0.32	0.24	0.27	−0.41
	(0.14)	(0.07)	(0.09)	(0.08)	(0.08)	(0.09)	(0.09)
	0	0	0.00	0	0.00	0	0
THA	0.35	0.28	0.72	−0.09	0.28	0.10	0.53
	(0.19)	(0.13)	(0.11)	(0.09)	(0.11)	(0.08)	(0.09)
	0.02	0.01	0	**0.15	0.01	**0.12	0
SMA	−0.10	0.30	−0.08	0.54	−0.40	0.45	0.07
	(0.18)	(0.13)	(0.13)	(0.07)	(0.10)	(0.08)	(0.07)
	**0.28	0.01	**0.24	0	0	0	**0.21

Note: ROIs in rows exert influence on ROIs in columns. The diagonal cells indicate autoregressive effects of ROIs on themselves. Nonsignificant paths at $\alpha = 0.05$ are marked by two asterisks on p-values.

INS, insula; DCG, median cingulate and paracingulate gyri; HIP, hippocampus; MOG, middle occipital gyrus; PCUN, precuneus; THA, thalamus; MTG, middle temporal gyrus.

10.4 Summary and Discussion

In this chapter, we discussed dynamic generalized structured component analysis for the analysis of multivariate time series data. This method extends conventional generalized structured component analysis by incorporating a multivariate autoregressive model to accommodate the dynamic nature of data taken over time. It is capable of examining both contemporaneous and time-lagged effects between latent variables, as well as direct and modulating effects of input variables on the latent variables and their connections.

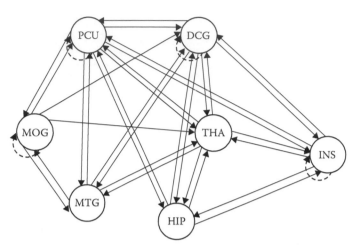

FIGURE 10.6
The resultant path diagram in the second empirical example.

With proper modifications of the conventional bootstrap method, it can also assess the reliability of parameter estimates despite the dependencies among observations over time. Two real data sets were analyzed. Through the exemplary analyses, it has been demonstrated that dynamic generalized structured component analysis allows specifications of complex bidirectional relations in the structural model and provides stable parameter estimates with no computational difficulties such as improper solutions and nonconvergence.

There have been two factor-based SEM methods, called unified SEM (Kim et al. 2007) and extended unified SEM (euSEM; Gates et al. 2011). These methods have features similar to dynamic generalized structured component analysis, incorporating time-lagged effects and stimulus effects in the structural model. In parameter estimation, however, they use a lagged covariance (correlation) matrix as input data and a maximum likelihood estimation method under the multivariate normality assumption. As this covariance matrix is large when there are many observed variables (voxels), latent variables are constructed outside the SEM methods themselves. The externally constructed latent variables are then treated as if they were the observed variables. A problem with this procedure is that the construction of the latent variables is carried out without any consideration of their relationships. Furthermore, asymptotic properties of the maximum likelihood estimators, developed under the assumption of statistical independence among observations, do not hold under correlated observations. In dynamic generalized structured component analysis, on the other hand, both measurement model (specifying the relationships between observed and latent variables) and the structural model (specifying the relationships among the latent variables) are simultaneously taken into account in a unified framework.

Dynamic generalized structured component analysis also employs a special bootstrap method to deal with time dependence among consecutive observations, thereby obtaining more accurate standard errors of parameter estimates in the analysis of fMRI data.

The model for dynamic generalized structured component analysis may be extended in a variety of ways to enhance its data-analytic capability. We briefly discuss two of them below, incorporating latent interactions among latent variables and simultaneous analysis of multiple-subject data.

One possible extension of dynamic generalized structured component analysis involves incorporation of latent interactions. A latent (linear by linear) interaction is defined as a product of interacting latent variables (Hwang, Ho and Lee 2010; Kenny and Judd 1984; see also Chapter 6 of this volume). This extension may be particularly useful for brain connectivity analysis to model changes in the magnitude of path coefficients between ROIs as a function of activities of different ROIs, thereby capturing important aspects of neuronal interactions. This kind of feature is required for modeling complex neurobiological processes, including top-down modulation, learning, and effects exerted by neuromodulatory transmitters (Stephan et al. 2008).

To incorporate latent interactions, we add the following term to the predictor side of the structural model in (Equation 10.6):

$$\sum_{l=0}^{L4}\sum_{s=1}^{p} S_l \mathrm{diag}(\gamma_s)\Gamma\mathbf{Q}_{ls}, \tag{10.12}$$

where $\mathbf{Q}_{ls}$'s ($l = 0, \dots, L_4$, where L_4 is the maximum time lag for the latent interactions; $s = 1, \dots, p$) are the matrices of path coefficients for the interaction terms (Zhou 2014). The matrix $\mathbf{Q}_{ls}$ has many structural zeros to avoid redundancies and logically impossible connections. For example, $q_{0s,ij} = 0$ for $i < s$ (to avoid redundancy between $\gamma_s\gamma_i$ and $\gamma_i\gamma_s$), $i = s$ (to avoid a quadratic term such as γ_s^2), $j = s$ (to avoid $\gamma_s\gamma_i \rightarrow \gamma_s$), and $j = i$ (to avoid $\gamma_s\gamma_i \rightarrow \gamma_i$), where $q_{0s,ij}$ is the ijth element of $\mathbf{Q}_{0s}$. The updating equations for the path coefficients and component weights have to take into account this additional term in a manner similar to other existing terms. Otherwise, the algorithm can proceed very much like that described in the appendix to this chapter.

As has been noted before, dynamic generalized structured component analysis is currently capable of analyzing only single-subject data at a time. However, the importance of simultaneous analysis of multiple-subject data is obvious because results obtained from a single subject may be peculiar to that subject and not generalizable to a population of subjects. There are several ways in which dynamic generalized structured component analysis can be extended to simultaneous analysis of multiple-subject data. Jung et al. (2012) suggested two such possibilities. One approach is based on the idea of multi-sample (multi-group) comparison (Hwang and Takane 2004; see also Section 3.3 of this volume), and the other based on multilevel analysis

(Hwang, Takane, and Malhotra 2007; Chapter 7). In the former, the same structural model is fitted to more than one subject simultaneously with some of the path coefficients assumed to be equal across subjects while others are assumed to vary over the subjects. In this way, both differences and commonalities among subjects can be simultaneously analyzed.

The second approach takes into account the hierarchical nature of observed data. That is, typical fMRI data are hierarchically structured in such a way that their individual-level measures (i.e., BOLD signals) are grouped within higher-level units (e.g., groups, subjects, and trials). In this approach, observed data are split into between-subjects and within-subjects data, each of which is separately modeled by SEM. This analysis allows us to investigate across-level interactions of explanatory variables for loadings and path coefficients in different levels. Multilevel dynamic generalized structured component analysis is an extension of original multilevel generalized structured component analysis (Hwang et al. 2007b) to accommodate multivariate time series data. To the best of our knowledge, neither of the two approaches suggested above has been implemented in practical forms.

A third approach to a simultaneous analysis of multiple-subject data has recently been proposed by Zhou (2014), in which a model for generalized (multiple-set) canonical correlation analysis (Carroll 1968) is used as the measurement model to extract variations in ROIs most representative of all subjects. This is in contrast with the principal component analysis-like measurement model in dynamic generalized structured component analysis. This is because there is only one matrix of BOLD signals per ROI in the single-subject case, whereas there are many such matrices in the multiple-subject case. In the latter, information from multiple subjects must be integrated into a quantity (a component) that captures the most representative variations over the subjects, before it is related to other ROIs' activations by a structural model similar to the one used in dynamic generalized structured component analysis. This new method is called dynamic generalized structured canonical correlation analysis (Zhou 2014) because it employs generalized (multiple-set) canonical correlation analysis for the information integration.

In generalized structured canonical correlation analysis, the usual bootstrap method may be used for reliability assessment by sampling subjects rather than a block of time points within subjects, assuming that the subjects in the original sample are randomly selected from a target population, and that there are enough subjects in the original sample.

Multiple subjects may constitute a relatively homogeneous group, or they may belong to distinct groups. The distinct groups may be intact groups (e.g., male vs. female, normal control vs. schizophrenic patients, etc.), or they may represent experimentally manipulated treatment groups. In either case, the multi-sample comparison approach, suggested above as a possible approach to simultaneous analysis of multiple-subject data, may be used to analyze both similarities and differences among the groups.

Technical details and examples of simultaneous analysis of multiple-subject data in a single and multiple groups by generalized structured canonical correlation analysis can be found in the study of Zhou (2014).

Appendix 10.1 Algorithm for Dynamic Generalized Structured Component Analysis

In this appendix, we provide a detailed explanation of the alternating LS algorithm for dynamic generalized structured component analysis. This algorithm is similar to the one described in Appendix 2.1. However, due to the additional terms in the structural model of dynamic generalized structured component analysis, the steps involved are somewhat more complicated than those in conventional generalized structured component analysis.

Step 1: Updating component loadings (c_i's) and path coefficients (A_l's, D_l's, and M_{lj}'s) given component weights: the component loadings are only in the measurement model, whereas the path coefficients are only in the structural model, so that they can be updated separately. Furthermore, the component loadings can be separately updated for each block of variables corresponding to a latent variable. Let $\varphi^{(M)}$ represent the portion of φ in Equation 10.11 pertaining to the measurement model. Then,

$$\varphi^{(M)} = SS(E^{(M)}) = SS(Z - ZWC) = \sum_{i=1}^{I} SS(Z_i - \gamma_i c_i') = \sum_{i=1}^{I} (vec(Z_i) - (I_{vi} \otimes \gamma_i)c_i)$$

(A10.1)

The c_i that minimizes $\varphi^{(M)}$ can be obtained by

$$c_i = (G_i'G_i)^{-1} G_i' vec(Z_i)$$

(A10.2)

for $i = 1, \ldots, p$, where $G_i = I_{vi} \otimes \gamma_i$. The path coefficients, on the other hand, can be updated as follows: let $\varphi^{(S)}$ denote the LS criterion pertaining to the structural model. Then,

$$\varphi^{(S)} = SS(E^{(S)}) = SS\{vec(\Gamma) - \sum_{l=0}^{L1} (I_p \otimes S_l\Gamma)vec(A_l)$$

$$-\sum_{l=0}^{L2} (I_p \otimes S_l U)vec(D_l) - \sum_{l=0}^{L3}\sum_{j=1}^{J} (I_p \otimes S_l diag(u_j)\Gamma)vec(M_{lj})\}.$$

(A10.3)

Let $\gamma = \text{vec}(\Gamma)$,

$$\mathbf{a}^* = \text{vec}(\left[\mathbf{A}_0, \quad \cdots, \quad \mathbf{A}_{L1}\right], \left[\mathbf{D}_0, \quad \cdots, \quad \mathbf{D}_{L2}\right], \left[\mathbf{M}_{01}, \quad \cdots, \quad \cdots \quad \mathbf{M}_{L3J}\right]), \quad (A10.4)$$

$$\mathbf{X}_1^* = [\mathbf{I}_p \otimes \mathbf{S}_0\Gamma, \ \dots, \mathbf{I}_p \otimes \mathbf{S}_{L1}\Gamma], \quad (A10.5)$$

$$\mathbf{X}_2^* = [\mathbf{I}_p \otimes \mathbf{S}_0\mathbf{U}, \ \dots, \mathbf{I}_p \otimes \mathbf{S}_{L2}\mathbf{U}], \quad (A10.6)$$

$$\mathbf{X}_3^* = [\mathbf{I}_p \otimes \mathbf{S}_0\text{diag}(\mathbf{u}_1)\Gamma, \ \dots, \dots, \mathbf{I}_p \otimes \mathbf{S}_{L3}\text{diag}(\mathbf{u}_J)\Gamma], \quad (A10.7)$$

and $\mathbf{X}^* = [\mathbf{X}_1^*, \mathbf{X}_2^*, \mathbf{X}_3^*]$. Let $\mathbf{a}$ denote the vector obtained by eliminating all 0 elements in $\mathbf{a}^*$, and let $\mathbf{X}$ denote the matrix obtained by eliminating the corresponding columns from $\mathbf{X}^*$. Then, $\varphi^{(S)}$ can be rewritten as

$$\varphi^{(S)} = \text{SS}(\gamma - \mathbf{X}\mathbf{a}). \quad (A10.8)$$

The LS estimate of $\mathbf{a}$ can be obtained by

$$\hat{\mathbf{a}} = (\mathbf{X}'\mathbf{X})^{-1}\mathbf{X}'\gamma. \quad (A10.9)$$

The updates of the original path coefficient matrices are constructed from the elements of this vector.

Step 2: Updating the component weights $\mathbf{w}$ given $\mathbf{c}_i$'s, $\mathbf{A}_l$'s, $\mathbf{D}_l$'s, and $\mathbf{M}_{lj}$'s: as in conventional generalized structured component analysis, $\mathbf{w}_i$ ($i = 1, \dots, p$) is updated for each block of variables sequentially. Let $\mathbf{W}_{(i)} = \mathbf{w}_i^* \mathbf{e}_i'$ and $\mathbf{W}_{(-i)} = \mathbf{W} - \mathbf{W}_{(i)}$, where $\mathbf{w}_i^*$ is the ith column vector of $\mathbf{W}$ and $\mathbf{e}_i$ is the p-element vector of zeros except the ith element, which is unity. (The $\mathbf{w}_i^*$ has $\mathbf{w}_i$ as its part, but is also appended by zeros because $\mathbf{W}$ is a block diagonal matrix with $\mathbf{w}_i$ as the ith block.) Using $\mathbf{W} = \mathbf{W}_{(i)} + \mathbf{W}_{(-i)}$, we separate the effect related to $\mathbf{w}_i$ from the rest. Note that the component weights are in both measurement and structural models. The $\varphi^{(M)}$ can be rewritten as

$$\varphi^{(M)} = \text{SS}(\mathbf{Z} - \mathbf{Z}\mathbf{W}\mathbf{C}) = \text{SS}(\mathbf{Z} - \mathbf{Z}\mathbf{W}_{(-i)}\mathbf{C} - \mathbf{Z}\mathbf{W}_{(-i)}\mathbf{C}) = \text{SS}(\mathbf{y}_1 - \mathbf{X}_1^*\mathbf{w}_i^*), \quad (A10.10)$$

where $\mathbf{y}_1 = \text{vec}(\mathbf{Z} - \mathbf{Z}\mathbf{W}_{(-i)}\mathbf{C})$ and $\mathbf{X}_1^* = \mathbf{C}'\mathbf{e}_i \otimes \mathbf{Z}$. The $\varphi^{(S)}$, on the other hand, can be rewritten as

$$\varphi^{(S)} = \text{SS}(\mathbf{y}_2 - \mathbf{X}_2^*\mathbf{w}_i^*), \quad (A10.11)$$

where

$$\mathbf{y}_2 = \text{vec}(\mathbf{ZW}_{(-i)} - \sum_{l=0}^{L1} \mathbf{S}_l\mathbf{ZW}_{(-i)}\mathbf{A}_l - \sum_{l=0}^{L2} \mathbf{S}_l\mathbf{UD}_l - \sum_{l=0}^{L3}\sum_{j=1}^{J} \mathbf{S}_l\text{diag}(\mathbf{u}_j)\mathbf{ZW}_{(-i)}\mathbf{M}_{lj}),$$

(A10.12)

and

$$\mathbf{X}_2^* = \sum_{l=0}^{L1}(\mathbf{A}_l'\mathbf{e}_i \otimes \mathbf{S}_l\mathbf{Z}) + \sum_{l=0}^{L2}(\mathbf{D}_l'\mathbf{e}_i \otimes \mathbf{S}_l\mathbf{U}) + \sum_{l=0}^{L3}\sum_{j=1}^{J}(\mathbf{M}_{lj}'\mathbf{e}_i \otimes \mathbf{S}_l\text{diag}(\mathbf{u}_j)\mathbf{Z}) - (\mathbf{e}_i \otimes \mathbf{Z}).$$

(A10.13)

Let

$$\mathbf{y} = \begin{pmatrix} \mathbf{y}_1 \\ \mathbf{y}_2 \end{pmatrix}, \text{ and } \mathbf{X}^* = \begin{bmatrix} \mathbf{X}_1^* \\ \mathbf{X}_2^* \end{bmatrix}.$$

As alluded to earlier, $\mathbf{w}_i$ is obtained from $\mathbf{w}_i^*$ by eliminating 0 elements. Let $\mathbf{X}$ denote the matrix obtained by eliminating the corresponding columns of $\mathbf{X}^*$. Then,

$$\varphi = \varphi^{(M)} + \varphi^{(S)} = \text{SS}(\mathbf{y} - \mathbf{Xw}_i),$$

(A10.14)

which should be minimized with respect to $\mathbf{w}_i$ under the normalization restriction that $\mathbf{w}_i'\mathbf{Z}_i'\mathbf{Z}_i\mathbf{w}_i = T$. This is a special kind of constrained LS problem proposed by ten Berge and Nevels (1977). Their algorithm is different from the weight updating scheme in conventional generalized structured component analysis, where $\mathbf{w}_i$ is first obtained by ordinary LS, which is then normalized to satisfy the normalization restriction. In dynamic generalized structured component analysis, the normalization restriction must be explicitly taken into account in the minimization of Equation A10.14. This is because $\mathbf{X}'\mathbf{X}$ with $\mathbf{X}$ defined above is not a constant multiple of $\mathbf{Z}_i'\mathbf{Z}_i$.

As before, the two steps described above are repeatedly applied until the change in $\mathbf{e}'\mathbf{e}$ from one iteration to the next gets smaller than a certain prescribed value, say 1.0×10^6.

11

Functional Generalized Structured Component Analysis

Functional data represent data collected in the form of curves, surfaces, images, or anything else varying over a continuum. The continuum can be time, spatial position, wavelength, probability, and so forth. Owing to the emergence of various novel measurement tools, such as eye-trackers, motion capture devices, and functional neuroimaging modalities (e.g., positron emission tomography, functional magnetic resonance imaging, electroencephalography, etc.), functional data become ubiquitous. A few examples of such data collected in psychology or other social sciences include data from motor control (e.g., Mattar and Ostry 2010; Olshen et al. 1989), emotional speech production (Lee, Bresch, and Narayanan 2006), musical cognition (e.g., Almansa and Delicado 2009; Vines et al. 2006; Vines, Nuzzo, and Levitin 2005), gaze-tracking (e.g., Jackson and Sirois 2009), functional neuro imaging (e.g., Hwang et al. 2012a; Tian 2010), online auction (e.g., Jank and Shmueli 2006; Reddy and Dass 2006), and intra-daily stock market (e.g., Alva, Romo, and Ruiz 2009).

Compared to conventional multivariate data, functional data may be characterized by high-frequency repeated measurements that reflect a smooth function, which is assumed to generate them. Although such smooth functions are the elements of analysis, in practice, the data observed are likely to be a sample of discrete observations measured repeatedly over occasions (Ramsay and Silverman 2005, p. 38). In this sense, at least at an initial measurement level, functional data may be comparable to traditional repeated measures or panel data. In the social sciences, two streams of statistical techniques have been considered for the analysis of repeated measures data: autoregressive models and latent trajectory models (Bollen and Curran 2004). Autoregressive models, which include univariate/multivariate autoregressive simplex models and dynamic factor models, focus on the relationship between two or more adjacent data points and aim to examine the effect of the value of a variable measured at an earlier occasion on the value at the current occasion. This variable can be either latent or observed (e.g., Browne and Nesselroade 2005; Kessler and Greeberg 1981; Nesselroade et al. 2001; Nesselroade and Molenaar 1999). On the other hand, latent trajectory models, which include latent growth curve models or hierarchical linear models, are used to discern intra-individual and inter-individual trajectories of change in repeated measures data over occasions and to examine the influences of covariates on certain characteristics (e.g., intercepts and slopes) of the trajectories (e.g., Browne and Du Toit 1991; Bollen and Curran 2006; Duncan, Duncan, and Strycker 2006; McArdle 1986; Meredith and Tisak 1990; Rao 1958; Raudenbush 2001; Raudenbush and Bryk 2002; Tucker 1966). More recently, latent difference score models were developed to explicitly take into account the amount of change occurred between two occasions (e.g., McArdle 2001, 2009; McArdle and Hamagami 2001).

These traditional techniques have typically focused on the analysis of repeated measures data that entail not more than 10 occasions (Walls and Schafer 2006). On the other hand, in principle, functional data are intrinsically infinite-dimensional. In practice, we usually have a finite number of discrete records of a function evaluated at different occasions. However, the number of occasions can often be huge, so that it exceeds the number of individuals. For example, as a prototype of functional data, Ramsay and Silverman (2005, Chapter 1) show the heights of 10 girls recorded at 31 ages in the Berkeley Growth Study (Tuddenham and Snyder 1954), in which the number of occasions exceeds that of girls. In such situations, the sample covariance matrix of the data will become singular, that is, nonpositive definite. Dynamic factor models, latent growth curve models, and latent difference score models generally estimate parameters via maximum likelihood or generalized least-squares, which requires the sample covariance matrix to be positive definite (Bollen 1989, p. 107; Wothke 1993, p. 263).

In addition, functional data often exhibit an intricate trajectory of change, which is difficult to describe satisfactorily by using a simple parametric model, such as a linear- or quadratic-trend latent growth curve model.

Owing to these distinctive characteristics, it may be less suitable to apply traditional statistical techniques for the analysis of functional data. Consequently, there has been a continuing need to analyze functional data effectively and gain insight into the data. *Functional Data Analysis*, the term coined by Ramsay and Dalzell (1991), is a branch of statistics, which aims to develop and apply statistical techniques for the analysis of functional data.

A vast literature on functional data analysis exists and it continues to grow rapidly. Ramsay and Silverman (1997, 2005) provide comprehensive discussions about various characteristics of functional data and functional approaches to multivariate statistical techniques. Ferraty and Vieu (2006) discuss complementary and extensive treatments on nonparametric techniques in functional data analysis. Moreover, many past special issues of statistics journals provide review articles on functional data analysis (e.g., Davidian, Lin, and Wang 2004; González-Manteiga and Vieu 2007; Rice 2004; Valderrama 2007). A description of more recent developments in functional data analysis can be found in Ferraty (2011), Ferraty, Mas, and Vieu (2007), Ferraty and Romain (2011), and Horváth and Kokoszka (2012). Moreover, Ramsay and Silverman (2002) provide various enlightening applications of functional data analysis in a wide range of disciplines. Computational software and support for functional data analysis is available in different platforms such as MATLAB, R, and S-PLUS from Ramsay's Functional Data Analysis website (http://www.psych.mcgill.ca/misc/fda/) (also see Ramsay, Hooker, and Graves 2009). Computational software for nonparametric functional data analysis is available from the Nonparametric Functional Data Analysis website (http://www.math.univ-toulouse.fr/staph/npfda/) developed under a research group known as STAPH (Groupe de Travail en Statistique Fonctionnelle et Opeatorielle) and from a group of Spanish researchers at the University of Santiago de Compostela (e.g., Febrero-Bande and Oviedo de la Fuente 2012).

A variety of techniques have been developed for the analysis of functional data, including functional principal component analysis, functional canonical correlation analysis, and functional linear models. Functional principal component analysis is a useful technique for identifying and exploring low-dimensional sources of representative variation in functional data (Rice and Silverman 1991; also see Ramsay and Silverman 2005, Chapters 8, 9, and 10). It technically aims to extract a series of components from a single set of functional data in such a way that the components are mutually orthogonal and successively account for the maximum variation in the data. A number of researchers have contributed to developing this technique (e.g., Besse and Ramsay 1986; Ramsay and Dalzell 1991; Rice and Silverman 1991), to investigating its properties and extensions (e.g., Cardot 2000; Girard 2000; Hall and Hosseini-Nasab 2006; James, Hastie, and Sugar 2000; Ocaña, Aguilera, and Valderrama 1999; Silverman 1995, 1996; Yao and Lee 2006; Yao, Müller, and Wang 2005), and to applying the technique to various types of functional data (e.g., Kneip and Utikal 2001; Viviani, Grön, and Spitzer 2005).

Functional canonical correlation analysis is employed for exploring the associations between a pair of functional data (e.g., He, Müller, and Wang 2003; Leurgans, Moyeed, and Silverman 1993; Ramsay and Silverman 2005, Chapter 11). For example, Leurgans, Moyeed, and Silverman (1993) have shown how variation in the knee angle functions of children was related to their hip angle functions. In functional canonical correlation analysis, a series of components, called canonical variates, are extracted from each set of functional data such that a component from one set is maximally correlated only with its corresponding component from the other set, while remaining orthogonal to all the other components. Recently, functional multiple-set canonical correlation analysis (Hwang et al. 2012) was developed that extended functional canonical correlation analysis to the analysis of more than two sets of functional data.

Functional canonical correlation analysis is used to investigate nondirectional relationships (i.e., correlations) between two sets of functional data. Thus, there is no distinction between dependent and predictor data. Functional linear models are developed to investigate the effects of predictor data on dependent data. They may be classified into four different types based on which data are considered functional. In the first type, also known as functional analysis of variance, dependent data are functional, whereas predictor data are multivariate (e.g., Gu 2002; Ramsay and Silverman 2005, Chapter 13). For example, Ramsay and Silverman (2005, Chapter 13) examined how 35 Canadian weather stations' (log) daily temperature functions over a year could be predicted by their multivariate, geographical characteristics. In the second type, dependent data are scalar, whereas predictor data are functional (e.g., Cardot, Ferraty, and Sarda 1999; Ramsay and Silverman 2005, Chapter 15; Wu, Fan, and Müller 2010). For example, we may study how a subject's rating on emotion such as happiness may change after listening to a musical piece in which perceived tension varies continuously. An emotion rating can then be a scalar dependent variable, while the continuous changes of musical tension can be considered functional (e.g., Vines et al. 2005). In the third type, both dependent and predictor data are functional, and the effect of predictor functions at occasion t on dependent functions at occasion t is to be examined. This type of model is also called the concurrent model (e.g., Ramsay and Silverman 2005, Chapter 14; Tan et al. 2012). For example, the effect of temperature functions on precipitation functions can be evaluated at the same day (Ramsay and Silverman 2005, Chapter 14). Finally, the fourth type is the most general one, in which both dependent and predictor data are functional and the effect of predictor functions evaluated at occasion t on dependent functions at occasion t' ($t \neq t'$) can be considered (e.g., Ramsay and Silverman 2005, Chapter 16; Yao, Müller, and Wang 2005). For example, the values of temperature functions evaluated at day t can be used to predict the values of precipitation functions evaluated at day t' (Ramsay and Silverman 2005, Chapter 16).

The second type of functional linear models has been extended in various ways. For example, it was extended to generalized functional linear models that could deal with a scalar dependent variable arising from different exponential-family distributions (e.g., James 2002; Müller and Stadtmüller 2005). It was also extended to functional linear mixed models that could accommodate both fixed and random effects (e.g., Morris and Carroll 2006). It was further generalized to generalized multilevel functional linear models that combined generalized functional linear models with functional linear mixed models (e.g., Crainiceanu, Staicu, and Di 2009). A mixture version of the second model was also developed that permitted model parameters to vary across heterogeneous subgroups of the population (e.g., Yao, Fu, and Lee 2011).

Despite the usefulness of the techniques discussed above, obviously, they were not developed for investigating path-analytic relationships between functional data and latent variables. In this chapter, we discuss how to specify and test structural equation models that involve functional data. Specifically, we focus on a functional extension of generalized structured component analysis, called *functional generalized structured component analysis* (Suk 2013; Suk and Hwang 2014). Functional generalized structured component analysis is one of the most recent developments in functional data analysis and generalized structured component analysis. We provide a technical account and empirical applications of functional generalized structured component analysis. We also discuss *functional extended redundancy analysis* (Hwang et al. 2012b) as a special case of functional generalized structured component analysis.

11.1 Functional Generalized Structured Component Analysis

11.1.1 Model Specification

As in generalized structured component analysis, functional generalized structured component analysis involves the specification of three sub-models: *measurement, structural,* and *weighted relation* models. Figure 11.1 displays a prototype model of functional generalized structured component analysis. In the prototype model, there are three smooth functions (x_1, x_2, and x_3), each of which is depicted as a curve. That is, in this model, a block of indicators per latent variable is replaced by a function, which will be called a *data function* hereafter. The prototype model also contains three latent variables (γ_1, γ_2, and γ_3), which are assumed to underlie the corresponding data functions.

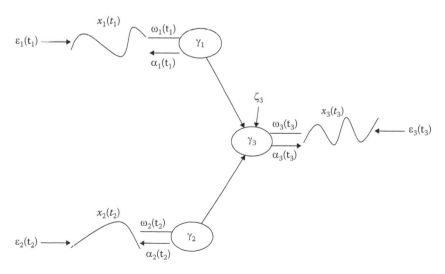

FIGURE 11.1
A prototype model of functional generalized structured component analysis.

Two latent variables (γ_1 and γ_2) are further specified to influence a latent variable (γ_3).

The measurement model of functional generalized structured component analysis expresses the relationships between data functions and latent variables. A data function is considered reflective if it is influenced by its latent variable, whereas it is considered formative if it forms its latent variable. As in generalized structured component analysis, the measurement model is specified only when there exist reflective data functions. The prototype model depicted in Figure 11.1 requires the specification of the measurement model because data functions are reflective. The measurement model for the prototype model is given as

$$x_1(t_1) = \gamma_1\alpha_1(t_1) + \varepsilon_1(t_1)$$

$$x_2(t_2) = \gamma_2\alpha_2(t_2) + \varepsilon_2(t_2)$$

$$x_3(t_3) = \gamma_3\alpha_3(t_3) + \varepsilon_3(t_3), \tag{11.1}$$

where $x_p(t_p)$ is the value of the pth data function x_p that is available for argument t_p in some finite interval T_p ($t_p \in T_p$) ($p = 1, 2, 3$), $\alpha_p(t_p)$ is the value of the pth loading function α_p evaluated at t_p, which relates the pth latent variable γ_p to x_p, and $\varepsilon_p(t_p)$ is the value of the pth residual function evaluated at t_p. In the prototype model, the data are functional, so that their corresponding loading and residual are functional as well. On the other hand, the scores of latent variables are scalar, which is equivalent to an inner product of two functions, as will be further discussed below. We can generally express the measurement model as

$$x_p(t_p) = \gamma_p\alpha_p(t_p) + \varepsilon_p(t_p). \tag{11.2}$$

The structural model is used to specify the relationships among latent variables. In functional generalized structured component analysis, all latent variable scores are scalar. This is the same as in generalized structured component analysis. Thus, the structural model for the prototype model is given as

$$\gamma_3 = \gamma_1 b_2 + \gamma_2 b_2 + \zeta_3, \tag{11.3}$$

where b's are path coefficients relating a latent variable to other latent variables, and ζ_3 is the residual for the endogenous latent variable. As in generalized structured component analysis, we can generally express the structural model in matrix notation, as follows.

$$\gamma = \mathbf{B}'\gamma + \zeta. \tag{11.4}$$

In functional generalized structured component analysis, a latent variable is defined as a component of a data function. The weighted relation model is used to express such a relationship between data functions and latent variables. Specifically, the weighted relation model for the prototype model is given as

$$\gamma_1 = \int_{T_1} x_1(t_1)\omega_1(t_1)dt_1$$

$$\gamma_2 = \int_{T_2} x_2(t_2)\omega_2(t_2)dt_2$$

$$\gamma_3 = \int_{T_3} x_3(t_3)\omega_3(t_3)dt_3, \tag{11.5}$$

where $\omega_p(t_p)$ denotes the value of the pth weight function ω_p evaluated at t_p, which is associated with the pth data function. In the weighted relation model, both x_p and ω_p are functions, so that the scalar latent variable score γ_p, which is an inner product of these functions, is defined as integration over T_p (Ramsay and Silverman 2005, p. 21). This is a main difference from generalized structured component analysis in which the inner product is defined as summation over the number of indicators. In functional generalized structured component analysis, an infinite-dimensional data function replaces a block of indicators for a latent variable in generalized structured component analysis. Thus, a block of weights for a latent variable is also replaced by an infinite-dimensional weight function. The value of a data function evaluated at t_p may be seen as a single indicator for a latent variable in generalized structured component analysis. The value of a weight function evaluated at t_p indicates the contribution of a data function to determining its latent variable at t_p. We can generally express the weighted relation model as

$$\gamma_p = \int_{T_p} x_p(t_p)\omega_p(t_p)dt_p. \tag{11.6}$$

11.1.2 Parameter Estimation

The unknown parameters of functional generalized structured component analysis include weight functions (ω_p), loading functions (α_p), and path coefficients (b's). To estimate the parameters, we seek to minimize the following penalized least-squares criterion

$$\phi = \sum_{p=1}^{P}\sum_{i=1}^{N}\int \left(x_{ip}(t_p) - \gamma_{ip}\alpha_p(t_p)\right)^2 dt_p + SS(\mathbf{\Gamma} - \mathbf{\Gamma}\mathbf{B}) + \lambda \sum_{p=1}^{P}\int [D^2\omega_p(t_p)]^2 dt_p$$

$$+ \rho \sum_{p=1}^{P}\int [D^2\alpha_p(t_p)]^2 dt_p, \tag{11.7}$$

subject to the standardization constraint on each latent variable,

$$\text{that is, } \sum_{i=1}^{N}\gamma_{ip}^{2} = N.$$

The first term of Equation 11.7 is the least-squares criterion for the measurement model, which is equal to the sum of integrated squared residuals over a sample of N data functions for each of P latent variables. The second term is the least-squares criterion for the structural model, which remains the same as that for the structural model of generalized structured component analysis. In the third and the fourth terms,

$$\sum_{p=1}^{P}\int \left[D^2\omega_p(t_p)\right]^2 dt_p \text{ and } \sum_{p=1}^{P}\int \left[D^2\alpha_p(t_p)\right]^2 dt_p$$

are called roughness penalties, which are equal to the sums of the integrated squared second derivative of ω_p and α_p over P latent variables, respectively; and λ and ρ are called smoothing parameters that are non-negative. In Equation 11.7, D^s denotes the derivative of order s. The interval T_p for integrals is omitted to make the notation concise. This will be the case for all the equations that follow.

The measurement model (Equation 11.2) involves reflective data functions, so that the least-squares criterion for the measurement model is defined as the sum of the integrated squared residuals. On the other hand, the structural model (Equation 11.4) does not entail such functions. Thus, no change needs to be made to the least-squares criterion for the structural model of generalized structured component analysis. Equation 11.7 is derived by adding the two roughness penalty terms into these least-squares criteria. Each roughness penalty takes into account the degree of a function's roughness or curvature. As will be further discussed in Section 11.1.3, a central property of a function is smoothness (Ramsay and Silverman 2005, p. 38). A function's curvature at argument t is reflected by its second derivative at t, as a straight line, which is deemed to be the smoothest curve, has no curvature, producing a zero second derivative. Thus, a function's roughness is usually assessed by its integrated squared second derivative (Ramsay and Silverman

2005, p. 84). In Equation 11.7, such roughness penalty terms are introduced to account for the degree of roughness in ω_p and α_p. When these functions are highly variable, their roughness penalty values tend to be large.

The smoothing parameters λ and ρ play a role in balancing out the relative importance of their roughness penalty terms in estimation of the weight and loading functions. If the smoothing parameter values become large, then a greater penalty is imposed on the roughness of the corresponding estimated functions. This helps keep the degree of each parameter function's roughness small and make it smoother. On the other hand, when the smoothing parameter values are set to zero, no penalty is placed on the roughness of the estimated functions, so that each function estimated becomes identical to its ordinary least-squares counterpart because the third and fourth terms of Equation 11.7 disappear.

We have considered all data functional thus far. However, in practice, we can often encounter situations in which some data are multivariate, consisting of indicators, whereas the other data are functional. For example, we may be interested in relating subjects' blood oxygen level-dependent (BOLD) signal changes in a brain region of interest to their cognitive ability. The BOLD signal changes recorded continuously over scans, called time courses, can be considered smooth functions (Tian 2010). On the other hand, cognitive ability can be measured based on a cognitive rating scale that consists of a number of items, which cannot be considered functional. Figure 11.2 displays a prototype model of such a situation, where data functions and indicators are mixed. In this prototype model, there are two data functions (x_1 and x_2) and two indicators (z_3 and z_4). Two latent variables underlying the corresponding

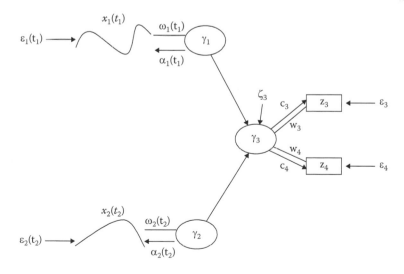

FIGURE 11.2
A prototype model of functional generalized structured component analysis, which involves both data functions and indicators.

data functions (γ_1 and γ_2) are to influence a latent variable underlying the two indicators (γ_3).

The measurement model for the prototype model is given as

$$x_1(t_1) = \gamma_1 \alpha_1(t_1) + \varepsilon_1(t_1)$$

$$x_2(t_2) = \gamma_2 \alpha_2(t_2) + \varepsilon_2(t_2)$$

$$z_3 = \gamma_3 c_3 + \varepsilon_3$$

$$z_4 = \gamma_3 c_4 + \varepsilon_4. \tag{11.8}$$

The last two equations are expressed in the same way as in the measurement model of generalized structured component analysis, because z_3 and z_4 are indicators, not functions. The structural model for the prototype model is given as

$$\gamma_3 = \gamma_1 b_2 + \gamma_2 b_2 + \zeta_3, \tag{11.9}$$

This is of the same form as Equation 11.3, because all latent variable scores remain scalar. On the other hand, the weighted relation model for the prototype model is given as

$$\gamma_1 = \int_{T_1} x_1(t_1) \omega_1(t_1) dt_1$$

$$\gamma_2 = \int_{T_2} x_2(t_2) \omega_2(t_2) dt_2$$

$$\gamma_3 = z_3 w_3 + z_4 w_4, \tag{11.10}$$

where w_3 and w_4 denote the scalar weights assigned to the two indicators. The last latent variable (γ_3) is thus defined as the weighted sum of the two indicators, as in generalized structured component analysis.

Functional generalized structured component analysis can easily accommodate such mixed cases by minimizing the following penalized least-squares criterion

$$\phi_1 = \sum_{p=1}^{P_1} \sum_{i=1}^{N} \int \left(x_{ip}(t_p) - \gamma_{ip} \alpha_p(t_p) \right)^2 dt_p + \sum_{k=1}^{P_2} \sum_{j=1}^{J_k} \sum_{i=1}^{N} (z_{ijk} - \gamma_{ik} c_{jk})^2 + \text{SS}(\mathbf{\Gamma} - \mathbf{\Gamma} \mathbf{B})$$

$$+ \lambda \sum_{p=1}^{P_1} \int [D^2 \omega_p(t_p)]^2 dt_p + \rho \sum_{p=1}^{P_1} \int [D^2 \alpha_p(t_p)]^2 dt_p, \tag{11.11}$$

where z_{ijk} is the ith scalar value of the jth indicator for the kth latent variable $(i = 1,..., N; j = 1,..., J_p; k = 1,..., P_2)$, c_{jk} is the scalar loading relating the kth latent variable to the jth indicator, and P_1 is the number of latent variables associated with data functions (x's), and P_2 is the number of latent variables associated with indicators (z's), so that $P = P_1 + P_2$. The second term of Equation 11.11 is equivalent to the least-squares criterion for the measurement model of generalized structured component analysis. Moreover, the latent variable γ_{ik} is a scalar value of the weighted composite of indicators, that is,

$$\gamma_{ik} = \sum_{j=1}^{J_k} z_{ijk} w_{jk},$$

as in generalized structured component analysis.

An alternating regularized least-squares algorithm (Hwang 2009) was developed to minimize Equations 11.7 and 11.11. As will be explained in more detail in the next section, in functional generalized structured component analysis, a function is represented as a linear combination of known basis functions (Hastie, Tibshirani, and Friedman 2009, Chapter 5; Ramsay and Silverman 2005, Chapter 3). The central idea of this so-called basis function expansion approach is that any infinite-dimensional function can be approximated to some arbitrary degree of precision by a linear combination of a finite number of known basis functions. The alternating regularized least-squares algorithm was developed in combination with the basis function expansion approach. We provide a description of the algorithm for functional generalized structured component analysis in the Appendix.

As in generalized structured component analysis, functional generalized structured component analysis provides the value of FIT for a given model. The FIT is given by

$$\text{FIT} = 1 - \frac{\sum_{p=1}^{P} \sum_{i=1}^{N} \int \left(x_{ip}(t_p) - \gamma_{ip}\alpha_p(t_p) \right)^2 dt_p + \text{SS}(\Gamma - \Gamma B)}{\sum_{p=1}^{P} \sum_{i=1}^{N} \int x_{ip}(t_p)^2 dt_p + \text{SS}(\Gamma)} \tag{11.12}$$

The first term in the denominator in Equation 11.12 indicates the total variability in data functions, whereas the second term represents the total variability in latent variables. Similarly, when indicators are included in the model, the FIT can be computed as

$$\text{FIT} = 1 - \frac{\sum_{p=1}^{P_1}\sum_{i=1}^{N}\int\left(x_{ip}(t_p) - \gamma_{ip}\alpha_p(t_p)\right)^2 dt_p + \sum_{k=1}^{P_2}\sum_{j=1}^{J_k}\sum_{i=1}^{N}(z_{ijk} - \gamma_{ik}c_{jk})^2 + \text{SS}(\mathbf{\Gamma} - \mathbf{\Gamma B})}{\sum_{p=1}^{P_1}\sum_{i=1}^{N}\int x_{ip}(t_p)^2 dt_p + \sum_{k=1}^{P_2}\sum_{j=1}^{J_k}\sum_{i=1}^{N} z_{ijk}^2 + \text{SS}(\mathbf{\Gamma})}$$

$$(11.13)$$

As in generalized structured component analysis, the bootstrap method is used to estimate the standard errors (SE) or confidence intervals (CI) of parameter estimates in functional generalized structured component analysis. Functional generalized structured component analysis provides the pointwise 95% CI of each estimated parameter function.

11.1.3 Additional Computational Considerations

11.1.3.1 A Basis Function Expansion Approach to Approximating Smooth Functions

The basic premise of functional data analysis is to treat data repeatedly measured at multiple occasions as a single entity or a function, rather than a collection of unconnected discrete observations. In other words, discrete observations or records taken over multiple occasions are assumed to be generated by an underlying smooth function. Being smooth means that the adjacent values of a function are connected and unlikely to be so different from each other (Ramsay and Silverman 2005, Chapter 3).

In practice, the data observed are typically discrete observations or records of z_{jp} for the pth latent variable ($j = 1,...,J_p$). We assume that a smooth function for the pth latent variable, denoted by $x_p(t_p)$, underlies each observed record. As stated earlier, this smooth function is called a data function in functional generalized structured component analysis. We generally assume a relationship between the discrete observations and their underlying smooth data function as follows.

$$z_{jp} = x(t_{jp}) + \varepsilon_{jp}, \tag{11.14}$$

where z_{jp} indicates a discrete observation value for the pth latent variable measured at the jth measurement occasion, $x(t_{jp})$ is the value of the underlying smooth function evaluated at the jth measurement occasion, and ε_{jp} is a noise or error for z_{jp} (Ramsay and Silverman 2005, p. 40).

Such a data function $x_p(t_p)$ can be approximated to some arbitrary degree of precision by a linear combination of a finite number of so-called basis functions, which are already known to reflect important characteristics of the data function (Ramsay and Silverman 2005, Chapter 3). In this basis function expansion approach, any function in a function space can be represented as a linear combination of basis functions that span the function space. This is comparable to that any vector in a vector space can be represented as a linear combination of basis vectors that span the vector space. Specifically, a data function can be generally expressed as basis function systems as follows.

$$x_p(t_p) = \sum_{l=1}^{L} u_{lp}\theta_{lp}(t_p) = \mathbf{\theta}_p(t_p)'\mathbf{u}_p, \tag{11.15}$$

where u_{lp} indicates the coefficient of the *l*th basis function, $\theta_{lp}(t_p)$ is the value of the *l*th basis function evaluated at t_p, $\mathbf{\theta}_p(t_p) = [\theta_{1p}(t_p), ..., \theta_{Lp}(t_p)]'$ is an *L* by 1 vector of the basis functions, and $\mathbf{u}_p = [u_{1p}, ..., u_{Lp}]'$ is an *L* by 1 vector of the basis function coefficients.

As described in Section 11.1.1, functional generalized structured component analysis involves two parameter functions (weight and loading functions), as well as data functions. Based on the basis function expansion approach, the weight and loading functions can be expressed as

$$\omega_p(t_p) = \sum_{l=1}^{L} y_{lp}\theta_{lp}(t_p) = \mathbf{y}_p'\mathbf{\theta}_p(t_p) \tag{11.16}$$

$$\alpha_p(t_p) = \sum_{l=1}^{L} a_{lp}\theta_{lp}(t_p) = \mathbf{a}_p'\mathbf{\theta}_p(t_p), \tag{11.17}$$

where y_{lp} indicates the coefficient of the *l*th basis function to approximate the *p*th weight function, $\mathbf{y}_p = [y_{1p}, ..., y_{Lp}]'$ is an *L* by 1 vector of the basis function coefficients, a_{lp} denotes the coefficient of the *l*th basis function to approximate the *p*th loading function, and $\mathbf{a}_p = [a_{1p}, ..., a_{Lp}]'$ is an *L* by 1 vector of these basis function coefficients.

The basis function expansion approach has advantages. For example, by representing a function as a weighted sum of basis functions, we can convert an infinite-dimensional functional problem into a finite-dimensional one that involves a vector of unknown basis function coefficients, as in conventional multivariate statistical techniques. This renders it possible to explain hundreds or thousands of data points with a manageable number of parameters. Moreover, the basis function expansion approach is highly flexible in the sense that functions of infinitely many different shapes can be represented by changing the values of basis function coefficients, rather than using infinitely many different models.

A wide array of basis functions is available for approximating functions. For example, Fourier and spline basis functions are some of the widely used basis functions (Ramsay and Silverman 2005, Chapter 3). Fourier basis functions are useful for representing stable periodic functions that do not involve strong local fluctuations. Spline basis functions are more appropriate for representing nonperiodic functions that tend to fluctuate locally.

Spline basis functions are mainly employed in functional generalized structured component analysis, because of their flexibility, popularity, and computational efficiency (Suk 2013). Spline functions are piecewise polynomial functions (see de Boor 2001; Gu 2002; Wahba 1990). To construct a spline function, it is necessary to divide the interval, over which a function is approximated, into the predetermined number (M) of subintervals. The values that separate these M subintervals are called breakpoints. Then, for each subinterval, a spline function is defined as a polynomial of order m. The order of a polynomial indicates the number of parameters required to determine the polynomial. For example, the order is two for a linear function, three for a quadratic function, four for a cubic function, and so on. A spline function must join at each breakpoint, and its derivatives up to order $m - 2$ must also match up at each breakpoint so as to be smooth. There are a number of different spline functions that satisfy these constraints, including P-spline, M-spline, and B-spline. For more detailed discussions on various types of spline functions, refer to de Boor (2001), Ramsay and Silverman (2005, Chapter 3), and Schumaker (2007).

Once the type and number of basis functions are determined, we need to estimate basis function coefficients, for example, θ_{lp}'s in Equation 11.15. We here focus on the estimation of the basis function coefficients for approximating a *data function*. A simple way is to estimate the basis function coefficients by minimizing the following ordinary least-squares criterion.

$$\phi_2 = \sum_{j=1}^{J_p} (z_{jp} - x_p(t_p))^2$$

$$= \sum_{j=1}^{J_p} (z_{jp} - \boldsymbol{\theta}_p(t_p)'\mathbf{u}_p)^2$$

$$= (\mathbf{z}_p - \boldsymbol{\Theta}_p\mathbf{u}_p)'(\mathbf{z}_p - \boldsymbol{\Theta}_p\mathbf{u}_p), \tag{11.18}$$

where $\mathbf{z}_p = [z_{1p}, \ldots, z_{Jp}]'$ is a J_p by 1 vector of discrete observations, and $\boldsymbol{\Theta}_p = [\boldsymbol{\theta}_p(t_{1p}), \ldots, \boldsymbol{\theta}_p(t_{Jp})]'$ is a J_p by L matrix of basis functions. The least-squares estimate of $\mathbf{u}_p$ is given as

$$\hat{\mathbf{u}}_p = (\boldsymbol{\Theta}_p'\boldsymbol{\Theta}_p)^{-1}\boldsymbol{\Theta}_p'\mathbf{z}_p. \tag{11.19}$$

After obtaining the least-squares estimates of the basis function coefficients, a data function can be approximated based on Equation 11.15. Although this least-squares approach is simple, it offers somewhat poor discontinuous control over the degree of smoothing (Ramsay and Silverman 2005, p. 81).

The so-called roughness penalty or regularization approach is often a more ideal way for approximating a data function while directly controlling the degree of smoothness. This approach aims to estimate the basis function coefficients by minimizing the following penalized least-squares criterion.

$$\phi_3 = \sum_{j=1}^{J_p} (z_{jp} - x_p(t_p))^2 + \tau \int \left(D^2 x_p(t_p) \right)^2 dt_p, \tag{11.20}$$

where

$$\int \left(D^2 x_p(t_p) \right)^2 dt_p$$

is a roughness penalty term for the data function, and τ is the smoothing parameter, as described in Section 11.1.2. The value of τ may be selected via cross validation or generalized cross-validation (Craven and Wahba 1979; Gu 2002; Ramsay and Silverman 2005, Chapter 5). Based on the basis function expansion approach, Equation 11.20 can be rewritten as

$$\phi_4 = \sum_{j=1}^{J_p} (z_{jp} - \boldsymbol{\theta}_p(t_p)' \mathbf{u}_p)^2 + \tau \int \left(D^2 \boldsymbol{\theta}_p(t_p)' \mathbf{u}_p \right)^2 dt_p$$

$$= (\mathbf{z}_p - \boldsymbol{\Theta}_p \mathbf{u}_p)'(\mathbf{z}_p - \boldsymbol{\Theta}_p \mathbf{u}_p) + \tau \int \left(D^2 \mathbf{u}_p' \boldsymbol{\theta}_p(t_p) D^2 \boldsymbol{\theta}_p(t_p)' \mathbf{u}_p \right) dt_p$$

$$= (\mathbf{z}_p - \boldsymbol{\Theta}_p \mathbf{u}_p)'(\mathbf{z}_p - \boldsymbol{\Theta}_p \mathbf{u}_p) + \tau \, \mathbf{u}_p' \left[\int D^2 \boldsymbol{\theta}_p(t_p) D^2 \boldsymbol{\theta}_p(t_p)' dt_p \right] \mathbf{u}_p$$

$$= (\mathbf{z}_p - \boldsymbol{\Theta}_p \mathbf{u}_p)'(\mathbf{z}_p - \boldsymbol{\Theta}_p \mathbf{u}_p) + \tau \, \mathbf{u}_p' \mathbf{R}_p \mathbf{u}_p, \tag{11.21}$$

where

$$\mathbf{R}_p = \int D^2 \boldsymbol{\theta}_p(t_p) D^2 \boldsymbol{\theta}_p(t_p)' dt_p,$$

which can be computed by using a numerical integration method such as the trapezoidal rule or Romberg integration (e.g., Press et al. 1999). Similar to solving the ridge regression problem (Hoerl and Kennard 1970), the penalized least-squares estimate of $\mathbf{u}_p$ is obtained as

$$\hat{\mathbf{u}}_p = (\boldsymbol{\Theta}_p' \boldsymbol{\Theta}_p + \tau \mathbf{R}_p)^{-1} \boldsymbol{\Theta}_p' \mathbf{z}_p. \tag{11.22}$$

When Equation 11.21 is minimized with no further assumptions made, it has been proven that the function $x_p(t_p)$ that minimizes this criterion is a cubic spline with breakpoints placed at every data point t_{jp}, which is called *cubic spline smoothing* (de Boor 2001, Wahba 1990; also see Ramsay and Silverman 2005, Chapter 5). Once the number of breakpoints and the order of polynomials are determined, the number of spline basis functions (L) can be selected automatically as follows: $L = M + m - 1$, where M is the number of breakpoints and m is the order of polynomials (see Ramsay and Silverman 2005, Chapter 3). In functional generalized structured component analysis, the number of basis functions for a data function is determined based on cubic spline smoothing, unless stated otherwise. As shown in Equations 11.16 and 11.17, the same set of basis functions are employed for approximating weight and loading functions. Minimizing the penalized least-squares criterion (Equation 11.7) indicates that the basis function coefficients for these parameter functions are estimated based on the roughness penalty approach.

11.1.3.2 Selection of Smoothing Parameters

The values of the smoothing parameters λ and ρ should be decided before applying the alternating regularized least-squares algorithm for estimating parameters. It is acceptable to choose the values of λ and ρ subjectively (Ramsay and Silverman 2005, p. 206). At the same time, they can be chosen in an automatic fashion based on K-fold cross validation (Hastie, Tibshirani, and Friedman 2009, p. 214). As described in Chapter 8, K-fold cross validation begins by dividing the entire set of data into K subgroups of approximately equal size. One of the K subgroups is used as a validation sample and the remaining subgroups are used as a calibration sample. The parameters are estimated based on the calibration sample under a given pair of smoothing parameters, and then the estimated parameters are applied to the validation sample for calculating prediction error. This procedure repeats K times with each of the K subgroups being used as a validation sample. Finally, the cross-validation estimate of prediction error is computed over the K validation samples.

Under a given pair of λ and ρ, the cross-validation estimate of prediction error, denoted by CV, can be calculated by

$$
\mathrm{CV}(\lambda,\rho) = \frac{1}{K}\sum_{k=1}^{K}\left[\sum_{p=1}^{P}\sum_{i=1}^{N}\int(x_{ip}^{(k)}(t_p) - \gamma_{ip}^{(k)}\alpha_p^{(-k)}(t_p))^2 dt_p + \mathrm{SS}(\mathbf{\Gamma}^{(k)} - \mathbf{\Gamma}^{(k)}\mathbf{B}^{(-k)})\right],
$$

(11.23)

where $x_{ip}^{(k)}(t_p)$ is the value of a data function evaluated at t_p, which belongs to the kth validation sample, $\gamma_{ip}^{(k)}$ is the scalar value of a latent variable obtained from the validation sample, $\alpha_p^{(-k)}(t_p)$ is a loading function estimated from the calibration sample, $\mathbf{\Gamma}^{(k)}$ is a matrix consisting of all latent variables obtained from the validation sample, and $\mathbf{B}^{(-k)}$ is a matrix of path coefficients estimated from the calibration sample. The pair of smoothing parameter values associated with the smallest cross-validation estimate of prediction error can be chosen as the final one.

11.2 A Related Method: Functional Extended Redundancy Analysis

There exists no functional extension of partial least squares path modeling, although a function version of partial least squares regression was proposed (Reiss and Ogden 2007). This suggests that no component-based approach to structural equation modeling is yet available that can be comparable to functional generalized structured component analysis in scope and applicability. Perhaps, the most relevant method to functional generalized structured component analysis is functional extended redundancy analysis (Hwang et al. 2012b). In functional extended redundancy analysis, a latent variable is obtained as a component of each of multiple sets of exogenous data. All latent variables are assumed to influence endogenous data. The exogenous and endogenous data can be either functional or multivariate. As the name suggests, functional extended redundancy analysis was proposed as a functional extension of extended redundancy analysis (Takane and Hwang 2005). As discussed in Chapter 2, extended redundancy analysis can be viewed as a special case of generalized structured component analysis. Thus, functional extended redundancy analysis can be considered a special case of functional generalized structured component analysis. In this section, we briefly describe this special case of functional generalized structured component analysis.

We begin by providing the general model of functional extended redundancy analysis. Figure 11.3 displays a prototype of the general model. The prototype model contains three exogenous smooth functions (x_1, x_2, and x_3) and one endogenous smooth function (x_4). A latent variable (i.e., γ_1, γ_2, or γ_3) is assumed to be formed as a weighted composite of each exogenous function, and is also assumed to have an impact on the endogenous function. Accordingly, the general model considers both sets of endogenous and exogenous data functional. Moreover, a formative relationship is assumed between a latent variable and its functional data. Also importantly, endogenous functional data are always observed, not latent.

As discussed in Chapter 2, extended redundancy analysis is a component-based method that is related closely to the measurement model of generalized structured component analysis. Likewise, functional extended redundancy analysis is composed of the measurement and weighted relation models only. Specifically, the measurement model for the prototype model can be given as

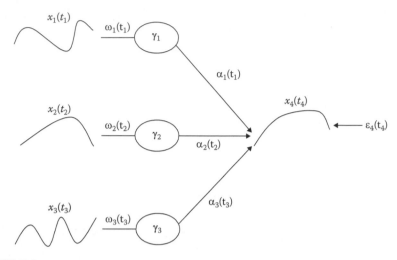

FIGURE 11.3
A prototype of the general model of functional extended redundancy analysis.

$$x_4\left(t_4\right) = \gamma_1\alpha_1\left(t_4\right) + \gamma_2\alpha_2\left(t_4\right) + \gamma_3\alpha_3\left(t_4\right) + \varepsilon_4\left(t_4\right). \tag{11.24}$$

The weighted relation model for the prototype model can be given as

$$\gamma_1 = \int x_1(t_1)\omega_1(t_1)dt_1$$

$$\gamma_2 = \int x_2(t_2)\omega_2(t_2)dt_2$$

$$\gamma_3 = \int x_3(t_3)\omega_3(t_3)dt_3 \tag{11.25}$$

As shown in the above sub-models, functional extended redundancy analysis aims to examine the effects of latent variables, that is, inner products of exogenous data and weight functions, on an observed endogenous data function. In this regard, functional extended redundancy analysis is viewed as a special case of functional generalized structured component analysis, which does not involve the structural model among latent variables.

We may also come across situations where only either endogenous or exogenous data are functional. This leads to two special cases of functional extended redundancy analysis. The first special case permits only exogenous data to be functional, while keeping endogenous data scalar or multivariate. Figure 11.4 displays a prototype model of the first special case. In this prototype model, there are three exogenous smooth functions (x_1, x_2, and x_3) and one endogenous indicator (z_4). Each latent variable is assumed to be formed as a weighted composite of an exogenous data function and also to influence the endogenous indicator.

The measurement model for the prototype model in Figure 11.4 can be given as

$$z_4 = \gamma_1 c_1 + \gamma_2 c_2 + \gamma_3 c_3 + \varepsilon_4. \tag{11.26}$$

In Equation 11.26, all loadings (c's) are scalar, which relate a latent variable to the endogenous indicator, and the residual (ε_4) is scalar as well. The weighted relation model for the prototype model is the same as Equation 11.25, because all exogenous data are functional.

In the second special case, only endogenous data are functional, while all exogenous data remain multivariate. Figure 11.5 depicts a prototype model of the second case. In this prototype model, each latent variable is defined as a weighted composite of two exogenous indicators and is assumed to influence an endogenous function (x_4).

The measurement model for the prototype model in Figure 11.5 can be given as

$$x_4(t_4) = \gamma_1 \alpha_1(t_1) + \gamma_2 \alpha_2(t_1) + \gamma_3 \alpha_3(t_1) + c_4(t_4). \tag{11.27}$$

The weighted relation model for the prototype model can be expressed as

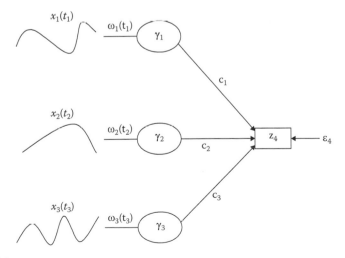

FIGURE 11.4
A prototype of the first special case of functional extended redundancy analysis.

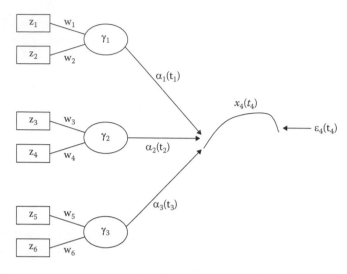

FIGURE 11.5
A prototype of the second special case of functional extended redundancy analysis.

$$\gamma_1 = z_1 w_1 + z_2 w_2$$

$$\gamma_2 = z_3 w_3 + z_4 w_4$$

$$\gamma_3 = z_5 w_5 + z_6 w_6. \tag{11.28}$$

To estimate parameter functions (ω_p and α_p) in the general model, we seek to minimize the following penalized least-squares criterion

$$\phi_5 = \sum_{i=1}^{N} \int \left(x_{ip}(t_p) - \sum_{k=1, k \neq p}^{P} \gamma_{ik}\, \alpha_k(t_p) \right)^2 dt_p + \lambda \sum_{k=1, k \neq p}^{P} \int [D^2 \omega_k(t_k)]^2 dt_k$$

$$+ \rho \sum_{k=1, k \neq p}^{P} \int [D^2 \alpha_k(t_p)]^2 dt_p, \tag{11.29}$$

with respect to ω_p and α_p, subject to

$$\sum_{i=1}^{N} \gamma_{ip}^{2} = N,$$

where λ and ρ indicate the smoothing parameters associated with the weight and loading functions, respectively. Equation 11.29 can reduce to the optimization criterion for each of the two special cases (see Hwang et al. 2012b).

An alternating regularized least-squares algorithm, which is similar to the algorithm for functional generalized structured component analysis, can be used to minimize Equation 11.29 in combination with a basis function expansion approach to approximating data and parameter functions. Moreover, virtually the same algorithm can be used for estimating parameters in the two special cases.

11.3 Examples

In this section, we apply functional generalized structured component analysis for the analysis of two real datasets. We will illustrate that functional generalized structured component analysis can deal with a situation where all data are functional, as well as where some data are functional while the others remain multivariate.

11.3.1 The Movie Box Office Revenue and Advertising Spending Data

The first example comes from a dataset that linked movie box office revenues to movie advertising spending. The box office revenues were collected from an on-line box office database available publicly (www.the-numbers.com), and the movie advertising spending data were provided by a commercial advertising consulting company. In this example, 152 movie titles released in the United States from 2006 to 2007 were measured on their weekly box office revenue for 10 consecutive weeks, as well as their weekly advertising spending on three different media (network televisions, newspapers, and national spot radios) for 15 consecutive weeks – 5 weeks prior to movie release and 10 weeks since release. Figure 11.6 displays the values of the box office revenues and advertising spending recorded for 152 movie titles.

We treated the box office revenues and weekly advertising spending records as data functions that varied continuously over time (i.e., weeks). Thus, this example considers data functions only. We hypothesized that a latent variable underlay each data function, and the latent variables underlying the three types of weekly advertising spending, labeled *TV advertising*, *newspaper advertising*, and *radio advertising*, had impacts on the latent variable underlying the weekly box office revenue, labeled *box office revenue*. Figure 11.7 displays the structural model specified for the data.

We used B-spline functions to approximate both data and parameter functions. We determined the number of basis functions, as described in Section 11.1.3.1. That is, we placed breakpoints at each data point and set the polynomial order at 4 (i.e., a cubic polynomial). Then, the number of basis functions was selected automatically. We applied five-fold cross validation for selecting the values of λ and ρ, considering 13 different values for each

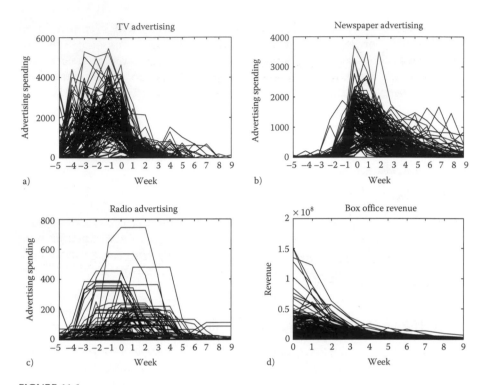

FIGURE 11.6
Weekly advertising spending on a) network televisions, b) newspapers, and c) national spot radios over 15 weeks and d) box office revenue over 10 weeks for 152 movie titles. Week zero denotes the movie release week.

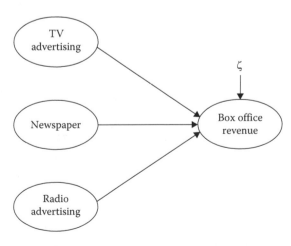

FIGURE 11.7
The structural model specified for the movie box office revenue and advertising spending data.

smoothing parameter $(10^{-10}, \ldots, 10^2)$. We found that the cross-validated prediction error estimate was minimized at $\lambda = 10$ and $\rho = 10^{-8}$. Thus, we used these values of λ and ρ for the example.

Given the values of the smoothing parameters, functional generalized structured component analysis was applied to fit the specified model to the data. It provided that FIT = 0.82 (SE = 0.02, 95% CI = 0.76–0.83), indicating that the model accounted for about 82% of the total variance of the data functions and latent variables. In addition, it provided that $\text{FIT}_M = 0.82$ (SE = 0.02, 95% CI = 0.77–0.94) and $\text{FIT}_S = 0.41$ (SE = 0.06, 95% CI = 0.33–0.54). We used 100 bootstrap samples to estimate SE or CI.

Figure 11.8 shows the estimated loading functions, along with their pointwise 95% CI. The solid and dashed lines indicate the estimated loading functions and pointwise 95% CI, respectively. In the figure, week zero denotes the week of movie release. The advertising spending on network televisions appeared to peak at around two weeks before movie release and end in the third week after the release. The advertising spending on newspapers tended to increase from the second week before the release

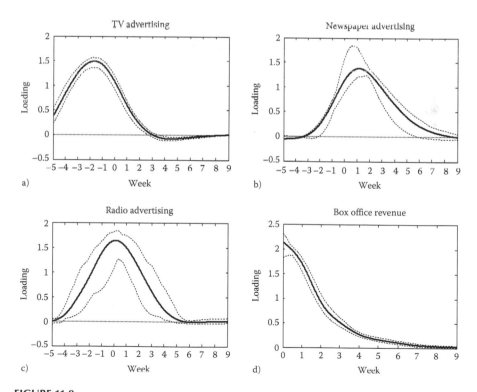

FIGURE 11.8
The estimated loading functions (solid lines) and their 95% pointwise confidence intervals (dashed lines) obtained from the movie box office revenue and advertising spending data. The horizontal dotted line indicates the zero line.

and to peak in the first week after the release. The advertising spending on national spot radios seemed to show a temporal pattern of change similar to that of the advertising spending on newspapers. Lastly, the weekly box office revenue was highest in the release week and then decreased somewhat sharply.

Figure 11.9 displays the estimated weight functions, along with their pointwise 95% CI. The estimated weight functions show which weeks play a more important role in determining the corresponding latent variables. In general, the estimated weight functions looked similar to the estimated loading functions depicted in Figure 11.8. Thus, they can be interpreted in a similar manner.

Table 11.1 presents the estimated path coefficients and their 95% CI. All exogenous latent variables (*TV advertising, newspaper advertising,* and *radio advertising*) had statistically significant and positive impacts on *box office revenue. Radio advertising* tended to show a smaller effect than *TV advertising* and *newspaper advertising.*

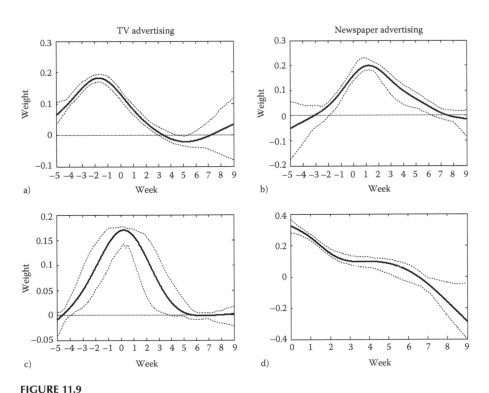

FIGURE 11.9

The estimated weight functions (solid lines) and their 95% pointwise confidence intervals (dashed lines) obtained from movie box office revenue and advertising spending data. The horizontal dotted line indicates the zero line.

TABLE 11.1

Estimates of Path Coefficients and Their 95% Confidence Intervals Obtained from the Movie Box Office Revenue and Advertising Spending Data

	Estimate	95% CI
TV advertising → box office revenue	0.33	0.11–0.52
Newspaper advertising → box office revenue	0.37	0.20–0.53
Radio advertising → box office revenue	0.14	0.01–0.29

CI, confidence interval.

11.3.2 The Gait Data

The second example is part of the gait data (Yogev et al. 2005), which is available on the PhysioNet website (http://www.physionet.org/physiobank/database/gaitpdb/; also see Goldberger et al. 2000). In this example, 23 patients diagnosed with idiopathic Parkinson's disease were measured on total force under the left foot (in Newtons) over 80 time points for 8 seconds, while they walked at their usual pace. In the original data, total force under the left foot was measured at 100 Hz per second, which resulted in a total of 800 time points for 8 seconds. We used only 80 time points by sampling observations at every 10 time points. Figure 11.10 displays the data functions of total force under the left foot for the patients. The patients began to be measured on their total force under the left foot at different phases of walking. Thus, these data functions were aligned to be in the same phase across the patients by using the continuous registration method (Ramsay, Hooker, Graves 2009, Chapter 8).

In addition, physical body size of the patients was measured based on two variables – height (in centimeters) and weight (in kilograms). Parkinson's disease severity was evaluated with two scale variables – the Hoehn and Yahr staging scale (HY) and the unified Parkinson's disease rating scale (UPDRS) variables. The former variable was used to rate six different levels of disability (Hoehn and Yahr 1967), whereas the latter was the sum of all item scores in the UPDRS. In both variables, a higher score indicated a more severe level of Parkinson's disease. Thus, the second example considers a data function and two sets of indicators at the same time. In addition, it can be seen as an example of high-frequency repeated measures data that involve a greater number of occasions than the number of individuals.

We hypothesized that a latent variable, labeled *total force*, underlay the data function, and it was affected by two other latent variables, labeled *body size* and *disease severity*, each of which was associated with two indicators. Figure 11.11 displays the structural model specified for the data.

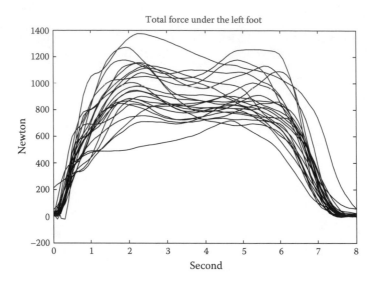

FIGURE 11.10
The data functions of total force under the left foot for 23 Parkinson's disease patients.

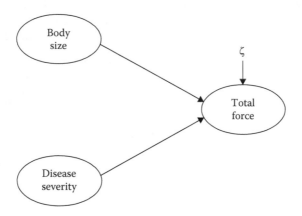

FIGURE 11.11
The structural model specified for the gait data.

We used B-spline functions to approximate both data and parameter functions. To reduce computational burden, we set the number of breakpoints at 50 and the polynomial order at 4. The breakpoints were equally spaced. We then determined the number of basis functions, as described in Section 11.1.3.1. We applied five fold cross validation for selecting the values of λ and ρ, considering 10 different candidates for each smoothing parameter (10^0,..., 10^9). We found that the minimum of the cross-validated prediction error estimate was achieved at $\lambda = 10^5$ and $\rho = 10^2$.

TABLE 11.2

Estimates of Loadings and Weights and Their 95% Confidence Intervals
Obtained from the Gait Data

Latent	Indicator	Weight		Loading	
		Estimate	95% CI	Estimate	95% CI
Body size	Height	0.45	0.32–0.58	0.90	082–0.96
	Weight	0.62	0.50–0.73	0.95	0.88–0.98
Disease severity	HY	0.57	−0.00–0.64	0.79	−0.04–0.93
	UPDRS	0.66	0.50–0.98	0.84	0.79–0.98

CI, confidence interval; HY, Hoehn and Yahr staging scale; UPDRS, unified Parkinson's
disease rating scale.

Under these values of λ and ρ, functional generalized structured component analysis was applied to fit the specified model to the data. It provided that FIT = 0.74 (SE = 0.05, 95% CI = 0.63–0.84), indicating that the model accounted for about 74% of the total variance of the data functions and latent variables. It also provided that FIT_M = 0.74 (SE = 0.05, 95% CI = 0.63–0.84) and FIT_S = 0.57 (SE = 0.13, 95% CI = 0.28–0.79).

Table 11.2 provides the estimated loadings and weights of the four indicators for the two exogenous latent variables – *body size* and *disease severity*. The loading and weight estimates of height and weight for *body size* were positive and statistically significant. The loading and weight estimates of UPDRS for *disease severity* were also positive and statistically significant. On the other hand, the weight estimate of HY was not statistically significant. This suggests that the Hoehn and Yahr scale value made little contribution to determining *disease severity*. The loading estimate of HY was not statistically significant, either.

Figure 11.12 shows the estimated loading and weight functions for *total force*, along with their pointwise 95% CI. The estimated loading function appeared to show an M-shape trajectory of change over 8 seconds. That is, it tended to increase sharply for the first two seconds, decrease gradually for the next two and a half seconds, increase again for the next two seconds, and then decrease for the remaining seconds. The estimated weight function showed a similar pattern of change, indicating that time points associated with high weight function values might be more crucial for determining *total force*.

Table 11.3 presents the estimated path coefficients and their 95% CI. *Body size* had a positive and statistically significant effect on *total force*. That is, the greater *body size*, the more force was exerted under the left foot. On the other hand, *disease severity* had a negative and statistically significant impact on *total force*. This suggests that patients in a more severe stage of Parkinson's disease tended to place a weaker level of total force under the left foot.

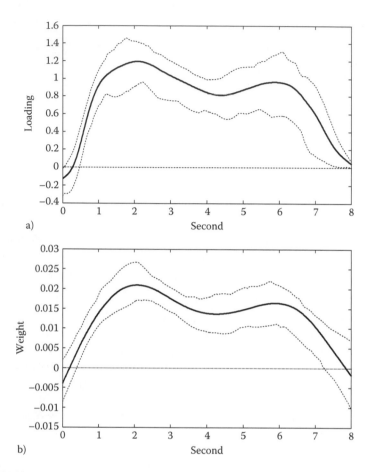

FIGURE 11.12

The estimated a) loading and b) weight functions for total force under the left foot (solid lines) and their 95% pointwise confidence intervals (dashed lines) obtained from the gait data. The horizontal dotted line indicates the zero line.

TABLE 11.3

Estimates of Path Coefficients and Their 95% Confidence Intervals Obtained from the Gait Data

	Estimate	95% CI
Body size → Total force	0.63	0.37–0.84
Disease severity → Total force	−0.45	−0.68 to −0.14

CI, confidence interval.

11.4 Summary

We discussed a functional extension of generalized structured component analysis, called functional generalized structured component analysis. In this extension, a block of indicators for a latent variable is replaced by a smooth data function. This leads to revising the measurement and weight relation models for accommodating such data functions. Specifically, a latent variable is defined as the inner product of data and weight functions in the weighted relation model. In addition, when data functions are reflective, their loadings and residuals also become functional in the measurement model. On the other hand, no change needs to be made to the structural model, because the individual scores of latent variables still remain scalar as in generalized structured component analysis. A penalized least-squares criterion was derived to estimate parameters, in which roughness penalty terms are combined with ordinary least-squares criteria in order to control for the degree of roughness embedded in parameter functions. An alternating regularized least-squares algorithm can be used to minimize the criterion. The algorithm repeats several steps until convergence. In particular, through the adoption of a basis function expansion approach to approximating functions, each step essentially reduces to solving a standard regularized least-squares regression problem.

As illustrated in Section 11.3, functional generalized structured component analysis can be of use in summarizing the relationships between data functions and latent variables and in understanding the associations between latent variables. In addition, it is flexible to accommodate both functional and multivariate data at the same time and can effectively handle situations where the number of occasions exceeds the number of individuals.

The collection of functional data will be more prevalent in a greater variety of scientific disciplines, along with technological advances in measurement. Functional generalized structured component analysis may be a useful tool for researchers who want to specify and test a structural equation model for such an emerging type of data. Moreover, it makes a technical contribution to generalized structured component analysis by expanding its scope and applicability beyond the conventional realm of multivariate statistics.

Appendix 11.1 The Alternating Regularized Least-Squares Algorithm for Functional Generalized Structured Component Analysis

Using the basis function expansion approach in Equations 11.15 and 11.16, the ith individual score of the pth latent variable in the weighted relation model ($i = 1,..., N$) can be rewritten as

$$\gamma_{ip} = \int x_{ip}(t_p)\omega_p(t_p)dt_p$$

$$= \int \mathbf{u}_{ip}'\boldsymbol{\theta}(t_p)\boldsymbol{\theta}(t_p)'\mathbf{y}_p dt_p$$

$$= \mathbf{u}_{ip}'\left(\int \boldsymbol{\theta}(t_p)\boldsymbol{\theta}(t_p)'dt_p\right)\mathbf{y}_p$$

$$= \mathbf{u}_{ip}'\mathbf{Q}_p\mathbf{y}_p, \tag{A11.1}$$

where

$$\mathbf{Q}_p = \int \boldsymbol{\theta}(t_p)\boldsymbol{\theta}(t_p)'dt_p.$$

We can express the simultaneous expansion of each set of all N data functions as

$$\mathbf{x}_p(t_p) = \mathbf{U}_p\boldsymbol{\theta}_p(t_p), \tag{A11.2}$$

where $\mathbf{U}_p = [\mathbf{u}_{ip},..., \mathbf{u}_{Np}]'$ denotes an N by L matrix of basis function coefficients for an N by 1 vector-valued data function, denoted by $\mathbf{x}_p(t_p) = [x_{1p}(t_p), ..., x_{Np}(t_p)]'$. Moreover, an N by 1 vector of the individual scores of the pth latent variable, denoted by γ_p, can be compactly expressed as

$$\gamma_p = \begin{bmatrix} \gamma_{1p} \\ \vdots \\ \gamma_{Np} \end{bmatrix} = \mathbf{U}_p\mathbf{Q}_p\mathbf{y}_p. \tag{A11.3}$$

Likewise, the measurement model (Equation 11.2) can be rewritten as

$$\mathbf{U}_p\boldsymbol{\theta}(t_p) = \mathbf{U}_p\mathbf{Q}_p\mathbf{y}_p\mathbf{a}_p'\boldsymbol{\theta}_p(t_p) + \boldsymbol{\varepsilon}_p(t_p), \tag{A11.4}$$

where $\boldsymbol{\varepsilon}_p(t_p) = [\varepsilon_{1p}(t_p), ..., \varepsilon_{Np}(t_p)]'$ denotes an N by 1 vector of the values of the residual function evaluated at t_p. Moreover, the two roughness penalty terms in Equation 11.7 can be re-expressed as

$$\int \left(D^2\omega_p(t_p)\right)^2 dt_p = \int \left(D^2\mathbf{y}_p'\boldsymbol{\theta}_p(t_p)D^2\boldsymbol{\theta}_p(t_p)'\mathbf{y}_p\right)dt_p$$

$$= \mathbf{y}_p'\left(\int D^2\boldsymbol{\theta}_p(t_p)D^2\boldsymbol{\theta}_p(t_p)'dt_p\right)\mathbf{y}_p$$

$$= \mathbf{y}_p'\mathbf{R}_p\mathbf{y}_p \tag{A11.5}$$

$$\int \left(D^2 \alpha_p(t_p) \right)^2 dt_p = \int \left(D^2 \mathbf{a}_p' \boldsymbol{\theta}_p(t_p) D^2 \boldsymbol{\theta}_p(t_p)' \mathbf{a}_p \right) dt_p$$

$$= \mathbf{a}_p' \mathbf{R}_p \mathbf{a}_p, \tag{A11.6}$$

where $\mathbf{R}_p = \int D^2 \boldsymbol{\theta}_p(t_p) D^2 \boldsymbol{\theta}_p(t_p)' \, dt_p$.

Consequently, the penalized least-squares criterion (Equation 11.7) can be re-expressed as

$$\phi = \sum_{p=1}^{P} \int \mathrm{SS}\left(\mathbf{U}_p \boldsymbol{\theta}(t_p) - \mathbf{U}_p \mathbf{Q}_p \mathbf{y}_p \mathbf{a}_p' \boldsymbol{\theta}_p(t_p) \right) dt_p + \mathrm{SS}(\boldsymbol{\Gamma} - \boldsymbol{\Gamma} \mathbf{B}) + \lambda \sum_{p=1}^{P} \mathbf{y}_p' \mathbf{R}_p \mathbf{y}_p$$

$$+ \rho \sum_{p=1}^{P} \mathbf{a}_p' \mathbf{R}_p \mathbf{a}_p. \tag{A11.7}$$

By using the basis function expansion approach, our problem is to minimize this criterion with respect to $\mathbf{y}_p$, $\mathbf{a}_p$, and $\mathbf{B}$. An alternating regularized least-squares algorithm can be used to minimize the criterion. The algorithm repeats the following three steps until convergence.

Step 1: Update $\mathbf{y}_p$ for fixed $\mathbf{a}_p$ and $\mathbf{B}$. Note that the first term of Equation A11.7 can be further written as

$$\phi^* = \sum_{p=1}^{P} \int \mathrm{SS}\left(\mathbf{U}_p \boldsymbol{\theta}_p(t_p) - \mathbf{U}_p \mathbf{Q}_p \mathbf{y}_p \mathbf{a}_p' \boldsymbol{\theta}_p(t_p) \right) dt_p$$

$$= \sum_{p=1}^{P} \mathrm{tr}\left[\int \left(\mathbf{U}_p \boldsymbol{\theta}_p(t_p) - \mathbf{U}_p \mathbf{Q}_p \mathbf{y}_p \mathbf{a}_p' \boldsymbol{\theta}_p(t_p) \right) \left(\mathbf{U}_p \boldsymbol{\theta}_p(t_p) - \mathbf{U}_p \mathbf{Q}_p \mathbf{y}_p \mathbf{a}_p' \boldsymbol{\theta}_p(t_p) \right)' dt_p \right]$$

$$= \sum_{p=1}^{P} \mathrm{tr}\left[\mathbf{U}_p \left(\int \boldsymbol{\theta}_p(t_p) \boldsymbol{\theta}_p(t_p)' dt_p \right) \mathbf{U}_p' \right] - 2\mathrm{tr}\left[\begin{array}{c} \mathbf{U}_p \left(\int \boldsymbol{\theta}_p(t_p) \boldsymbol{\theta}_p(t_p)' dt_p \right) \\ \times \mathbf{a}_p \mathbf{y}_p' \mathbf{Q}_p \mathbf{U}_p' \end{array} \right]$$

$$+ \mathrm{tr}\left[\mathbf{U}_p \mathbf{Q}_p \mathbf{y}_p \mathbf{a}_p' \left(\int \boldsymbol{\theta}_p(t_p) \boldsymbol{\theta}_p(t_p)' dt_p \right) \mathbf{a}_p \mathbf{y}_p' \mathbf{Q}_p \mathbf{U}_p' \right]$$

$$= \sum_{p=1}^{P} \mathrm{tr}\left[\mathbf{U}_p \mathbf{Q}_p \mathbf{U}_p' \right] - 2\mathrm{tr}\left[\mathbf{U}_p \mathbf{Q}_p \mathbf{a}_p \mathbf{y}_p' \mathbf{Q}_p \mathbf{U}_p' \right] + \mathrm{tr}\left[\mathbf{U}_p \mathbf{Q}_p \mathbf{y}_p \mathbf{a}_p' \mathbf{Q}_p \mathbf{a}_p \mathbf{y}_p' \mathbf{Q}_p \mathbf{U}_p' \right]$$

$$\tag{A11.8}$$

Let $\mathbf{e}_k$ denote a P by 1 vector of zeros except for the pth element being unity. Let $\mathbf{\Gamma}^{(-p)}$ denote $\mathbf{\Gamma}$ whose pth column contains all zeros. Then, the second term of Equation A11.7 can be rewritten as

$$\phi^{**} = SS(\mathbf{\Gamma} - \mathbf{\Gamma B})$$

$$= SS(\mathbf{\Gamma}(\mathbf{I} - \mathbf{B}))$$

$$= SS((\mathbf{\Gamma}^{(-p)} + \mathbf{\gamma}_p \mathbf{e}_p{}')(\mathbf{I} - \mathbf{B}))$$

$$= SS((\mathbf{\Gamma}^{(-p)} + \mathbf{U}_p\mathbf{Q}_p\mathbf{y}_p\mathbf{e}_p{}')(\mathbf{I} - \mathbf{B}))$$

$$= \mathrm{tr}\left[\mathbf{\Gamma}^{(-p)}(\mathbf{I} - \mathbf{B})(\mathbf{I} - \mathbf{B})'\mathbf{\Gamma}^{(-p)\,'}\right] - 2\mathrm{tr}\left[\mathbf{\Gamma}^{(-p)}(\mathbf{I} - \mathbf{B})(\mathbf{I} - \mathbf{B})'\mathbf{e}_p\mathbf{y}_p{}'\mathbf{Q}_p\mathbf{U}_p{}'\right]$$

$$+ \mathrm{tr}\left[\mathbf{U}_p\mathbf{Q}_p\mathbf{y}_p\mathbf{e}_p{}'(\mathbf{I} - \mathbf{B})(\mathbf{I} - \mathbf{B})'\mathbf{e}_p\mathbf{y}_p{}'\mathbf{Q}_p\mathbf{U}_p{}'\right]. \tag{A11.9}$$

By solving

$$\frac{\partial \phi}{\partial \mathbf{y}_p} = \frac{\partial \phi^*}{\partial \mathbf{y}_p} + \frac{\partial \phi^{**}}{\partial \mathbf{y}_p} + \frac{\partial \lambda \mathbf{y}_p{}'\mathbf{R}_p\mathbf{y}_p}{\partial \mathbf{y}_p} = \mathbf{0},$$

the estimate of $\mathbf{y}_p$ can be obtained as

$$\breve{\mathbf{y}}_p = \left(\mathbf{a}_p{}'\mathbf{Q}_p\mathbf{a}_p\mathbf{Q}_p\mathbf{U}_p{}'\mathbf{U}_p\mathbf{Q}_p + \mathbf{e}_p{}'(\mathbf{I} - \mathbf{B})(\mathbf{I} - \mathbf{B})'\mathbf{e}_p\mathbf{Q}_p\mathbf{U}_p{}'\mathbf{U}_p\mathbf{Q}_p + \lambda\mathbf{R}_p\right)^{-1}$$

$$\times \left(\mathbf{Q}_p\mathbf{U}_p{}'\mathbf{U}_p\mathbf{a}_p + \mathbf{Q}_p\mathbf{U}_p{}'\mathbf{\Gamma}^{(-p)}(\mathbf{I} - \mathbf{B})(\mathbf{I} - \mathbf{B})'\mathbf{e}_p\right). \tag{A11.10}$$

The updated $\mathbf{y}_p$ is subsequently multiplied by

$$\sqrt{\frac{N}{\mathbf{y}_p{}'\mathbf{Q}_p\mathbf{U}_p{}'\mathbf{U}_p\mathbf{Q}_p\mathbf{y}_p}}$$

in order to satisfy the standardization constraint imposed on the pth latent variable.

Step 2: Update $\mathbf{a}_p$ for fixed $\mathbf{y}_p$ and $\mathbf{B}$. By solving

$$\frac{\partial \phi}{\partial \mathbf{a}_p} = \frac{\partial \phi^*}{\partial \mathbf{a}_p} + \frac{\partial \rho \mathbf{a}_p{}'\mathbf{R}_p\mathbf{a}_p}{\partial \mathbf{a}_p} = \mathbf{0},$$

the estimate of $\mathbf{a}_p$ is obtained as

$$\hat{\mathbf{a}}_p = \left(\mathbf{y}_p{}'\mathbf{Q}_p\mathbf{U}_p{}'\mathbf{U}_p\mathbf{Q}_p\mathbf{y}_p\mathbf{Q}_p + \rho\mathbf{R}_p\right)^{-1}\left(\mathbf{Q}_p\mathbf{U}_p{}'\mathbf{U}_p\mathbf{Q}_p\mathbf{y}_p\right). \tag{A11.11}$$

Step 3: Update **B** for fixed $\mathbf{y}_p$ and $\mathbf{a}_p$. This step reduces to minimize the second term of Equation A11.7 only, which can be re-expressed as

$$\phi^{**} = SS(\boldsymbol{\Gamma} - \boldsymbol{\Gamma}\mathbf{B})$$

$$= SS(\mathrm{vec}(\boldsymbol{\Gamma}) - (\mathbf{I} \otimes \boldsymbol{\Gamma})\mathrm{vec}(\mathbf{B}))$$

$$= SS(\mathrm{vec}(\boldsymbol{\Gamma}) - \boldsymbol{\Omega}\mathbf{b}), \tag{A11.12}$$

where **b** denotes a vector formed by eliminating all zero elements in vec(**B**), and $\boldsymbol{\Omega}$ is a matrix formed by removing the columns of $\mathbf{I} \otimes \boldsymbol{\Gamma}$ corresponding to the zero elements in vec(**B**). The estimate of **b** is given as

$$\hat{\mathbf{b}} = (\boldsymbol{\Omega}'\boldsymbol{\Omega})^{-1}\boldsymbol{\Omega}'\mathrm{vec}(\boldsymbol{\Gamma}). \tag{A11.13}$$

The updated **B** is reconstructed from $\hat{\mathbf{b}}$.

References

Abela, J.R.Z., Ho, M.R., Webb, C.A., & McWhinnie, C.M. (2008). Materialistic values and well-being: a multi-wave longitudinal analysis. Paper submitted for publication.

Addinsoft (2009). *XLSTAT 2009*. France: Addinsoft.

Agarwal, R. & Karahanna, E. (2000). Time flies when you're having fun: cognitive absorption and beliefs about information technology usage. *MIS Quarterly*, 24, 665–694.

Aiken, L.S. & West, S.G. (1991). *Multiple Regression: Testing and Interpreting Interactions.* Newbury Park, CA: Sage Publications.

Allen, N.J. & Meyer, J.P. (1990). The measurement and antecedents of affective, continuance and normative commitment to the organization. *Journal of Occupational Psychology*, 63, 1–18.

Almansa, J. & Delicado, P. (2009). Analysing musical performance through functional data analysis: rhythmic structure in Schumann's Traumerei. *Connection Science*, 21, 207–225.

Alwin, D.F. & Hauser, R.M. (1975). The decomposition of effects in path analysis. *American Sociological Review*, 40, 37–47.

Algina, J. & Moulder, B.C. (2001). A note on estimating the Jöreskog-Yang model for latent variable interaction using LSIRESL 8.3. *Structural Equation Modeling*, 8, 40 52.

Alva, K., Romo, J., & Ruiz, E. (2009). Modelling intra-daily volatility by functional data analysis: an empirical application to the Spanish stock market. Working paper 09-28. *Statistics and Econometrics Series 9*. Universidad Carlos III de Madrid.

Amemiya, T. (1985). *Advanced Econometrics*. Cambridge, MA: Harvard University Press.

Anderson, E.W. & Fornell, C. (2000). Foundations of the American customer satisfaction index. *Total Quality Management*, 11, S869–S882.

Anderson, E.W., Fornell, C., & Lehman, D.R. (1994). Customer satisfaction, market share, and profitability: findings from Sweden. *Journal of Marketing*, 58, 53–66.

Anderson, E.W., Fornell, C., & Mazvancheryl, S.K. (2004). Customer satisfaction and shareholder value. *Journal of Marketing*, 68, 172–185.

Anderson, E.W., Fornell, C., & Rust, R.T. (1997). Customer satisfaction, productivity, and profitability: differences between goods and services. *Marketing Science*, 16, 129–45.

Anderson, T.W. (1951). Estimating linear restrictions on regression coefficients for multivariate normal distributions. *Annals of Mathematical Statistics*, 22, 327–351.

Arabie, P., Carroll, J.D., DeSarbo, W.S., & Wind, J. (1981). Overlapping clustering: a new method for product positioning. *Journal of Marketing Research*, 18, 310–317.

Arabie, P. & Hubert, L. (1994). Cluster analysis in marketing research. In R.P. Bagozzi (Ed.), *Advanced Methods of Marketing Research* (pp. 160–189). Oxford: Blackwell.

Bagozzi, R.P. (1982). A field investigation of causal relations among cognition, affect, intensions, and behavior. *Journal of Marketing Research*, 19, 562–584.

Bagozzi, R.P. (1994). Structural equation models in marketing research: basic princi-ples. In R.P. Bagozzi (Ed.), *Principles of Marketing Research* (pp. 317–385). Oxford: Blackwell.

Bagozzi, R.P., Baumgartner, H., & Yi, Y. (1992). State versus action orientation and the theory of reasoned action: an application to coupon usage. *Journal of Consumer Research*, 18, 505–518.

Baker, M.J. (1988). *Marketing Strategy and Management*. New York: Macmillan Education.

Baron, R.M. & Kenny, D.A. (1986). The moderator-mediator variable distinction in social psychological research: conceptual, strategic, and statistical consider-ations. *Journal of Personality and Social Psychology*, 51, 1173–1182.

Bauer, D.J. (2003). Estimating multilevel linear models as structural equation models. *Journal of Educational and Behavioral Statistics*, 28, 135–167.

Bauer, D.J. & Curran, P.J. (2005). Probing interactions in fixed and multilevel regression: inferential and graphical techniques. *Multivariate Behavioral Research*, 35, 373–400.

Baumann, K., Albert, H., & von Korff, M. (2002). A systematic evaluation of the ben-efits and hazards of variable selection in latent variable regression. Part I. Search algorithm, theory, and simulations. *Journal of Chemometrics*, 16, 339–350.

Beaton, A.E. & Tukey, J.W. (1974). The fitting of power series, meaning polynomials, illustrated on band-spectroscopic data. *Technometrics*, 16, 147–185.

Belsley, D.A., Kuh, E., & Welsch, R.E. (1980). *Regression Diagnostics: Identifying Influential Data and Sources of Collinearity*. New York: John Wiley & Sons.

Bentler, P.M. & Freeman, E.H. (1983). Tests for stability in linear structural equation systems. *Psychometrika*, 48, 143–145.

Bentler, P.M. & Weeks, D.G. (1980). Linear structural equations with latent variables. *Psychometrika*, 45, 3, 289–308.

Benzécri, J.P. (1973). *Analyse des Données*. Paris: Dunod.

Bergami, M. & Bagozzi, R.P. (2000). Self-categorization, affective commitment and group self-esteem as distinct aspects of social identity in the organization. *British Journal of Social Psychology*, 39, 555–577.

Berk, R.A. (2008). *Statistical Learning from a Regression Perspective*. New York: Springer.

Bernhardt, K.L., Donthu, N., & Kennett, P.A. (2000). A longitudinal analysis of satis-faction and profitability. *Journal of Business Research*, 47, 161–171.

Besse, P. & Ramsay, J.O. (1986). Principal components analysis of sampled functions. *Psychometrika*, 51, 285–311.

Bezdek, J.C. (1974a). Numerical taxonomy with fuzzy sets. *Journal of Mathematical Biology*, 1, 57–71.

Bezdek, J.C. (1974b). Cluster validity with fuzzy set. *Journal of Cybernetics*, 3, 58–72.

Bezdek, J.C. (1981). *Pattern Recognition with Fuzzy Objective Function Algorithms*. New York: Plenum Press.

Bezdek, J.C., Coray, C., Gunderson, R., & Watson, J. (1981). Detection and charac-teristics of cluster substructure. II. Fuzzy c-varieties and convex combinations thereof. *SIAM Journal on Applied Mathematics*, 40, 358–372.

Blankstein, K.R. & Flett, G.L. (1993). *Development of the General Hassles Scale for Students*. Unpublished manuscript, University of Toronto at Mississauga.

Bock, R.D. (1989). *Multilevel Analysis of Educational Data*. San Diego, CA: Academic Press.

Böckenholt, U. & Takane, Y. (1994). Linear constraints in correspondence analysis. In M.J. Greenacre & J. Blasius (Eds.), *Correspondence Analysis in Social Sciences* (pp. 112–127). London: Academic Press.

Bollen, K.A. (1987). Total, direct, and indirect effects in structural equation models. In C.C. Clogg (Ed.), *Sociological Methodology* (pp. 37–69). Washington, D.C.: American Sociological Association.

Bollen, K.A. (1989). *Structural Equations with Latent Variables*. New York: Wiley.

Bollen, K.A. & Curran, P.J. (2004). Autoregressive latent trajectory (ALT) models: a synthesis of two traditions. *Sociological Methods & Research*, 32, 336–383.

Bollen, K.A. & Curran, P.J. (2006). *Latent Curve Models: A Structural Equation Perspective*. New Jersey: Wiley.

Bollen, K.A., Kirby, J.B., Curran, P.J., Paxton, P.M., & Chen, F. (2007). Latent variable models under misspecification: two-stage least squares (2SLS) and maximum likelihood (ML) estimators. *Sociological Methods and Research*, 36, 46–86.

Browne, M.W. (1984). The decomposition of multitrait-multimethod matrices. *British Journal of Mathematical and Statistical Psychology*, 37, 1–21.

Browne, M.W. & Cudeck, R. (1993). Alternative ways to assessing model fit. In K.A. Bollen & J.S. Long (Eds.), *Testing Structural Equation Models* (pp. 136–162). Newbury Park, CA: Sage Publications.

Browne, M.W. & Du Toit, S.H.C. (1991). Models for learning data. In L.M. Collins & J.L. Horn (Eds.), *Best Methods for the Analysis of Change: Recent Advances, Unanswered Questions, Future Directions* (pp. 47–68). Washington, DC: American Psychological Association.

Browne, M. & Nesselroade, J.R. (2005). Representing psychological processes with dynamic factor models: some promising uses and extensions of autoregressive moving average time series models. In A. Maydeu-Olivares & J.J. McArdle (Eds.), *Contemporary Advances in Psychometrics* (pp. 415–452). Mahwah, NJ: Erlbaum.

Bryk, A.S. & Raudenbush, S.W. (1992). *Hierarchical Linear Models: Applications and Data Analysis Methods*. Newbury Park: Sage Publications.

Busemeyer, J.R. & Jones, L.E. (1983). Analysis of multiplicative combination rules when the causal variables are measured with error. *Psychological Bulletin*, 93, 549–562.

Campbell, D.T. & Fiske, D.W. (1959). Convergent and discriminant validation by the multitrait-multimethod matrix. *Psychological Bulletin*, 56, 81–105.

Cardot, H. (2000). Nonparametric estimation of smoothed principal components analysis of sampled noisy functions. *Nonparametric Statistics*, 12, 503–538.

Cardot, H., Ferraty, F., & Sarda, P. (1999). Functional linear model. *Statistics and Probability Letters*, 45, 11–22.

Carroll, J.D. (1968). A generalization of canonical correlation analysis to two or more sets of variables. *Proceedings of the 76th Annual Convention of the American Psychological Association*, 227–228.

Chatterjee, S. & Price, B. (1977). *Regression Analysis by Example*. New York: John Wiley and Sons.

Chin, W.W. (1998). The partial least squares approach for structural equation modeling. In G.A. Marcoulides (Ed.), *Modern Methods for Business Research* (pp. 295–358). Mahwah, NJ: Lawrence Erlbaum Associates.

Chin, W.W. (2001). *PLS-Graph User's Guide Version 3.0*. Soft Modeling Inc.

Chin, W.W., Marcolin, B.L., & Newsted, P.R. (1996). A partial least squares latent variable approach for measuring interaction effects: results from a Monte Carlo simulation study and voice mail emotion/adoption study. *Proceedings of the Seventeenth International Conference on Information Systems*, J.I. DeGross, S. Jarvenpaa and A. Srinivasan, (Eds.), Cleveland: Association for Information Systems.

Chin, W.W., Marcolin, B.L., & Newsted, P.R. (2003). A partial least squares latent variable modeling approach for measuring interaction effects: results from a Monte Carlo simulation study and an electronic-mail emotion/adoption study. *Information Systems Research*, 14, 189–217.

Clarke, K.A. (2005). The phantom menace: omitted variable bias in econometric research. *Conflict Management and Peace Science*, 22, 341–352.

Cleary, P.D. & Kessler, R.C. (1982). The estimation and interpretation of modifier effects. *Journal of Health and Social Behavior*, 23, 159–169.

Cohen, J. & Cohen, P. (1983). *Applied Multiple Regression/Correlation Analyses for the Behavioural Sciences*. Hillsdale, NJ: Lawrence Erlbaum Associates.

Cohen, J., Cohen, P., West, S.G., & Aiken, L.S. (2003). *Applied Multiple Regression/ Correlation Analysis for the Behavioral Sciences*. (3rd Edition). Hillsdale, NJ: Erlbaum.

Cook, R.D. & Weisberg, S. (1982). *Residuals and Influence in Regression*. New York: Chapman and Hall.

Coolen, H. & de Leeuw, J. (1987). Least squares path analysis with optimal scaling. Paper presented at the fifth international symposium of data analysis and informatics. Versailles, France.

Coppi, R., Gil, M.A., & Kiers, H.A.L. (2006). The fuzzy approach to statistical analysis. *Computational Statistics & Data Analysis*, 51, 1–14.

Crainiceanu, C.M., Staicu, A.M., & Di, C. (2009). Generalized multilevel functional regression. *Journal of the American Statistical Association*, 104, 1550–1561.

Craven, P. & Wahba, G. (1979). Smoothing noisy data with spline functions. *Numerische Mathematik*, 31, 377–403.

Cronbach, L.J. (1951). Coefficient alpha and the internal structure of tests. *Psychometrika*, 16, 297–334.

Croon, M. (2002). Using predicted latent scores in general latent structure models. In G.A. Marcoulides & I. Moustaki (Eds.), *Latent Variable and Latent Structure Models* (pp. 195–223). Mahwah, NJ: Lawrence Erlbaum Associates.

Curran, P.J. (1998). Introduction to hierarchical linear models of individual growth: an applied example using the SAS data system. Paper presented at the first international institute on developmental science, University of North Carolina, Chapel Hill.

Curran, P.J. (2003). Have multilevel models been structural equation models all along? *Multivariate Behavioral Research*, 38, 529–569.

Davidian, M., Lin, X., & Wang, J.-L. (2004). Introduction: emerging issues in longitudinal and functional data analysis. *Statistica Sinica*, 14, 613–614.

Davies, P.T. & Tso, M.K.-S. (1982). Procedures for reduced-rank regression. *Applied Statistics*, 31, 244–255.

de Boor, C. (2001). *A Practical Guide to Splines*. New York: Springer.

de Jong, S. & Kiers, H.A.L. (1992). Principal covariates regression. *Chemometrics and Intelligent Laboratory Systems*, 14, 155–164

de Leeuw, J., Young, F.W., & Takane, Y. (1976). Additive structure in qualitative data: an alternating least squares method with optimal scaling features. *Psychometrika*, 41, 471–503.

Dempster, A.P., Laird, N.M., & Rubin, D.B. (1977). Maximum likelihood from incomplete data via the EM-algorithm. *Journal of the Royal Statistical Society*, Series B, 39, 1–38.

DeSarbo, W. S., & Cron, W. L. (1988). A conditional mixture maximum likelihood methodology for clusterwise linear regression. *Journal of Classification*, 5, 249–289.

Diamantopoulos, A. & Winklhofer, H.M. (2001). Index construction with formative indicators: an alternative to scale development. *Journal of Marketing Research*, 38, 269–277.

Dijkstra, T.K. (1981). Latent Variables in Linear Stochastic Models: Reflections on Maximum Likelihood and Partial Least Squares Methods. PhD Thesis. University of Groningen, Groningen, the Netherlands.

Dijkstra, T.K. (2010). Latent variables and indices: Herman Wold's basic design and partial least squares. In V. Esposito Vinzi, W.W. Chin, J. Henseler, & H. Wang (Eds.), *Handbook of Partial Least Squares: Concepts, Methods and Applications* (pp. 23–46). Berlin: Springer-Verlag.

Dillon, W.R., Madden, T.J., Kirmani, A., & Mukherjee, S. (2001). Understanding what's in a brand rating: a model for assessing brand and attitude effects and their relationship to brand equity. *Journal of Marketing Research*, 38, 415–429.

Duncan, O.D. (1975). *Introduction to Structural Equation Models*. New York: Academic Press.

Duncan, T.E., Duncan, S.C., Alpert, A., Hops, H., Stoolmiller, M., & Muthén, B. (1997). Latent variable modeling of longitudinal and multilevel substance use data. *Multivariate Behavioral Research*, 32, 275–318.

Duncan, T.D., Duncan, S.C., & Strycker, L.A. (2006). *An Introduction to Latent Variable Growth Curve Modeling: Concepts, Issues, and Applications*. Mahwah NJ: Lawrence Erlbaum Associates.

Dunn, J.C. (1974). A fuzzy relative of the ISODATA process and its use in detecting compact well-separated clusters. *Journal of Cybernetics*, 3, 32–57.

D'Urso, P., Massari, R., & Santoro, A (2010). A class of fuzzy clusterwise regression models. *Information Sciences*, 180, 4737–4762.

D'Urso, P. & Santoro, A. (2006). Fuzzy clusterwise linear regression analysis with symmetrical fuzzy output variable. *Computational Statistics & Data Analysis*, 51, 287–313.

Edwards, J.R. (1995). Alternatives to difference scores as dependent variables in the study of congruence in organizational research. *Organizational Behavior and Human Decision Processes*, 64, 307–324.

Edwards, J.R. & Lambert, L.S. (2007). Methods for integrating moderation and mediation: a general analytical framework using moderated path analysis. *Psychological Methods*, 12, 1–22.

Efron, B. (1979). Bootstrap methods: another look at the Jackknife. *Annals of Statistics*, 7, 1–26.

Efron, B. (1982). *The Jackknife, the Bootstrap and Other Resampling Plans*. Philadelphia: SIAM.

Efron, B., Hastie, T., Johnstone, I., & Tibshirani, R. (2004). Least angle regression. *Annals of Statistics*, 32, 407–499.

Egri, K. & Ralston, D.A. (2004). Generation cohorts and personal values: a comparison of China and the United States. *Organization Science*, 15, 210–220.

Escoufier, Y. (1987). The duality diagram: a means for better practical applications. In P. Legendre & L. Legendre (Eds.), *Development in Numerical Ecology* (pp. 139–156). Berlin: Springer.

Esposito Vinzi, V. (2008). The contribution of PLS regression to PLS path modeling: formative measurement model and causality network in the structural model. Paper presented at Joint Statistical Meetings, Denver, Colorado, USA.

Esposito Vinzi, V. (2009). PLS path modeling and PLS regression: a joint partial least squares component-based approach to structural equation modeling. Paper presented at the 11th biennial conference of the International Federation of Classification Societies. University of Technology, Dresden, Germany.

Esposito Vinzi, V., Trinchera, L., & Amato, S. (2010). PLS Path Modeling: recent developments and open issues for model assessment and improvement. In V. Esposito Vinzi, W.W. Chin, J. Henseler, & H. Wang (Eds.). *Handbook of Partial Least Squares: Concepts, Methods and Applications* (pp. 47–82). Berlin: Springer-Verlag.

Esposito Vinzi, V. & Russolillo, G. (2010). Partial least squares path modeling and regression. In E. Wegman, Y. Said, & D. Scott (Eds.), *Wiley Interdisciplinary Reviews: Computational Statistics*. New York: Wiley.

Esposito Vinzi, V., Trinchera, L., Squillacciotti, S., & Tenenhaus, M. (2008). REBUS-PLS: A response-based procedure for detecting unit segments in PLS path modeling. *Applied Stochastic Models in Business and Industry*, 24, 439–458.

Everitt, B.S., Landau, S., & Leese, M. (2001). *Cluster Analysis* (4th Edition). London: Arnold Press.

Febrero-Bande, M. & Oviedo de la Fuente, M. (2012). Statistical computing in functional data analysis: the R package fda.usc. *Journal of Statistical Software*, 51, 1–28.

Ferraty, F. (2011). *Recent Advances in Functional Data Analysis and Related Topics*. New York: Springer.

Ferraty, F., Mas, A., & Vieu, P. (2007). Nonparametric regression on functional data: inference and practical. *Australian & New Zealand Journal of Statistics*, 49, 267–286.

Ferraty, F. & Romain, Y. (2011). *The Oxford Handbook of Functional Data Analysis*. Oxford: University Press.

Ferraty, F. & Vieu, P. (2006). *Nonparametric Functional Data Analysis: Theory and Practice*. New York: Springer.

Finney, J.M. (1972). Indirect effects in path analysis. *Sociological Methods and Research*, 1, 175–186.

Fornell, C. (1992). A national customer satisfaction barometer, the Swedish experience. *Journal of Marketing*, 56, 6–21.

Fornell, C., Barclay, D.W., & Rhee, B. (1988). A model and simple iterative algorithm for redundancy analysis. *Multivariate Behavioral Research*, 23, 349–360.

Fornell, C. & Bookstein, F.L. (1982). Two structural equation models: LISREL and PLS applied to consumer exit-voice theory. *Journal of Marketing Research*, 19, 440–452.

Fornell, C. & Cha, J. (1994). Partial least squares. In R. Bagozzi (Ed.), *Advanced Methods of Marketing Research* (pp. 52–78). Oxford, UK: Blackwell Publishers.

Fornell, C., Johnson, M.D., Anderson, E.W., Cha, J., & Bryant, B.E. (1996). The American customer satisfaction index: nature, purpose, and findings. *Journal of Marketing*, 60, 7–18.

Fornell, C. & Larcker, D.F. (1981). Structural equation models with unobservable variables and measurement error: algebra and statistics. *Journal of Marketing Research*, 18, 328–388.

Fornell, C. & Wernerfelt, W. (1987). Defensive marketing strategy by customer complaint management: a theoretical analysis. *Journal of Marketing Research*, 24, 337–346.

Fox, J. (1980). Effect analysis in structural equation models. *Sociological Methods and Research*, 9, 3–28.

Frank, I.E. & Friedman, J.H. (1993). A statistical view of chemometrics regression tools. *Technometrics*, 35 109–148.

Frank, R.E., Massy, W.F., & Wind, Y. (1972). *Marketing Segmentation*. Englewood Cliffs, NJ: Prentice Hall.

Friedman, J., Hastie, T., Höfling, H., & Tibshirani, R. (2007). Pathwise coordinate optimization. *The Annals of Applied Statistics*, 1, 302–332.

Friston, K.J. (1994). Functional and effective connectivity in neuroimaging: a synthesis. *Human Brain Mapping*, 2, 56–78.

Friston, K. J., Ashburner, J., Kiebel, S. J., Nichols, T. E., & Penny, W. D. (2007). *Statistical Parametric Mapping: The Analysis of Functional Brain Images*. London: Academic Press.

Gabriel, K.R. & Zamir, S. (1979). Low rank approximation of matrices by least squares with any choice of weights. *Technometrics*, 21, 489–498.

Gates, K.M., Molenaar, P.C., Hillary, F.G., & Slobounov, S. (2011). Extended unified SEM approach for modeling event-related fMRI data. *Neuroimage*, 54, 1151–1158.

Gifi, A. (1990). *Nonlinear Multivariate Analysis*. Chichester: Wiley.

Girard, S. (2000). A nonlinear PCA based on manifold approximation. *Computational Statistics*, 15, 145–167.

Glang, M. (1988). Maximierung der Summe erklärter Varianzen in linearrekursiven Strukturgleichungsmodellen mit multiple Indikatoren: Eine Alternative zum Schätzmodus B des Partial-Least-Squares-Verfahren, Dissertation zur Erlangung der Würde des Doktors der Philosophie de Universität Hamburg, Hamburg.

Goldberger, A.L., Amaral, L.A.N., Glass, L., Hausdorff, J.M., Ivanov, P.C., Mark, R.G., Mietus, J.E., Moody, G.B., Peng, C.K., & Stanley, H.E. (2000). PhysioBank, PhysioToolkit, and PhysioNet: components of a new research resource for complex physiologic signals. *Circulation*, 101, e215–e220.

Goldstein, H.I. (1987). *Multilevel Models in Educational and Social Research*. London: Oxford University.

González-Manteiga, W. & Vieu, P. (2007). Statistics for functional data. *Computational Statistics & Data Analysis*, 51, 4788–4792.

Gordon, A.D. (1999). *Classification*. London: Chapman & Hall/CRC.

Götz, O., Liehr-Gobbers, K., Krafft, M. (2010). Evaluation of structural equation models using the partial least squares (PLS) approach. In V. Esposito Vinzi, W. W. Chin, J. Henseler, H. Wang (Eds.). *Handbook of Partial Least Squares: Concepts, Methods and Applications* (pp. 691–711). Berlin: Springer-Verlag.

Grady, C.L., Springer, M.V., Hongwanishkul, D., McIntosh, A.R., Winocur, G. (2006). Age-related changes in brain across the adult lifespan. *Journal of Cognitive Neuroscience*, 16, 227–241.

Graff, J. & Schmidt, P. (1982). A general model for decomposition of effects. In K.G. Jöreskog & H. Wold (Eds.), *System under Indirect Observation: Causality, Structure, and Prediction* (pp. 131–148). Amersterdam: North Holland.

Grapentine, T. (2000). Path analysis vs. structural equation modeling. *Marketing Research*, 12, 12–20.

Greenacre, M.J. (1984). *Theory and Applications of Correspondence Analysis*. London: Academic Press.

Grewal, R., Cote, J.A., & Baumgartner, H. (2004). Multicollinearity and measurement error in structural equation models: implications for theory testing. *Marketing Science*, 23, 519–529.

Griep, M.I., Wakeling, I.N., Vankeerberghen, P., & Massart, D.L. (1995). Comparison of semirobust and robust partial least squares procedures. *Chemometrics and Intelligent Laboratory Systems*, 29, 37–50.

Grizzle, J.E. & Allen, D.M. (1969). Analysis of growth and dose response curves. *Biometrics*, 25, 357–381.

Gu, C. (2002). *Smooth Spline ANOVA Models*. New York: Springer.

Guttman, L. (1953). Image theory for the structure of quantitative variables. *Psychometrika*, 9, 277–296.

Hahn, C., Johnson, D.M., Herrmann, A., & Huber, F. (2002). Capturing customer heterogeneity using a finite mixture PLS approach. *Schmalenbach Business Review*, 54, 243–269.

Hall, P. & Hosseini-Nasab, M. (2006). On properties of functional principal components analysis. *Journal of the Royal Statistical Society. Series B (Methodological)*, 68, 109–126.

Hanafi, M. (2007). PLS path modeling: computation of latent variables with the estimation mode B. *Computational Statistics*, 22, 275–292.

Harris, R. (1989). A canonical cautionary. *Multivariate Behavioral Research*, 24, 17–39.

Hastie, T., Tibshirani, R., & Friedman, J. (2009). *The Elements of Statistical Learning: Data Mining, Inference, and Prediction* (2nd Edition). New York: Springer.

Hathaway, R.J. & Bezdek, J.C. (1993). Switching regression models and fuzzy clutering. *IEEE Transactions on Fuzzy Systems*, 1, 195–204.

Hayashi, C. (1993). *Theory and Methods of Quantification*. Tokyo: Asakura Publishing Co. (in Japanese).

He, G., Müller, H.-G., & Wang, J.-L. (2003). Functional canonical analysis for square integrable stochastic processes. *Journal of Multivariate Analysis*, 85, 54–77.

Heiser, W.J. & Groenen, P.J.F. (1997). Cluster differences scaling with a within-clusters loss component and a fuzzy successive approximation strategy to avoid local minima. *Psychometrika*, 62, 63–83.

Henseler, J. (2010). On the convergence of the partial least squares path modeling algorithm. *Computational Statistics*, 25, 107–120.

Henseler, J. (2012). Why generalized structured component analysis is not universally preferable to structural equation modeling. *Journal of the Academy of Marketing Science*, 40, 402–413.

Henseler, J. & Chin, W.W. (2010). A comparison of approaches for the analysis of interaction effects between latent variables using partial least squares path modeling. *Structural Equation Modeling*, 17, 82–109.

Henseler, J. & Fassott, G. (2010). Testing moderating effects in PLS path models: an illustration of available procedures. In V. Esposito Vinzi, W.W. Chin, J. Henseler, H. Wang (Eds.). *Handbook of Partial Least Squares: Concepts, Methods and Applications* (pp. 713–735). Berlin: Springer-Verlag.

Henseler, J., Ringle, C.M., & Sinkovics, R.R. (2009). The use of partial least squares path modeling in international marketing. *Advances in International Marketing*, 20, 277–320.

Henseler, J. & Sarstedt, M. (2013). Goodness-of-fit indices for partial least squares path modeling. *Computational Statistics*, 28, 565–580.

Hoehn, M.M. & Yahr, M.D. (1967). Parkinsonism: onset, progression and mortality. *Neurology*, 17, 427–442.

Hoerl, A.E. & Kennard, R.W. (1970). Ridge regression: application to nonorthogonal problems. *Technometrics*, 12, 69–82.

Hoffman, D.L. & Novak, T.P. (1996). Marketing in hypermedia computer-mediated environments: conceptual foundations. *Journal of Marketing*, 60, 50–68.

Horst, P. (1936). Obtaining a composite measure from a number of different measures of the same attribute. *Psychometrika*, 1, 53–60.

Horst, P. (1961). Relations among m sets of variables. *Psychometrika*, 26, 129–149.

Horváth, L. & Kokoszka, P. (2012). *Inference for Functional Data with Applications*. New York: Springer.

Höskuldsson, A. (1988). PLS regression methods. *Journal of Chemometrics*, 2, 211–228.

Hotelling, H. (1933). Analysis of a complex of statistical variables into principal components. *Journal of Educational Psychology*, 24, 417–441 and 498–520.

Hotelling, H. (1936). Relations between two sets of variates. *Biometrika*, 28, 321–377.

Hox, J.J. (1995). *Applied Multilevel Analysis*. Amsterdam: T-T Publikaties.

Hruschka, H. (1986). Market definition and segmentation using fuzzy clustering methods. *International Journal of Research in Marketing*, 3, 117–134.

Hu, L.T. & Bentler, P.M. (1998). Fit indices in covariance structure modeling: sensitivity to underparameterized model misspecification. *Psychological Methods*, 3, 424–453.

Hu, L.T. & Bentler, P.M. (1999). Cutoff criteria for fit indices in covariance structure analysis: conventional criteria versus new alternatives. *Structural Equation Modeling*, 6, 1–55.

Huettel, S.A., Song, A.W., & McCarthy, G. (2004). *Functional Magnetic Resonance Imaging*. Sunderland, MA: Sinauer Associates.

Hwang, H. (2008). VisualGSCA 1.0 - A graphical user interface software program for generalized structured component analysis. In K. Shigemasu, A. Okada, T. Imaizumi, & T. Hoshino (Eds.). *New Trends in Psychometrics* (pp. 111–120). Tokyo: University Academic Press.

Hwang, H. (2009). Regularized generalized structured component analysis. *Psychometrika*, 74, 514–530.

Hwang, H., DeSarbo, S.W., & Takane, Y. (2007). Fuzzy clusterwise generalized structured component analysis. *Psychometrika*, 72, 181–198.

Hwang, H., Ho, R.M., & Lee, J. (2010). Generalized structured component analysis with latent interactions. *Psychometrika*, 75, 228–242.

Hwang, H., Jung, K., Takane, Y., & Woodward, T.S. (2012a). Functional multiple-set canonical correlation analysis. *Psychometrika*, 77, 48–64.

Hwang, H., Kim, Y., & Tomiuk, M.A. (2005). Latent growth curve modeling of the relationships among revenue, loyalty, and customer satisfaction by generalized structured component analysis. *Asia Pacific Advances in Consumer Research*, Vol. 6, 215–217.

Hwang, H., Malhotra, N.K., Kim, Y., Tomiuk, M.A., & Hong, S. (2010a). A comparative study on parameter recovery of three approaches to structural equation modeling. *Journal of Marketing Research*, 47, 699–712.

Hwang, H., Malhotra, N.K., Kim, Y., Tomiuk, M.A., & Hong, S. (2010b). Web errata for "A comparative study on parameter recovery of three approaches to structural equation modeling (Hwang, Malhotra, Kim, Tomiuk, & Hong, 2010, vol XLVII, August, pp. 699–712)." Available at http://www.marketingpower.com/AboutAMA/Documents/JMR_Web_Appendix/2010.4/erratum_comparative_study_on_parameter_recovery.pdf.

Hwang, H., Suk, H.W., Lee, J-H., Moskowitz, D.S., & Lim, J. (2012b). Functional extended redundancy analysis. *Psychometrika*, 77, 524–542.

Hwang, H., Suk, H.W., Takane, Y., Lee, J-H., & Lim, J. (in press). Generalized functional extended redundancy analysis. *Psychometrika*.

Hwang, H. & Takane, Y. (2004a). Generalized structured component analysis. *Psychometrika*, 69, 81–99.

Hwang, H. & Takane, Y. (2004b). A multivariate reduced-rank growth curve model with unbalanced data. *Psychometrika*, 69, 65–79.

Hwang, H. & Takane, Y. (2010). Nonlinear generalized structured component analysis. *Behaviormetrika*, 37, 1–14.

Hwang, H., Takane, Y., & Malhotra, N. (2007). Multilevel generalized structured component analysis. *Behaviormetrika*, 34, 95–109.

Ittner, C.D. & Larcker, D.F. (1998). Are nonfinacial measures leading indicators of financial performance? An analysis of customer satisfaction. *Journal of Accounting Research*, 36, Supplement, 1–35.

Izenman, A.J. (1975). Reduced-rank regression for the multivariate linear model. *Journal of Multivariate Analysis*, 5, 248–264.

Jaccard, J. & Wan, C.H. (1995). *LISREL Approaches to Interaction Effects in Multiple Regression*. Thousand Oaks, CA: Sage.

Jackson, I. & Sirois, S. (2009). Infant cognition: going full factorial with pupil dilation. *Developmental Science*, 12, 670–679.

Jagpal, H.S. (1982). Multicollinearity in structural equation models with unobserved variables. *Journal of Marketing Research*, 19, 431–439.

James, G.M. (2002). Generalized linear models with functional predictors. *Journal of the Royal Statistical Society*, Series B, 63, 533–550.

James, G.M., Hastie, T.J., & Sugar, C.A. (2000). Principal component models for sparse functional data. *Biometrika*, 87, 587–602.

Jank, W. & Shmueli, G. (2006). Functional data analysis in electronic commerce research. *Statistical Science*, 21, 155–166.

Jedidi, K., Jagpal, H.S., & DeSarbo, W.S. (1997). Finite-mixture structural equation models for response-based segmentation and unobserved heterogeneity. *Marketing Science*, 16, 39–59.

Johnson, M.D., Gustafsson, A., Andreassen, T.W., Lervik, L., & Cha, J. (2001). The evolution and future of national customer satisfaction index models. *Journal of Economic Psychology*, 22, 217–245.

Johnson, P.O. & Neyman, J. (1936). Tests of certain linear hypotheses and their applications to some eduacational problems. *Statistical Research Memoris*, 1, 57–93.

Jolliffe, I.T. (2002). *Principal Component Analysis* (2nd Edition). New York: Springer-Verlag.

Jöreskog, K.G. (1970). A general method for analysis of covariance structures. *Biometrika*, 57, 409–426.

Jöreskog, K.G. (1973). A generating method for estimating a linear structural equation system. In A.S. Goldberger & O.D. Duncan (Eds.), *Structural Equation Models in the Social Sciences* (pp. 85–112). New York: Academic Press.

Jöreskog, K.G. & Goldberger, A.S. (1975). Estimation of a model with multiple indicators and multiple causes of a single latent variable. *Journal of the American Statistical Association*, 10, 631–639.

Jöreskog, K.G. & Sörbom, D. (1986). *LISREL VI: Analysis of Linear Structural Relationships by Maximum Likelihood and Least Squares Methods*. Mooresville, IN: Scientific Software, Inc.

Jöreskog, K.G. & Wold, H. (1982). The ML and PLS techniques for modeling with latent variables: historical and comparative aspects. In H. Wold and K.G. Jöreskog (Eds.), *Systems under Indirect Observation: Causality, Structure, Prediction, Part I* (pp. 263–270). Amsterdam: North Holland.

Jöreskog, K.G. & Yang, F. (1996). Non-linear structural equation models: the Kenny-Judd model with interaction effects. In G.A. Marcoulides & R.P. Shumacker (Eds.), *Advanced Structural Equation Modeling: Issues and Techniques* (pp. 57–88). Mahwah, NJ: Lawrence Erlbaum Associates.

Jung, K., Takane, Y., Hwang, H., & Woodward, T.S. (2012). Dynamic GSCA (Generalized Structured Component Analysis) with applications to the analysis of effective connectivity in functional neuroimaging data. *Psychometrika, 77*, 827–848.

Kaiser, H.F. (1960). The application of electronic computers to factor analysis. *Educational and Psychological Measurement, 20*, 141–151.

Kano, Y. (2002). Variable selection for structural models. *Journal of Statistical Planning and Inference, 108*, 173–187.

Kaplan, D. (1994). Estimator conditioning diagnostics for covariance structure models. *Sociological Methods Research, 23*, 200–229.

Kasser, T. & Ryan, R.M. (1993). A dark side of the American dream: correlates of financial success as a central life aspiration. *Journal of Personality and Social Psychology, 65*, 410–422.

Kasser, T. & Ryan, R.M. (1996). Further examining the American dream: differential correlates of intrinsic and extrinsic goals. *Personality and Social Psychology Bulletin, 22*, 280–287.

Kempler, B. (1971). Stimulus correlates of area judgments: a psychological developmental study. *Developmental Psychology, 4*, 158–163.

Kenny, D.A. (2011). SEM: Miscellaneous variables. http://davidakenny.net/cm/mvar.htm

Kenny, D.A. & Judd, C.M. (1984). Estimating the non-linear and interactive effects of latent variables. *Psychological Bulletin, 96*, 201–210.

Kessler, R.C. & Greenberg, D.F. (1981). *Linear Panel Analysis*. New York: Academic Press.

Khatri, C.G. (1966). A note on a MANOVA model applied to problems in growth curves. *Annals of the Institute of Statistical Mathematics, 18*, 75–86.

Kiers, H.A.L., Takane, Y., & ten Berge, J.M.F. (1996). The analysis of multitrait-multimethod matrices via constrained components analysis. *Psychometrika, 61*, 601–628.

Kiers, H.A.L. & ten Berge, J.M.F. (1989). Alternating least squares algorithms for simultaneous components analysis with equal component weight matrices in two or more populations. *Psychometrika, 54*, 467–473.

Kim, J., Zhu, W., Chang, L., Bentler, P.M., & Ernst, T. (2007). Unified structural equation modeling approach for the analysis of multisubject multivariate functional MRI data. *Human Brain Mapping, 28*, 85–93.

Kline, R.B. (1998). *Principles and Practice of Structural Equation Modeling* (1st Edition). New York: Guilford Press.

Kneip, A. & Utikal, K.J. (2001). Inference for density families using functional principal component analysis. *Journal of the American Statistical Association, 96*, 519–532.

Kruskal, J.B. (1964a). Multidimensional scaling by optimizing goodness of it to a nonmetric hypothesis. *Psychometrika, 29*, 1–29.

Kruskal, J.B. (1964b). Nonmetric multidimensional scaling: a numerical method. *Psychometrika, 29*, 115–129.

Lahiri, S.N. (2003). *Resampling Methods for Dependent Data*. New York: Springer-Verlag.

Laid, N.M. & Ware, J.H. (1982). Random-effects models for longitudinal data. *Biometrics, 38*, 963–974.

Lambert, Z.V., Wildt, A.R., & Durand, R.M. (1988). Redundancy analysis: an alternative to canonical correlation and multivariate multiple regression in exploring interest associations. *Psychological Bulletin*, 104, 282–289.

Lazarsfeld, P.F. (1955). Interpretation of statistical relations as a research operation. In P.F. Lazarsfeld, M. Rosenberg (Eds.), *The Language of Social Research: A Reader in the Methodology of Social Research* (pp. 115–125). Glencoe, IL: Free Press.

Lazarsfeld, P.F. & Henry, N.W. (1968). *Latent Structure Analysis*. New York: Houghton Mifflin.

Lebart, L., Morineau, A., & Warwick, K.M. (1984). *Multivariate Descriptive Statistical Analysis*. New York: Wiley.

Le Cessie, S. & Van Houwelingen, J.C. (1992). Ridge estimators in logistic regression. *Applied Statistics*, 41, 191–201.

Lee, A. & Silvapulle, M. (1988). Ridge estimation in logistic regression. Communications in Statistics. *Simulation and Computation*, 17, 1231–1257.

Lee, S., Bresch, E., & Narayanan, S. (2006). An exploratory study of emotional speech production using functional data analysis techniques. *Proceedings of the 7th International Seminar on Speech Production*. Ubatuba, Brazil: Center for Research on Speech, Acoustics, Language and Music.

Leurgans, S.E., Moyeed, R.A., & Silverman, B.W. (1993). Canonical correlation analysis when the data are curves. *Journal of the Royal Statistical Society. Series B (Methodological)*, 55, 725–740.

Lewis-Beck, M.S. & Mohr, L.B. (1976). Evaluating effects of independent variables. *Political Methodology*, 3, 27–47.

Li, F., Duncan, T.E., & Hops, H. (2001). Examining developmental trajectories in adolescent alcohol use using piecewise growth mixture modeling analysis. *Journal of Studies on Alcohol*, 62, 199–201.

Li, R-P. & Mukaidono, M. (1995). A maximum entropy approach to fuzzy clustering. *Proceedings of the 4th IEEE International Conference on Fuzzy Systems*, pp. 2227–2232, Yokohama, Japan.

Little, T.D., Bovaird, J.A., & Widaman, K.F. (2006). On the merits of orthogonalizing powered and product terms: implications for modeling interactions among latent variables. *Structural Equation Modeling*, 13, 497–519.

Lohmöller, J.B. (1984). *LVPLS Program Manual Version 1.6*. Zentralarchiv fr Empirische Sozialforschung. Koln: Universitat zu Koln.

Lohmöller, J.B. (1989). *Latent Variable Path Modeling with Partial Least Squares*. New York: Springer-Verlag.

Loudon, D. & Della Bitta, A.J. (1984). *Consumer Behavior. Concepts and Applications*. London: McGraw-Hill.

Lu, I.R.R., Kwan, E., Thomas, R., & Cedzynski, M. (2011). Two new methods for estimating structural equation models: an illustration and a comparison with two established methods. *International Journal of Research in Marketing*, 28, 258–268.

MacCallum, R.C., Kim, C., Malarkey, W.B., & Kiecolt-Glaser, J.K. (1997). Studying multivariate change using multilevel models and latent curve models. *Multivariate Behavioral Research*, 32, 215–253.

Mael, F.A. (1988). Organizational identification: construct redefinition and a field application with organizational alumni. Unpublished doctoral thesis, Wayne State University, Detroit.

Magidson, J. & Vermunt, J.K. (2004). Latent Class Models. In D. Kaplan (Ed.), *Handbook of Quantitative Methodology for the Social Sciences*. Newbury Park, CA: Sage Publications.

Mallows, C.L. (1973). Some comments on Cp. *Technometrics*, 15, 661–675.

Marsh, H.W., Dowson, M., Pietsch, J., & Walker, R. (2004). Why multicollinearity matters: a reexamination of relations between self-efficacy, self-concept, and achievement. *Journal of Educational Psychology*, 96, 518–522.

Marsh, H.W., Wen, Z., & Hau, K.-T. (2004). Structural equation models of latent interactions: evaluation of alternative estimation strategies and indicator construction. *Psychological Methods*, 9, 275–300.

Mattar, A.A.G. & Ostry, D.J. (2010). Generalization of dynamics learning across changes in movement amplitude. *The Journal of Neurophysiology*, 104, 426–438.

McArdle, J.J. (1986). Latent growth within behavior genetic models. *Behavior Genetics*, 16, 163–200.

McArdle, J.J. (2001). A latent difference score approach to longitudinal dynamic structural analyses. In R. Cudeck, S. Du Toit & D. Sorbom (Eds.), *Structural Equation Modeling: Present and Future* (pp. 342–380). Lincolnwood, IL: Scientific Software International.

McArdle, J.J. (2009). Latent variable modeling of differences and changes with longitudinal data. *Annual Review of Psychology*, 60, 577–605.

McArdle, J.J. & Hamagami, F. (2001). Latent difference score structural models for linear dynamic analyses with incomplete longitudinal data. In L.M. Collins & A.G. Sayer (Eds.), *New Methods for the Analysis of Change* (pp. 137–176). Washington, DC: American Psychological Association.

McArdle, J.J. & McDonald, R.P. (1984). Some algebraic properties of the reticular action model for moment structures. *British Journal of Mathematical and Statistical Psychology*, 37, 234–251.

McBratney, A.B. & Moore, A.W. (1985). Application of fuzzy sets to climatic classification. *Agricultural and Forest Meteorology*, 35, 165–185.

McDonald, R.P. (1996). Path analysis with composite variables. *Multivariate Behavioral Research*, 31, 239–270.

McDonald, R.P. & Ho, M-H.R. (2002). Principles and practice in reporting structural equation analyses. *Psychological Methods*, 7, 64–82.

McLachlan, G.J. & Peel, D. (2000). *Finite Mixture Models*. New York: John Wiley.

Meredith, W. & Millsap, R.E. (1985). On component analysis. *Psychometrika*, 50, 495–507.

Meredith, W. & Tisak, J. (1990). Latent curve analysis. *Psychometrika*, 55, 107–122.

Millsap, R.E. & Meredith, W. (1988). Component analysis in cross-sectional and longitudinal data. *Psychometrika*, 53, 123–134.

Miyamoto, S. (1998). An overview and new methods in fuzzy clustering. *Proceedings of the Second International Conference on Knowledge-Based Intelligent Electronic Systems*, pp. 21–23, Adelaide, Australia.

Miyamoto, S. & Mukaidono, M. (1997). Fuzzy c-means as a regularization and maximum entropy approach. *Proceedings of the 7th International Fuzzy Systems Association World Congress*, pp. 86–92, Prague, Czech Republic.

Moffitt, T.E. (1993). Adolescent-limited and life-course-persistent antisocial behavior: a developmental taxonomy. *Psychological Review*, 100, 674–701.

Mood, A.M., Graybill, F.A., & Boes, D.C. (1974). *Introduction to the Theory of Statistics*. New York: McGraw-Hill.

Morris, J.S. & Carroll, R.J. (2006). Wavelet-based functional mixed models. *Journal of the Royal Statistical Society, Series B*, 68, 179–199.

Müller, H-G. & Stadtmüller, U. (2005). Generalized functional linear models. *Annals of Statistics*, 33, 774–805.

Muthén, B.O. (1989). Latent variable modeling in heterogeneous populations. *Psychometrika*, 54, 557–585.

Myers, R. (1990). *Classical and Modern Regression with Applications* (2nd Edition). Boston, MA: Duxbury.

Nelson, E.C., Rust, R.T., Zahorik, A., Rose, R.L., Batalden, P., & Siemanski, B.A. (1992). Do patient perceptions of quality relate to hospital financial performance? *Journal of Health Care Marketing*, 12, 6–13.

Nesselroade, J.R., McArdle, J.J., Aggen, S.H., & Meyers, J. (2001). Dynamic factor analysis models for multivariate time series analysis. In D.M. Moskowitz & S.L. Hershberger (Eds.), *Modeling Intraindividual Variability with Repeated Measures Data: Advances and Techniques* (pp. 235–265). Mahwah, NJ: Erlbaum.

Nesselroade, J.R. & Molenaar, P.C.M. (1999). Pooling lagged covariance structures based on short, multivariate time series for dynamic factor analysis. In R.H. Hoyle (Ed.), *Statistical Strategies for Small Sample Research* (pp. 223–250). Thousand Oaks, CA: Sage Publications.

Ng, S. (2012). Variable selection in predictive regressions. Unpublished manuscript. Columbia University.

Nishisato, S. (1980). *Analysis of Categorical Data: Dual Scaling and Its Applications*. Toronto: University of Toronto Press.

Novak, T.P., Hoffman, D.L., & Yung, Y-F. (2000). Measuring the customer experience in online environments: a structural modeling approach. *Marketing Science*, 19, 22–42.

Nunnally, J. (1978). *Psychometric Theory*. New York: McGraw-Hill.

Ocaña, F.A., Aguilera, A.M., & Valderrama, M.J. (1999). Functional principal components analysis by choice of norm. *Journal of Multivariate Analysis*, 71, 262–276.

Ohtani, K. (2000). Bootstrapping R^2 and adjusted R^2 in regression analysis. *Economic Modelling*, 17, 473–483.

Okeke, F. & Karnieli, A. (2006). Linear mixture model approach for selecting fuzzy exponent value in fuzzy c-means algorithm. *Ecological Informatics*, 1, 117–124.

Oliver, R.L. (1996). *Satisfaction: A Behavioral Perspective on the Consumer*. Boston, MA: Irwin-McGraw-Hill.

Oliver, R.L. (1999). Whence consumer loyalty. *Journal of Marketing*, 63, 33–44.

Olshen, R.A., Biden, E.N., Wyatt, M.P., & Sutherland, D.H. (1989). Gait analysis and the bootstrap. *Annals of Statistics*, 17, 1419–1440.

Osborne, M.R., Presnell, B., & Thurlach, B.A. (2000). On the LASSO and its dual. *Journal of Computational and Graphical Statistics*, 9, 319–337.

Paxton, P., Curran, P.J., Bollen, K.A., Kirby, J., & Chen, F. (2001). Monte Carlo experiments: design and implementation. *Structural Equation Modeling*, 8, 287–312.

Pearson, K. (1901). On lines and planes of closest fit to systems of points in space. *Philosophical Magazine*, 2, 559–572.

Pedhazur, E.J. (1982). *Multiple Regression in Behavioral Research: Explanation and Prediction* (2nd Edition). New York: Holt, Rinehart & Winston.

Potthoff, R.F. & Roy, S.N. (1964). A generalized multivariate analysis of variance model useful especially for growth curve problems. *Biometrika*, 51, 313–326.

Preacher, K.J., Curran, P.J., & Bauer, D.J. (2006). Computational tools for probing interations in multiple linear regression, multilevel modeling, and latent curve analysis. *Journal of Educational and Behavioral Statistics, 31,* 437–448.

Preacher, K.J., Rucker, D.D., & Hayes, A.F. (2007). Addressing moderated mediation hypotheses: theory, methods, and prescriptions. *Multivariate Behavioral Research,* 42, 185–227.

Preacher, K.J., Wichman, A.L., MacCallum, R.C., & Briggs, N.E. (2008). *Latent Growth Curve Modeling.* Thousand Oaks, CA: Sage Publications.

Press, W.H., Teukolsky, S.A., Vetterling, W.T., & Flannery, B.P. (1999). *Numerical Recipes in C. The Art of Scientific Computing* (2nd Edition). New York: Cambridge University Press.

Radloff, L.S. (1977). The CES-D scale: a self-report depression scale for research in the general population. *Applied Psychological Measurement, 1,* 385–401.

Ramsay, J.O. & Dalzell, C.J. (1991). Some tools for functional data analysis (with discussion). *Journal of the Royal Statistical Society B, 53,* 539–572.

Ramsay, J.O., Hooker, G., & Graves, S. (2009). *Functional Data Analysis with R and MATLAB.* New York: Springer.

Ramsay, J.O. & Silverman, B.W. (1997). *Functional Data Analysis* (1st Edition). New York: Springer.

Ramsay, J.O. & Silverman, B.W. (2002). *Applied Functional Data Analysis: Methods and Case Studies.* New York: Springer.

Ramsay, J.O. & Silverman, B.W. (2005). *Functional Data Analysis* (2nd Edition). New York: Springer.

Rao, C.R. (1958). Some statistical methods for the comparison of growth curves. *Biometrics,* 110, 49–58.

Rao, C.R. (1964). The use and interpretation of principal component analysis in applied research. *Sankhya,* A, 26, 329–358.

Rao, C.R. (1965). The theory of least squares when the parameters are stochastic and its application to the analysis of growth curves. *Biometrika,* 52, 447–458.

Rao, C. R. (1980). Matrix approximations and reduction of dimensionality in multivariate statistical analysis. In P. R. Krishnaiah (Ed.), *Multivariate Analysis* (pp. 3–22). Amsterdam: North Holland.

Raudenbush, S.W. (2001). Comparing personal trajectories and drawing causal inferences from longitudinal data. *Annual Review of Psychology,* 52, 501–525.

Raudenbush, S.W. & Bryk, A.S. (2002). *Hierarchical Linear Models: Applications and Data Analysis Methods.* (2nd Edition). Thousand Oaks, CA: Sage Publications.

Reddy, S.K. & Dass, M. (2006). Modeling on-line art auction dynamics using functional data analysis. *Statistical Science,* 21, 179–193.

Reichheld, F.F. & Sasser, W.E. (1990). Zero defections: quality comes to services. *Harvard Business Review,* 68, 105–111.

Reinartz, W., Krafft, M., & Hoyer, W.D. (2004). The customer relationship management process: its measurement and impact on performance. *Journal of Marketing Research,* 41, 293–305.

Reinsel, G.C. & Velu, R.P. (1998). *Multivariate Reduced-Rank Regression.* New York: Springer-Verlag.

Reiss, P.T. & Ogden, T.R. (2007). Functional principal component regression and functional partial least squares. *Journal of the American Statistical Association,* 102, 984–996.

Rice, J.A. (2004). Functional and longitudinal data analysis: perspectives on smoothing. *Statistica Sinica*, 14, 631–647.

Rice, J.A. & Silverman, B.W. (1991). Estimating the mean and covariance structure non-parametrically when the data are curves. *The Journal of the Royal Statistical Society B*, 53, 233–243.

Richins, M.L. & Dawson, S. (1992). A Consumer values orientation for materialism and its measurement: scale development and validation. *Journal of Consumer Research*, 19, 303–316.

Ridgon, E.E., Schumacker, R.E., & Worthe, W. (1998). A comparative review of interaction and nonlinear modeling. In R.E. Schumaker & G.A. Marcoulides (Eds.), *Interaction and Nonlinear Effects in Structural Equation Modeling* (pp. 1–16). Mahwah, NJ: Lawrence Erlbaum Associates.

Ringle, C.M., Wende, S., & Will, A. (2005). *SmartPLS 2.0.* Hamburg, Germany.

Rogosa, D. (1981). On the relationship between the Johnson-Neyman region of significance and statistical tests of parallel within-group regressions. *Educational and Psychological Measurement*, 41, 73–84.

Roubens, M. (1982). Fuzzy clustering algorithms and their cluster validity. *European Journal of Operational Research*, 10, 294–301.

Rousseeuw, P.J. & Leroy, A.M. (1987). *Robust Regression and Outlier Detection.* New York: Wiley.

Rovine, M.J. & Molenaar, P.C.M. (2000). A structural equation modeling approach to a multilevel random coefficients model. *Multivariate Behavioral Research*, 35, 51–88.

Rubin, D.B. (1987). *Multiple Imputation for Nonresponse in Surveys.* New York: John Wiley & Sons.

Rust, R.T. & Zahorik, A.J. (1993). Customer satisfaction, customer retention, and market Share. *Journal of Retailing*, 69, 193–215.

Schafer, J.L. (1997). *Analysis of Incomplete Multivariate Data.* London: Chapman & Hall/CRC.

Schmitt, N. & Stults, D.M. (1986). Methodology review: analysis of multitrait-multimethod matrices. *Applied Psychological Measurement*, 10, 1–22.

Schneider, B. (1991). Service quality and profits: can you have your cake and eat it too. *Human Resource Planning*, 14, 151–157.

Schumacker, R.E. & Marcoulides, G.A. (1998). *Interaction and Nonlinear Effects in Structural Equation Modeling.* Mahwah, NJ: Lawrence Erlbaum Associates.

Schumaker, L.L. (2007). *Spline Functions: Basic Theory.* Cambridge: Cambridge University Press.

Searle, S.R. (1971). *Linear Models.* New York: Wiley.

Seber, G.A.F. (1984). *Multivariate Observations.* New York: Wiley.

Segi, M. (1979). *Age-adjusted Death Rates for Cancer for Selected Sites (A-classification) in 51 Countries in 1974 (1979).* Nagoya, Japan: Segi Institute of Cancer Epidemiology.

Sharma, S., Durand, R.M., & Gur-Arie, O. (1981). Identification and analysis of moderator variables. *Journal of Marketing Research*, 18, 291–300.

Sharma, S., Mukherjee, S., Kumar, A., & Dillon, W.R. (2005). A simulation study to investigate the use of cutoff values for assessing model fit in covariance structure models. *Journal of Business Research*, 58, 935–943.

Sheldon, K.M. & Elliot, A.J. (1999). Goal striving, need satisfaction, and longitudinal well-being: the self-concordance model. *Journal of Personality and Social Psychology*, 76, 482-497.

Sheldon, K.M. & Kasser, T. (1998). Pursuing personal goals: skills enable progress but not all progress is beneficial. *Personality and Social Psychology Bulletin*, 24, 1319–1331.

Silverman, B.W. (1995). Incorporating parametric effects into functional principal components analysis. *Journal of the Royal Statistical Society. Series B (Methodological)*, 57, 673–689.

Silverman, B.W. (1996). Smoothed functional principal components analysis by choice of norm. *The Annals of Statistics*, 24, 1–24.

Singh, A. (1996). Outliers and robust procedures in some chemometric applications. *Chemometrics and Intelligent Laboratory Systems*, 33, 75–100.

Smith, S., Miller, K.L., Salimi-Khorshid, G., Webster, M., Beckman, C.F., Nichols, T.E., Ramsey, J.D., & Woolrich, M.W. (2010). Network modeling methods for fMRI. *Neuroimage*, 54, 875–891.

Snijders, T.A.B. & Bosker, R.J. (1999). *Multilevel Analysis: An Introduction to Basic and Advanced Multilevel Modeling*. London: Sage Publications.

Sörbom, D. (2001). Karl Jöreskog and LISREL: a personal history. In R. Cudeck, S. du Toit, & S. Sörbom (Eds.), *Structural Equation Modeling: Present and Future* (pp. 3–10). Lincoln Wood, IL: Scientific Software.

Spearman, C. (1913). Correlations of sums or differences. *British Journal of Psychology*, 5, 417–426.

Stephan, K.E., Kasper, L., Harrison, L.M., Daunizeau, J., den Ouden, H.E.M., Breakspear, M., & Friston, K.J. (2008). Nonlinear dynamic causal models for fMRI. *Neuroimage*, 42, 649–662.

Stevens, J.P. (2009). *Applied Multivariate Statistics for the Social Sciences* (5th Edition). New York: Routledge.

Suk, H. (2013). Functional generalized structured component analysis. Unpublished doctoral thesis. Department of Psychology. McGill University.

Suk, H. & Hwang, H. (2014). Functional generalized structured component analysis. Paper submitted for publication.

Tabachnick, B.G. & Fidell, L.S. (2012). *Using Multivariate Statistics* (6th Edition). New York: Pearson.

Takane, Y. & de Leeuw, J. (1987). On the relationship between item response theory and factor analysis of discretized variables. *Psychometrika*, 52, 393–408.

Takane, Y. & Hunter, M.A. (2001). Constrained principal component analysis: a comprehensive theory. *Applicable Algebra in Engineering, Communication, and Computing*, 12, 391–419.

Takane, Y. & Hwang, H. (2005). An extended redundancy analysis and its applications to two practical examples. *Computational Statistics and Data Analysis*, 49, 785–808.

Takane, Y. & Hwang, H. (2007). Regularized linear and kernel redundancy analysis. *Computational Statistics and Data Analysis*, 52, 394–405.

Takane, Y., Kiers, H., & de Leeuw, J. (1995). Component analysis with different sets of constraints on different dimensions. *Psychometrika*, 60, 259–280.

Takane, Y. & Shibayama, T. (1991). Principal component analysis with external information on both subjects and variables. *Psychometrika*, 56, 97–120.

Takane, Y., Yanai, H., & Mayekawa, S. (1991). Relationships among several methods of linearly constrained correspondence analysis. *Psychometrika*, 56, 667–684.

Takane, Y., Young, F.W., & de Leeuw, J. (1980). An individual differences additive model: an alternating least squares method with optimal scaling features. *Psychometrika*, 45, 183–209.

Tan, X., Shiyko, M.P., Li, R., Li, Y., & Dierker, L. (2012). A time-varying effect model for intensive longitudinal data. *Psychological Methods*, 17, 61–77.

Temme, D., Kreis, H., & Hildebrandt, L. (2006). PLS path modeling – A software review. SFB 649 discussion paper. Humboldt University, Berlin, Germany.

ten Berge, J.M.F. (1993). *Least Squares Optimization in Multivariate Analysis*. Leiden: DSWO Press.

ten Berge, J.M.F. & Nevels, K. (1977). A general solution to Mosier's oblique procrustes problem. *Psychometrika*, 42, 593–600.

Tenenhaus, A. & Tenenhaus, M. (2011). Regularized generalized canonical correlation analysis. *Psychometrika*, 76, 257–284.

Tenenhaus, M. (2008a). Component-based SEM. Comparisons between various methods. Paper presented at the Joint Statistical Meetings, Denver, Colorado, USA.

Tenenhaus, M. (2008b). Component-based structural equation modelling. *Total Quality Management and Business Excellence*, 19, 871–886.

Tenenhaus, M., Amato, S., & Esposito Vinzi, V. (2004). A global goodness-of-fit index for PLS structural equation modelling. *Proceedings of the XLII SIS Scientific Meeting*, Vol. Contributed Papers, CLEUP, Padova, 739–742.

Tenenhaus, M., Esposito Vinzi, V., Chateline, Y.-M., & Lauro, C. (2005). PLS path modeling. *Computational Statistics and Data Analysis*, 48, 159–205.

Thorndike, R.L., Hagen, E.P., & Sattler, J.M. (1986). *The Stanford-Binet Intelligence Scale, Fourth Edition: Guide for Administering and Scoring* (4th Edition). Chicago: Riverside.

Tian, T.S. (2010). Functional data analysis in brain imaging studies. *Frontiers in Quantitative Psychology and Measurement*, 1, 1–11.

Tibshirani, R. (1996). Regression shrinkage and selection via the lasso. *Journal of the Royal Statistical Society. Series B (Methodological)*, 58, 267–288.

Tornow, W.W. & Wiley, J.W. (1991). Service quality and management practices: a look at employee attitudes, customer satisfaction, and bottom line consequences. *Human Resource Planning*, 14, 105–115.

Trinchera, L., Squillacciotti, S., & Esposito Vinzi, V. (2006). PLS typological path modeling: a model-based approach to classification. *Proceedings of Knowledge Extraction and Modeling*. Island of Capri, Italy.

Tucker, L.R. (1966). Learning theory and multivariate experiment: illustration by determination of parameters of generalized learning curves. In R.B. Cattell (Ed.), *The Handbook of Multivariate Experimental Psychology* (pp. 476–501). Chicago, IL: Rand McNally.

Tuddenham, R.D. & Snyder, M.M. (1954). Physical growth of California boys and girls from birth to eighteen years. University of California Publications in Child Development, 1, 183–364.

Valderrama, M.J. (2007). An overview to modelling functional data. *Computational Statistics*, 22, 331–334.

van den Wollenberg, A.L. (1977). Redundancy analysis: an alternative for canonical analysis. *Psychometrika*, 42, 207–219.

van der Leeden, R. (1990). *Reduced Rank Regression with Structured Residuals*. Leiden: DSWO Press.

Velu, R.P. (1991). Reduced rank models with two sets of regressors. *Applied Statistics*, 40, 159–170.

Viertl, R. (2011). *Statistical Methods for Fuzzy Data*. New York: John Wiley & Sons.

Vines, B.W., Krumhansl, C.L., Wanderley, M.M., Dalca, I.M., & Levitin, D.J. (2005). Dimensions of emotion in expressive musical performance. *Annals of the New York Academy of Sciences*, 1060, 462–466.

Vines, B.W., Krumhansl, C.L., Wanderley, M.M., & Levitin, D.J. (2006). Cross-modal interactions in the perception of musical performance. *Cognition*, 101, 80–113.

Vines, B.W., Nuzzo, R.L., & Levitin, D.J. (2005). Quantifying and analyzing musical dynamics: differential calculus, physics and functional data techniques. *Music Perception*, 23, 137–152.

Vinod, H.D. (1978). A survey of ridge regression and related techniques for improvements over ordinary least squares. *The Review of Economics and Statistics*, 60, 121–131.

Vinod, H.D. & Ullah, A. (1981). *Recent Advances in Regression Methods*. New York: Marcel Dekker.

Viviani, R., Grön, G., & Spitzer, M. (2005). Functional principal component analysis of fMRI data. *Human Brain Mapping*, 24, 109–129.

Vonesh, E.F. & Carter, R.L. (1987). Efficient inference for random-coefficient growth curve models with unbalanced data. *Biometrics*, 43, 617–628.

Wahba, G. (1990). *Spline Models for Observational Data*. Philadelphia: Society for Industrial and Applied Mathematics.

Wakeling, I.N. & Macfie, H.J. (1992). A robust PLS procedure. *Journal of Chemometrics*, 6, 189–198.

Walczak, B. & Massart, D.L. (1995). Robust principal components regression as a detection tool for outliers. *Chemometrics and Intelligent Laboratory Systems*, 27, 41–54.

Walls, T.A. & Schafer, J.L. (2006). *Models for Intensive Longitudinal Data*. New York: Oxford University Press.

Wang, L., Li, Y., Metzak, P. He, Y., & Woodward, T. S. (2010). Age-related changes in topological patterns of large-scale brain functional networks during memory encoding and recognition. *NeuroImage*, 50, 862–872.

Watson, D. & Clark, L.A. (1991). The mood and anxiety symptom questionnaire. Unpublished manuscript, University of Iowa, Iowa City, IA.

Wedel, M. & Kamakura, W.A. (1998). *Market Segmentation: Conceptual and Methodological Foundations*. Boston: Kluwer Academic Publishers.

Wedel, M. & Steenkamp, J-B.E.M. (1989). Fuzzy clusterwise regression approach to benefit segmentation. *International Journal of Research in Marketing*, 6, 241–258.

Wedel, M. & Steenkamp, J-B.E.M. (1991). A clusterwise regression method for simultaneous fuzzy market structuring and benefit segmentation. *Journal of Marketing Research*, 28, 385–396.

Werts, C.E. & Linn, R.L. (1970). Path analysis: psychological examples. *Psychological Bulletin*, 74, 193–212.

Werts, C.E., Linn, R.L., & Jöreskog, K.G. (1974). Intraclass reliability estimates: testing structural assumptions. *Educational and Psychological Measurement*, 34, 25–33.

Widaman, K.F. (1985). Hierarchically nested covariance structure models for multitrait-multimethod data. *Applied Psychological Measurement*, 9, 1–26.

Wiesner, M. & Windle, M. (2004). Assessing covariates of adolescent delinquency trajectories. *Journal of Youth and Adolescence*, 33, 431–442.

Wiley, J.W. (1991). Customer satisfaction: a supportive work environment and its financial costs. *Human Resource Planning*, 117–127.

Willett, J.B. & Bub, K.L. (2005). Structural equation modeling: latent growth curve analysis. In Brian S. Everitt & David C. Howell (Eds.), *Encyclopedia of Statistics in Behavioral Science* (pp. 772–779). Chichester, UK: John Wiley & Sons.

Wold, H. (1966). Estimation of principal components and related methods by itera-tive least squares. In P.R. Krishnaiah (Ed.), *Multivariate Analysis* (pp. 391–420). New York: Academic Press.

Wold, H. (1973). Nonlinear iterative partial least squares (NIPALS) modeling: some current developments. In P.R. Krishnaiah (Ed.), *Multivariate Analysis* (pp. 383–487). New York: Academic Press.

Wold, H. (1982). Soft modeling: the basic design and some extensions. In K.G. Jöreskog & H. Wold (Eds.), *Systems under Indirect Observations II* (pp. 1–54). Amsterdam: North Holland.

Wothke, W. (1993). Nonpositive definite matrices in structural modeling. In K.A. Bollen & J.S. Long (Eds.), *Testing Structural Equation Models* (pp. 256–293). Newbury Park, CA: Sage Publications.

Wright, S. (1934). The method of path coefficients. *Annals of Mathematical Statistics, 5,* 161–215.

Wu, Y., Fan, J., & Müller, H-G. (2010). Varying-coefficient functional linear regression. *Bernoulli, 16,* 730–758.

Yang Jonsson, F. (1998). Modeling interaction and nonlinear effects: a step-by-step LISREL example. In R.E. Schumaker & G.A. Marcoulides (Eds.), *Interaction and Nonlinear Effects in Structural Equation Modeling* (pp. 17–42). Mahwah, NJ: Lawrence Erlbaum Associates.

Yates, F. (1933). The analysis of replicated experiments when the field results are incomplete. *Empire Journal of Experimental Agriculture, 1,* 129–142.

Yao, F., Fu, Y., & Lee, T.C.M. (2011). Functional mixture regression. *Biostatistics, 12,* 341–353.

Yao, F. & Lee, T.C.M. (2006). Penalized spline models for functional principal com-ponent analysis. *Journal of the Royal Statistical Society. Series B (Methodological), 68,* 3–25.

Yao, F., Müller, H.-G., & Wang, J.-L. (2005). Functional data analysis for sparse longi-tudinal data. *Journal of the American Statistical Association, 100,* 577–590.

Yogev, G., Giladi, N., Peretz, C., Springer, S., Simon, E.S., & Hausdorff, J.M. (2005). Dual tasking, gait rhythmicity, and Parkinson's disease: which aspects of gait are attention demanding? *European Journal of Neuroscience, 22,* 1248–1256.

Young, F.W. (1981). Quantitative analysis of qualitative data. *Psychometrika, 46,* 347–388.

Young, F.W., de Leeuw, J., & Takane, Y. (1976). Regression with qualitative and quan-titative variables: an alternating least squares method with optimal scaling fea-tures. *Psychometrika, 41,* 505–529.

Yuan, M. & Lin, Y. (2006). Model selection and estimation in regression with grouped variables. *Journal of the Royal Statistical Society. Series B (Methodological), 68,* 49–67.

Zadeh, L.A. (1965). Fuzzy sets. *Information and Control, 8,* 338–353.

Zahorik, A.J. & Rust, R.T. (1992). Modeling the impact of service quality profitability: a review. In T.A. Swartz, D.E., Bowen, & S.W. Brown (Eds.), *Advances in Services Marketing and Management* (pp. 247–276). Greenwich, CT: JAI Press.

Zhou, L. (2014). Dynamic generalized (multiple-set) structured canonical correlation analysis (Dynamic GCANO): a structural equation model for simultaneous analysis of multiple-subject effective connectivity in functional neuroimaging studies. Unpublished Doctoral Dissertation, Department of Psychology, McGill University.

Zou, H. & Hastie, T. (2005). Regularization and variable selection via the elastic net. *Journal of the Royal Statistical Society. Series B (Methodological), 67,* 301–320.

Index

A

ACSI, *see* American Customer
    Satisfaction Index
Advertising spending data, 289–293
Algorithm, dynamic generalized
    structured component
    analysis, 265–267
Alternating coordinate-descent
    algorithm
    for lasso generalized structured
        component analysis, 240–245
Alternating least-squares algorithm
    for estimating missing observations,
        125–126
    for fuzzy clusterwise generalized
        structured component
        analysis, 143–145
    for generalized structured
        component analysis with latent
        interactions, 189–191
    for two-level multilevel generalized
        structured component
        analysis, 208–210
Alternating regularized least-squares
    algorithm
    for functional generalized
        structured component
        analysis, 297–301
    for regularized generalized
        structured component
        analysis, 227–229
American Customer Satisfaction Index
    (ACSI)
    company-level data, 221, 235–237
    confidence intervals, 202, 203
    consumer complaints, 39
    consumer-level data, 87–89
    consumer loyalty, 40
    correlation residuals, 37, 38
    customer expectations, 38, 40
    customer satisfaction, 41
    generalized structured component
        analysis, 32–40, 67–70

    group-dependent path coefficients
        for, 205–207
    measurement model, 33
    multilevel generalized structured
        component analysis, 201–207
    partial least squares path modeling
        for, 68
    path coefficients, 40, 43, 69
    path diagram of, 33
    perceived quality, 38
    perceived value, 38
    real data, 67–70
    R-squared values of, 40, 68
    structural model for, 35
    two-level, 201
    weighted relation model, 35
    weights and loadings and, 39, 42
American customer satisfaction index
    model
    correlation matrix of, 224
    generalized structured component
        analysis, 225
    path diagram of, 223
Analysis of variance (ANOVA)
    model, 198
ANOVA model, *see* Analysis of variance
    model
Approximating smooth functions,
    280–284
Aspirations Index (ASPQ), 184
ASPQ, *see* Aspirations Index
Autoregressive models, 269
Auxiliary data, 256

B

Bidirectionally connected structural
    model, 258
Block bootstrap method, 256
Blood oxygen level-dependent (BOLD)
    signals, 247, 248, 277
BOLD signals, *see* Blood oxygen
    level-dependent signals
Bootstrap method, 95, 255–256

Printed and bound by CPI Group (UK) Ltd, Croydon, CR0 4YY

24/10/2024

01778281-0010